T0325452

FOSSILIZATION

FOSSILIZATION

Understanding the Material Nature
of Ancient Plants and Animals

EDITED BY

CAROLE T. GEE,

VICTORIA E. MCCOY,

AND P. MARTIN SANDER

JOHNS HOPKINS UNIVERSITY PRESS

BALTIMORE

This book has been brought to publication with the generous assistance of Darlene Bookoff.

Johns Hopkins University Press
2715 North Charles Street
Baltimore, Maryland 21218-4363
www.press.jhu.edu

Library of Congress Cataloging-in-Publication Data

Names: Gee, Carole T., editor. | McCoy, Victoria E., editor | Sander, Martin P., editor
Title: Fossilization : understanding the material nature of ancient plants and animals /
    edited by Carole T. Gee, Victoria E. McCoy, and P. Martin Sander.
Description: Baltimore : Johns Hopkins University Press, 2021. |
    Includes bibliographical references and index.
Identifiers: LCCN 2020015932 | ISBN 9781421440217 (hardcover) |
    ISBN 9781421440224 (ebook)
Subjects: LCSH: Fossilization. | Taphonomy.
Classification: LCC QE721.2.F6 F67 2021 | DDC 560/.41—dc23
LC record available at https://lccn.loc.gov/2020015932

A catalog record for this book is available from the British Library.

*Special discounts are available for bulk purchases of this book. For more information,
please contact Special Sales at specialsales@jh.edu.*

Johns Hopkins University Press uses environmentally friendly book materials,
including recycled text paper that is composed of at least 30 percent post-consumer
waste, whenever possible.

Contents

Preface

If I handed you, or any person really, a terrific-looking fossil, it is a sure bet that your first reaction will be a spontaneous, "Oh, what is it?" The next question is most likely, "How old is it?" This may be followed with a more contemplative query: "How did it get this way?"

Well, scientists are people, too, and we react in the same way and ask the same series of questions. Yet, in the case of paleontologists, geochemists, and biochemists at the University of Bonn in Germany, the last query has turned into a quest by becoming the goal of a collaborative research program on fossilization. Of a researcher's list of interrogatives consisting of the five Ws and one H—who, what, where, when, why, and how—the H is usually the toughest question to answer. But this is exactly what we want to tackle in this book: How do organisms fossilize? And, after they have been fossilized, what are they made of now?

Paleontologists, geochemists, and biochemists—as well as mineralogists and microbiologists—are so highly specialized today that many want to stay in their own lane and keep moving on as fast as possible. Yet, approaching questions of fossilization with modern methods requires interdisciplinary work and coordinated lane-changing into shared avenues of research. This means that paleontologists must seek out and work with others in disciplines that might seem quite distant academically, but actually hold the key to understanding to long-perplexing questions. We need to take time to learn each other's jargon and methodologies, as well as to start reaching for the same goals. Finally, in this modern day and age, it seems that the digital revolution is pushing us to come together and move forward in unison with joint high-resolution studies to understand the processes of fossilization and the substance preservation of ancient plants and animals.

Alternatively, we can also consider this on the singleton level. I like to think of a collaborative relationship between a paleontologist and geochemist as being the perfect research marriage between geoscientists: paleontologists have the material and questions, while geochemists have the methods, machines, and know-how to resolve issues of fossilization. A simple case in point: If I give a geochemist a piece of 150-million-year-old wood and ask "How did it get this way?" the geochemist will answer that it was preserved by silica-rich water that infilled the wood's cells and permeated the cell walls to form a rock-hard plant fossil. Not only that, but the geochemist will also tell me where the silica entered the tree, what kind of silica it is, and how many waves of silica-rich fluid had swept through it. Actually, after so many years of research, the general process of silicification of wood is fairly well worked out, so what we want to know now are the fine details, such as how the aqueous silica interacts with the molecules of the wood to enclose and protect the organic material of the cell walls, resulting in fossil wood with the perfect preservation of minute structures on the subcellular level. Hence, we are currently at the point at which we are concerned with processes happening on the nanoscale and chemical level that we can only discern with high-resolution analysis and purposeful experimentation.

Where do the biochemists fit into this relationship? Well, when a fossil "turns into stone," it is actually not completely replaced by inorganic minerals. A piece of rock-hard silicified wood needs to retain some organic matter of its former self in order to be excellently preserved. Indeed, some fossils never become mineralized at all, such as the bone cells of dinosaurs and soft tissues of insects found later in this book. There are even some remains of ancient life that never get embedded into mineral-rich rock, but instead are embedded in organic substances like resin. This is precisely when it gets interesting for biological chemists. Microbiologists, on the other hand, become intellectually engaged at the very onset of fossilization, when an animal gets deposited on the bottom of a water body and is covered with bacteria, for it is this microbial biofilm that facilitates and enhances the preservation of the carcass. In fact, it seems that nearly all scientists can enjoy a piece of the fossilization research pie, for the history of life relates to many of us.

The greatest sponsor of scientists and scientific research in Germany is the German Research Foundation, Deutsche Forschungsgemeinschaft (DFG), to whom we are indebted for funding our coordinated program "The Limits of the Fossil Record: Analytical and Experimental Approaches to Fossilization," a consortium of nine research projects on fossilization processes and the material nature of fossils. This book as a whole is contribution number 14 of the DFG Research Unit 2685.

Special thanks should also go to the 15 experts from all over the world who reviewed the scientific chapters in this book. We appreciate their helpful comments, the probing questions, and—well, yes—even the well-deserved critical remarks. Some specialists even stepped up to review more than one chapter of the book. Closer to home, we thank Aowei Xie and Mariah How-ell in Bonn for their assistance with image editing and proofreading.

Finally, let me express our collective gratitude to Tiffany Gasbarrini, Senior Acquisitions Editor, Life Sciences, Mathematics, and Physics, and Esther P. Rodriguez, Editorial Assistant, as well as the rest of the team at Johns Hopkins University Press, for the editing, design, production, and marketing of this book, as well as to Carrie Love for her meticulous and thoughtful copyedit-ing. On behalf of my co-editors, Tory McCoy and Martin Sander, as well as all chapter authors and our colleagues in the Research Unit 2685, I am

Yours in fossilization,
Carole Gee

FOSSILIZATION

CHAPTER **1**

Introduction to the Limits of the Fossil Record

P. MARTIN SANDER AND CAROLE T. GEE

Fossils are the only evidence of life that vanished from this
planet before the rise of humans. Yet little is known about how
fossils form.
MARY H. SCHWEITZER

The history of life on earth is documented primarily by the fossil record.
Researchers interpret and investigate this unique record through the science
of paleontology. But does the fossil record truly reflect the history of life?
This is still undoubtedly one of the most fundamental issues in the study of
ancient life. Nearly all research fields in paleontology, from the micro- to
macroevolutionary level, are affected in a complex manner by the loss of
information during the process of fossilization. The limits imposed by fossilization conceal data on animal behavior and physiology, and the lack of
preserved soft parts in animals leaves us with only a subset of the anatomical evidence once present in the living organism. There is also the vast
amount of biological information lost at the molecular level. The fossil record places distinct restrictions on our understanding of the evolution of
organisms and the history of life on earth, and thus we must find new ways
to transcend those limits.

As paleontologists, we commonly employ methods of inference based on
extant organisms, but we first have to ask ourselves how to extract as much
information as possible from the tangible remains of ancient life, our fossils.
Traditionally, paleontology has been limited to studying the morphology of
mineralized, sclerotized, and lignified parts and tissues of organisms, such as
wood, shells, bones, and teeth. Only in rare and famous instances of "soft
tissue preservation" have we gained a glimpse into the other aspects of the

Table 1.1.
Examples of two common fossils—dinosaur bone and silicified wood—and what is preserved at the different levels of organismal integration, from morphology to molecules

	Dinosaur	Tree
Organism (Morphology/Anatomy)	Bone	Tree trunk
Tissue (Histology)	Bone tissue	Wood
Cell (Cytology)	Osteocytes	Wood cells
Molecule (Biochemistry)	Bioapatite, collagen, lipids, other structural protein compounds (osteopontin, osteocalcin, etc.)	Lignin, cellulose, hemicellulose, proteins, resins, secondary plant compounds, chromophores

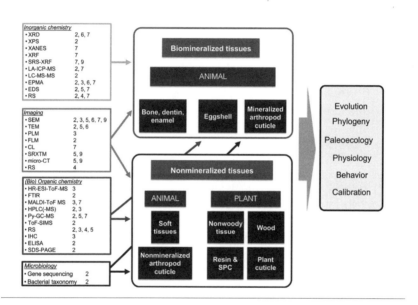

Figure 1.1. Analytical methods and approaches to fossilization research in regard to fossil tissues. Numbers in boxes with analytical methods refer to the chapters in this book. Abbreviations: CL = cathodoluminescence microscopy and spectroscopy; EDS = energy-dispersive spectroscopy; ELISA = enzyme-linked immunosorbent assay; EPMA = electron probe microanalysis; FLM = fluorescence light microscopy; FTIR = Fourier transformed infrared spectroscopy; HPLC-MS = high-performance liquid chromatography–mass spectrometry; HR-ESI-ToF-MS = high-resolution electrospray ionization time-of-flight mass spectrometry; IHC = immunohistochemistry; LA-ICP-MS = laser ablation inductively coupled plasma mass spectrometry; LC-MS-MS = reversed-phase microcapillary liquid chromatography tandem mass spectrometry; MALDI-ToF MS = matrix-assisted laser desorption/ionization–time-of-flight mass spectrometry; micro-CT = high-resolution X-ray microcomputed tomography; PLM = polarized light microscopy; Py-GC-MS = pyrolysis–gas chromatography–mass spectrometry; RS= Raman spectroscopy; SDS-PAGE = sodium dodecyl sulfate–polyacrylamide gel electrophoresis; SEM = scanning electron microscopy; SPC = secondary plant compounds; SRS-XRF = synchrotron radiation scanning X-ray fluorescence; SRXTM = synchrotron radiation X-ray tomographic microscopy; TEM = transmission electron microscopy; ToF-SIMS = time-of-flight secondary ion mass spectrometry; XANES = X-ray absorption near-edge structure spectroscopy; XPS = X-ray photoelectron spectroscopy; XRD = X-ray diffraction; XRF = X-ray fluorescence.

biology of ancient organisms. Hence, the material nature of fossils has remained poorly understood (Schweitzer 2011; Cunningham et al. 2014). We feel that, for much too long, paleontology has neglected potential information at histological, cytological, geochemical, and biomolecular levels (table 1.1). This is about to change as a result of the rapid development of diverse analytical technologies (fig. 1.1) that were not originally designed for studying the fossil record greatly expanding the horizons of paleontological research and redefining the fossil record (Briggs and Summons 2014; Cunningham et al. 2014).

In this book, which arose from a research unit funded by the German Research Foundation called "The Limits of the Fossil Record: Analytical and Experimental Approaches to Fossilization," our interest is on substance preservation in fossils as well as the taphonomic, microbiological, mineralogical, and geochemical pathways involved in the preservation processes (Gee, Preface). By substance preservation in fossils, we mean the materials found in a fossil today. This would include original biological substances, such as those in the blood cells of dinosaurs or in the lung tissue of insects embedded in amber. However, this would also include the mineral or organic compounds that make up the fossil now, such as those occurring in silicified wood or in the amber itself. The study of fossilization also embraces how these materials have changed from those of the once-living organism. In this book, we focus on the remains of whole organisms and their fossilized tissues, not on isolated organic molecules (biomarkers).

Why Study Fossilization?

A deeper understanding of the substance preservation of fossils offers innumerable benefits for reconstructing the history of life on earth (Schweitzer 2011). Quantitative analysis of large-scale patterns in the fossil record requires an understanding of any bias in the record that is associated with the transition of organic matter and biominerals from the living environment (the biosphere) to the geological context (the lithosphere). The same understanding is needed for paleoecological studies at the community level. Understanding the biology of ancient organisms requires equal understanding of the nature of their fossils. In the case of fossil bone, a multitude of questions arises from the observation that the process of mineralization results in the replacement of the organic material (collagen) by a mineral without obliterating the microstructure of the bone at the tissue level (table 1.1). The problem of bone fossilization has become even more intriguing with the discovery that organic compounds other than collagen and tissues may be preserved over hundreds of millions of years, such as the remains of bone cells and blood vessels (Schweitzer et al. 2009, 2013, 2014, 2016; Cleland et al. 2015; Geisler and Menneken, chap. 4; Wiersma et al., chap. 2).

The relevance of our approach, as laid out in this book, and the need for a coordinated multidisciplinary research effort are highlighted by recent debates on the color of dinosaur feathers (Vinther et al. 2008, 2010; Moyer et al. 2014; Vinther 2015; Dance 2016; McNamara et al. 2016a, b; Pinheiro et al. 2019) and on the preservation of chitin (e.g., Cody et al. 2011; Ehrlich et al. 2013; Ehrlich 2014), which is the primary component of arthropod cuticles. Fossil wood, the most abundant plant tissue by volume in the geological record, can be represented by coal, in which case little anatomical information remains, or by silicified wood (see Gee and Liesegang, chap. 6; Liesegang et al., chap. 7), which offers varying degrees of structural preservation (table 1.1). However, little is known about wood silicification pathways in autochthonous trees or allochthonous wood (Hellawell et al. 2015; Gee and Liesegang, chap. 6).

The ultimate goal of fossilization research is to answer questions of evolutionary biology. Reconstruction of skin pigments, eggshell color, and secondary plant compounds is relevant for existing paleobiological hypotheses, for example, about color vision, signaling (Koschowitz et al. 2014; Dance 2016; Roy et al. 2019), and reproductive biology in dinosaurs (Wiemann et al. 2017, 2018; Yang and Canoville, chap. 3). Detection of secondary plant compounds in fossils could provide information about the origin and coevolution of specific plant–animal interactions (McCoy et al., chap. 8). The detection of structural proteins such as keratin, chitin, and collagen may be useful in molecular phylogeny reconstruction, as well as to address questions of physiology and taphonomy (Service 2017; Wiemann et al. 2018).

Why Study Fossilization Now?

Despite the long interest in the subject of fossilization and the rich tradition from classical (Abel 1911; Weigelt 1927; Voigt 1935, 1937) to more recent studies (e.g., Pawlicki et al. 1966; Pawlicki 1975; Wuttke 1983; Wuttke and Reisdorf 2012), a comprehensive understanding of the substance preservation of fossils was far beyond the reach of paleontological investigations. This has changed only recently, underscoring the topicality of the fossilization research program. Topicality thus derives from the interplay of (1) new conceptual frameworks (fig. 1.2), (2) new paleontological questions, (3) new

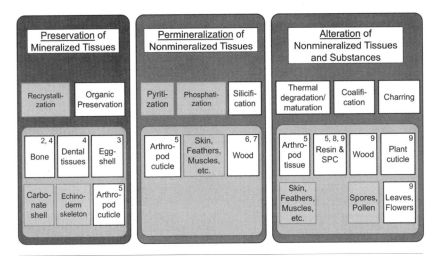

Figure 1.2. Processes of fossilization and their relationship to biological tissues and substances. The white boxes outlined in black indicate topics covered or touched on by the chapters in this book. Numbers in tissue and substance boxes correspond to book chapters. Abbreviation: SPC = secondary plant compounds.

sophisticated, high-resolution analytical tools (fig. 1.1), (4) the discovery of evermore striking fossils, and (5) the discovery of exceptional fossil deposits. Driven by these five factors, recent years have seen remarkable progress in understanding the biology of fossil organisms, such as the evolution of the dinosaur integument and its color (Schweitzer 2011; Roy et al. 2019). Furthermore, this research has raised many new questions and brought old ones into sharper focus.

The new field of "molecular paleontology" is a case in point, although this is often misunderstood to be synonymous with ancient DNA research. Molecular paleontology comprises the study of fossils at the molecular and atomic levels, particularly the organic compounds that are at the heart of most biominerals, such as bone proteins, arthropod cuticle chitin, and eggshell pigments (fig. 1.2). These compounds clearly have a preservation potential over deep time that is measured in hundreds of millions of years (e.g., Ehrlich et al. 2013; Schweitzer et al. 2016; Wiemann et al. 2017, 2018; McCoy et al. 2019; Pinheiro et al. 2019; Rogers et al. 2019), not just the last 700,000 years (Demarchi et al. 2016) that are currently covered by ancient DNA research.

The preservation of organic molecules, on one hand, depends on associated cells and tissues, but on the other hand, also on the physicochemical parameters the organism has been exposed to after death (e.g., temperature;

Demarchi et al. 2016), which commonly involve microbial activity (Briggs and McMahon 2016; Iniesto et al. 2016; McMahon et al. 2016; Barthel et al., chap. 5). The phenomenon of exceptional fossil preservation, especially the preservation of soft tissues, is one of the most controversial questions in modern paleontology.

Molecular paleontology also emerges as a powerful research field because of the advances in analytical chemistry with ever higher spatial resolution and more sophisticated analytical approaches that open up new avenues to our understanding of the fossil record (Demarchi et al. 2016; Schweitzer et al. 2016; Wiemann et al. 2017, 2018; Barbi et al. 2019; McCoy et al. 2019). This is highlighted, for instance, by Ehrlich et al. (2013), who were able to detect chitin in a 505-million-year-old demosponge fossil from the Cambrian Burgess Shale by the identification of its hydrolysate monomer, *D*-glucosamine, using a wide range of sophisticated analytical techniques. Biological methods, such as immunohistochemistry, are also a promising avenue of research, as shown by the detection of proteins in Mesozoic bone using antibodies (Schweitzer et al., 2013, 2016; Cleland et al., 2015; Boatman et al. 2019).

Similarly, careful extraction and biogeochemical identification of the pigment in a strikingly pink Jurassic calcareous red alga, *Solenopora jurassica*, suggests that the rosy color does not come directly from the fossil alga but likely from an ancient bacterium in the fossil plant (Wolkenstein et al. 2015). The pink color derives from borolithochromes, unusual boron-containing pigments, that are linked to a living *Clostridium* bacterium (see Gee and McCoy, chap. 9).

For these reasons, the study of fossilization benefits from a tightly integrated approach that involves the study of the fossil record with state-of-the-art analytical tools (fig. 1.1) that can deliver information on the molecular scale on fossil composition and structure, as well as on the timing of mineralization events (e.g., Geisler and Menneken, chap. 4; Gee and Liesegang, chap. 6; Liesegang et al., chap. 7). In addition, fossilization and decomposition experiments are increasingly being carried out in parallel under controlled physicochemical conditions (McCoy et al. 2019; Gee and Liesegang, chap. 6). The experimental results are providing the basis for understanding the data obtained from the fossil record. Experimental studies are also required to answer questions about the behavior of trace elements and stable isotopes during fossilization, which are used as proxies for diet, provenance, taphonomy, and paleoenvironment (e.g., Tütken et al., 2004; Koch, 2007; Herwartz et al., 2011; Tütken et al. 2011; Liesegang et al., chap. 7) and, in the case of uranium, for dating fossils (Balter et al., 2008; Fassett et al. 2011). In this book, we explore the limits of preservation, especially at the histological, cytological, and molecular levels. All chapters contain a review of the state-of-the-art, while two of them (chapters 2 and 4) also offer new data from original research. Fossils from conservation deposits such as the Hunsrück Slate in western Germany play an important role in this research program (Geisler and Menneken, chap. 4), but fossils from other deposits are also of compelling interest if they offer suborganismal preservation potential, such as insects from global amber deposits (Barthel et al., chap. 5) or most fossil bone (Wiersma et al., chap. 2).

Processes of Fossilization

Research on the topic of fossilization could be organized by taxonomic group, analytical methods, lagerstätten type, or kind of fossilization process, and each of these organization schemes has its own heuristic value. However, we prefer a framework based on biological tissue types (including their component cells and chemical substances) and their specific fossilization processes (fig. 1.2). The commonly recognized processes of fossilization—*silicification* (e.g., of wood), *pyritization* (e.g., of arthropods), *phosphatization* (e.g., of arthropods and vertebrate soft tissues), *permineralization* (e.g., of bone), and *alteration of organics* (e.g., amber, vertebrate integument, coalification of plant material)—are key processes that need to be studied analytically and experimentally. Our rationale is that the chemistry of a specific living tissue and the postmortem environment are commonly linked to a fossilization process. For example, wood generally silicifies or coalifies but does not commonly phosphatize, whereas bone does not pyritize or coalify, but rather it fossilizes by a more complex process of replacement and mineralization. Also, since certain tissue types represent key evolutionary innovations of certain organisms (e.g., bone in vertebrates), the other potential organizing principles such as taxonomic group and phylogenetic heritage will automatically be reflected in this organizing principle. We recognize three process clusters (fig. 1.2), the understanding of most of which requires a combination of analytical and experimental approaches.

Preservation of Mineralized Tissue

Bone, in particular, represents the greatest challenge to the understanding of preservation of mineralized tissue because of its high content of organics (ca. 30% vs. 70% bone apatite) in the composite bone material, the incorporation of neurovascular bundles and living cells into the tissue (Schweitzer

2011; Schweitzer et al. 2014; Cleland et al. 2015; Wiersma et al., chap. 2), and the unique processes of biological resorption and redeposition in bone (Padian and Lamm 2013).

The more highly mineralized vertebrate hard tissues of dentin, enamel, and eggshell also present questions about the preservation of the organic template and associated organics, as well as about the changes in microstructure and chemistry often associated with complex mineral replacement reactions, some of which are microbially mediated (e.g., Iniesto et al. 2016). The (bio)mineral replacement is characterized by the retention of fine morphological features, which is commonly explained as occurring by diffusion-controlled processes. However, recent experimental studies have revealed that many replacement reactions are better explained by a process that involves the congruent dissolution of the parent (bio)mineral, coupled in space and time to the precipitation of a new phase or phases (e.g., Geisler et al. 2005; Kasioptas et al. 2010).

In general, bone mineralization and replacement reactions during fossilization are only understood in a descriptive sense at the tissue level but not at the molecular scale, where the reactions are also intimately linked to the organic content (Schweitzer et al. 2007, 2014; Cleland et al. 2015). To further complicate matters, the full complexity of extant bone at the nanostructural level remains to be understood (Schwarcz et al. 2017; Reznikov et al. 2018). The lack of understanding of the importance of the organic phase in fossilization also applies to well-preserved arthropod cuticles and mollusk shells, particularly in the preservation of nacre, also known as mother-of-pearl (Olson et al. 2012). A particularly interesting case is represented by organic compound preservation in a crystalline matrix such as in organic purple quinone pigments in Jurassic crinoids (Wolkenstein et al. 2006; Wolkenstein 2015) and in the organic phase of fossil ostrich shell (Demarchi et al. 2016). This area of research is also the focus of the chapters by Wiersma et al. (chap. 2) and Yang and Canoville (chap. 3).

Permineralization of Nonmineralized Tissues

What traditionally has been called "soft tissue preservation" (Schweitzer 2011) can, in fact, be divided into two processes: (1) diagenetic mineralization of tissues, and (2) chemical alteration of tissues without mineralization. Mineralization is a common mode of wood fossilization (for instance, through silicification; e.g., Ballhaus et al. 2012; Läbe et al. 2012; Gee and Liesegang, chap. 6; Liesegang et al., chap. 7), but it is rarer in vertebrate integument and muscles (through phosphatization; e.g., Schweitzer 2011) and in arthropod bodies (through phosphatization, silicification, and pyritization; e.g., Brock et al. 2006; Laflamme et al. 2014). All diagenetic mineralization processes of nonmineralized animal tissues have an extremely rapid onset after the death of the organism, from only hours to days, with the process possibly completed within a nongeological time frame (Briggs and Kear 1993; Briggs 2003; Gee and Liesegang, chap. 6; Liesegang et al., chap. 7). Some permineralization processes have a crucial microbial component that remains poorly characterized (Briggs and McMahon 2016; Iniesto et al. 2016).

Alteration of Nonmineralized Tissues and Organic Substances

The process of preservation by alteration is commonplace and reasonably well understood in plant fossils. It consists of a gradual increase in compounds rich in carbon (cyclic aromatic compounds) and degradation or loss of other components (Taylor et al. 2009). Microbial activity plays only a limited role in the process of plant fossilization by alteration. The alteration of nonmineralized animal tissues, however, has received far less attention. It only takes place under exceptional circumstances (Schweitzer 2011; Briggs and Summons 2014) and forms the basis of the concept of "conservation deposit" that was first proposed by Seilacher (Briggs and Summons 2014). In fact, alteration as a mode of soft tissue preservation in animals is sometimes questioned altogether, and any preserved soft tissue is attributed to the process of templating, for example, by bacteria.

Increasing evidence from refined chemical analyses of fossils from new lagerstätten suggests that alteration is a process that needs to be investigated because it has great information potential (Schweitzer 2011). Preservation of chloroplasts (e.g., Schoenhut et al. 2004; Yang et al. 2005) and green color in Eocene leaves (Gee and McCoy, chap. 9) are other examples of remarkable, but poorly understood, preservation by alteration. Alteration may even include other cell contents such as secondary plant metabolites (McCoy et al., chap. 8). The classical altered plant compound actually is amber, a material that has been extensively researched. However, the fossil inclusions in amber, some of which preserve altered nonmineralized tissue (e.g., Rust et al., 2010; Barthel et al., chap. 5) or even seemingly unaltered nonmineralized tissue (e.g., resin bodies in leaves and flowers; Taylor 1988), have received insufficient attention from the perspective of fossilization. Microbiological processes are extremely relevant to understanding these processes of alteration (see discussion in Vinther et al. 2008; Colleary et al. 2015; Lindgren et al. 2015; Barthel et al., chap. 5).

Perspectives

This brief synopsis and review, together with the other chapters of this book, offer some perspectives on future research on fossilization. First, the focus has to be on understanding simpler processes, such as silicification using high-resolution analytical and advanced experimental approaches, followed by a shift to more complex types of fossilization and a search for common principles in the mineralization and preservation of organic compounds. Furthermore, standard research protocols across the disciplines and groups of organisms are needed to provide a deeper understanding of the origin and nature of exceptional fossil deposits and to develop a synthetic view of fossilization (see also McCoy, chap. 10). This agenda must go beyond classical taphonomy and aim at discovering basic principles that underlie the preservation of organisms and their parts. This goal cannot be reached without the close collaboration of paleontologists, microbiologists, organic chemists, mineralogists, geochemists, pharmacists, and petrologists, as represented by the integrated group of contributors to this book. The results of this research program will have profound applications in the fields of evolutionary biology, phylogeny, calibration of molecular clocks, paleoecology, physiology, and organismic behavior.

ACKNOWLEDGMENTS

This introductory chapter benefited greatly from the discussions and flow of ideas among members of Deutsche Forschungsgemeinschaft (DFG, German Research Foundation) Research Unit 2685, beginning with its preparatory phase and continuing on to today. This work is supported by the DFG, Project number 396710782 to P. Martin Sander and Project number 39676817 to Carole T. Gee (both University of Bonn). This is contribution number 15 of the DFG Research Unit 2685, "The Limits of the Fossil Record: Analytical and Experimental Approaches to Fossilization."

WORKS CITED

Abel, O. 1911. *Grundzüge der Paläobiologie der Wirbeltiere.* E. Schweizerbart'sche Verlagsbuchhandlung (Erwin Nägele), Stuttgart.
Ballhaus, C., Gee, C.T., Bockrath, C., Greef, K., Mansfeldt, T., and Rhede, D. 2012. The silicification of trees in volcanic ash—An experimental study. *Geochimica et Cosmochimica Acta*, 84: 62–74.
Balter, V., Blichert-Toft, J., Braga, J., Telouk, P., Thackeray, F., and Albarède, F. 2008. U–Pb dating of fossil enamel from the Swartkrans Pleistocene hominid site, South Africa. *Earth and Planetary Science Letters*, 267: 236–246.

Barbi, M., Bell, P.R., Fanti, F., Dynes, J.J., Kolaceke, A., Buttigieg, J., Coulson, I.M., and Currie, P.J. 2019. Integumentary structure and composition in an exceptionally well-preserved hadrosaur (Dinosauria: Ornithischia). *PeerJ*, 7: e7875.

Boatman, E.M., Goodwin, M.B., Holman, H.-Y.N., Fakra, S., Zheng, W., Gronsky, R., and Schweitzer, M.H. 2019. Mechanisms of soft tissue and protein preservation in *Tyrannosaurus rex*. *Scientific Reports*, 9: 15678.

Briggs, D.E.G. 2003. The role of decay and mineralization in the preservation of soft-bodied fossils. *Annual Review of Earth and Planetary Sciences*, 31: 275–301.

Briggs, D.E.G., and Kear, A.J. 1993. Fossilization of soft tissue in the laboratory. *Science*, 259: 1439–1442.

Briggs, D.E.G., and McMahon, S. 2016. The role of experiments in investigating the taphonomy of exceptional preservation. *Palaeontology*, 59: 1–11.

Briggs, D.E.G., and Summons, R.E. 2014. Ancient biomolecules: Their origins, fossilization, and role in revealing the history of life. *BioEssays*, 36: 482–490.

Brock, F., Parkes, R.J., and Briggs, D.E.G. 2006. Experimental pyrite formation associated with decay of plant material. *Palaios*, 21: 499–506.

Cleland, T.P., Schroeter, E.R., Zamdborg, L., Zheng, W., Lee, J.E., Tran, J.C., Bern, M., Duncan, M.B., Lebleu, V.S., Ahlf, D.R., Thomas, P.M., Kalluri, R., Kelleher, N.L., and Schweitzer, M.H. 2015. Mass spectrometry and antibody-based characterization of blood vessels from *Brachylophosaurus canadensis*. *Journal of Proteome Research*, 14: 5252–5262.

Cody, G.D., Gupta, N.S., Briggs, D.E.G., Kilcoyne, A.L.D., Summons, R.E., Kenig, F., Plotnick, R.E., and Scott, A.C. 2011. Molecular signature of chitin-protein complex in Paleozoic arthropods. *Geology*, 39: 255–258.

Colleary, C., Dolocan, A., Gardner, J., Singh, S., Wuttke, M., Rabenstein, R., Habersetzer, J., Schaal, S., Feseha, M., Clemens, M., Jacobs, B.F., Currano, E.D., Jacobs, L.L., Sylvestersen, R.L., Gabbott, S.E., and Vinther, J. 2015. Chemical, experimental, and morphological evidence for diagenetically altered melanin in exceptionally preserved fossils. *Proceedings of the National Academy of Sciences of the United States of America*, 112: 12592–12597.

Cunningham, J.A., Donoghue, P.C.J., and Bengtson, S. 2014. Distinguishing biology from geology in soft-tissue preservation, pp. 275–288. In: Laflamme, M., Schiffbauer, J.D., and Darroch, S.A.F., eds., *Reading and Writing of the Fossil Record: Preservational Pathways to Exceptional Fossilization. The Paleontological Society Papers*, vol. 20. The Paleontological Society Short Course, October 18, 2014.

Dance, A. 2016. New feature: Prehistoric animals, in living color. *Proceedings of the National Academy of Sciences of the United States of America*, 113: 8552–8556.

Demarchi, B., Hall, S., Roncal-Herrero, T., Freeman, C.L., Woolley, J., Crisp, M.K., Wilson, J., Fotakis, A., Fischer, R., Kessler, B.M., Rakownikow Jersie-Christensen, R., Olsen, J.V., Haile, J., Thomas, J., Marean, C.W., Parkington, J., Presslee, S., Lee-Thorp, J., Ditchfield, P., Hamilton, J.F., Ward, M.W., Wang, C.M., Shaw, M.D., Harrison, T., Domínguez-Rodrigo, M., MacPhee, R.D.E., Kwekason, A., Ecker, M., Kolska Horwitz, L., Chazan, M., Kröger, R., Thomas-Oates, J., Harding, J.H., Cappellini, E., Penkman, K., and Collins, M.J. 2016. Protein sequences bound to mineral surfaces persist into deep time. *eLife* 5: e17092.

Ehrlich, H. 2014. Identification of chitin in 200-million-year-old gastropod egg capsules. *Paleobiology*, 40: 529–540.

Ehrlich, H., Rigby, J.K., Botting, J.P., Tsurkan, M., Werner, C., Schwille, P., Petrášek, Z., Pisera, A., Simon, P., Sivkov, V.N., Vyalikh, D.V., Molodtsov, S.L., Kurek, D., Kammer, M., Hunoldt, S., Born, R., Stawski, D., Steinhof, A., Bazhenov, V.V., and Geisler, T. 2013. Discovery of 505-million-years old chitin in the basal demosponge *Vauxia gracilenta. Scientific Reports*, 3: 3497.

Fassett, J.E., Heaman, L.M., and Simonetti, A. 2011. Direct U–Pb dating of Cretaceous and Paleocene dinosaur bones, San Juan Basin, New Mexico. *Geology*, 39: 159–162.

Geisler, T., Pöml, P., Stephan, T., Janssen, A., and Putnis, A. 2005. Experimental observation of an interface-controlled pseudomorphic replacement reaction in a natural crystalline pyrochlore. *American Mineralogist* 90: 1683–1687.

Hellawell, J., Ballhaus, C., Gee, C.T., Mustoe, G.E., Nagel, T.J., Wirth, R., Rethemeyer, J., Tomaschek, F., Geisler, T., Greef, K., and Mansfeldt, T. 2015. Incipient silicification of recent conifer wood at a Yellowstone hot spring. *Geochimica et Cosmochimica Acta*, 149: 79–87.

Herwartz, D., Tütken, T., Münker, C., Jochum, K.P., Stoll, B., and Sander, P.M. 2011. Timescales and mechanisms of REE and Hf uptake in fossil bones. *Geochimica et Cosmochimica Acta*, 75: 82–105.

Iniesto, M., Buscalioni, A.D., Guerrero, M.C., Benzerara, K., Moreira, D., and López-Archilla, A.I. 2016. Involvement of microbial mats in early fossilization by decay delay and formation of impressions and replicas of vertebrates and invertebrates. *Scientific Reports*, 6: 25716.

Kasioptas, A., Geisler, T., Perdikour, C., Putnis, C.V., and Putnis, A. 2010. Crystal growth of apatite by replacement of an aragonite precursor. *Journal of Crystal Growth*, 312: 2431–2440.

Koch, P.L. 2007. Isotopic study of the biology of modern and fossil vertebrates, pp. 99–154. In: Minchener, R., and Lajtha, K., eds., *Stable Isotopes in Ecology and Environmental Science*, 2nd ed. Oxford University Press, Oxford.

Koschowitz, M.-C., Fischer, C., and Sander, P.M. 2014. Beyond the rainbow. Dinosaur color vision may have been key to the evolution of bird feathers. *Science*, 346: 1416–1418.

Läbe, S., Gee, C.T., Ballhaus, C., and Nagel, T. 2012. Experimental silicification of the tree fern *Dicksonia antarctica* at high temperature with silica-enriched H_2O vapor. *Palaios*, 27: 835–841.

Laflamme, M., Schiffbauer, J.D., and Darroch, S.A.F., eds. 2014. Reading and writing of the fossil record: Preservational pathways to exceptional fossilization. Paleontological Society Short Course, October 18, 2014.

Lindgren, J., Moyer, A., Schweitzer, M.H., Sjövall, P., Uvdal, P., Nilsson, D.E., Heimdal, J., Engdahl, A., Gren, J.A., Schultz, B.P., and Kear, B.P. 2015. Interpreting melanin-based coloration through deep time: A critical review. *Proceedings of the Royal Society B*, 282: 20150614.

McCoy, V.E., Gabbott, S.E., Penkman, K., Collins, M.J., Presslee, S., Holt, J., Grossman, H., Wang, B., Solórzano Kraemer, M.M., Delclòs, X., and Peñalver, E. 2019. Ancient amino acids from fossil feathers in amber. *Scientific Reports*, 9: 6420.

McMahon, S., Anderson, R.P., Saupe, E.E., and Briggs, D.E.G. 2016. Experimental evidence that clay inhibits bacterial decomposers: Implications for preservation of organic fossils. *Geology*, 44: 867–870.

McNamara, M.E., Orr, P.J., Kearns, S.L., Alcalá, L., Anadón, P., and Peñalver, E. 2016a. Reconstructing carotenoid-based and structural coloration in fossil skin. *Current Biology*, 26: 1075–1082.

McNamara, M.E., van Dongen, B.E., Lockyer, N.P., Bull, I.D., and Orr, P.J. 2016b. Fossilization of melanosomes via sulfurization. *Palaeontology*, 59: 337–350.

Moyer, A.E., Zheng, W., Johnson, E.A., Lamanna, M.C., Li, D.-Q., Lacovara, K.J., and Schweitzer, M.H. 2014. Melanosomes or microbes: Testing an alternative hypothesis for the origin of microbodies in fossil feathers. *Scientific Reports*, 4: 4233.

Olson, I.C., Kozdon, R., Valley, J.W., and Gilbert, P.U.P.A. 2012. Mollusk shell nacre ultrastructure correlates with environmental temperature and pressure. *Journal of the American Chemical Society*, 134: 7351–7358.

Padian, K., and Lamm, E.-T., eds. 2013. *Bone Histology of Fossil Tetrapods. Advancing Methods, Analysis, and Interpretation*. University of California Press, Berkeley.

Pawlicki, R. 1975. Studies of the fossil dinosaur bone in the scanning electron microscope. *Zeitschrift für mikroskopisch-anatomische Forschung*, 2: 393–398.

Pawlicki, R., Korbel, A., and Kubiak, H. 1966. Cells, collagen fibrils and vessels in dinosaur bone. *Nature Communications*, 211: 655–657.

Pinheiro, F.L., Prado, G., Ito, S., Simon, J.D., Wakamatsu, K., Anelli, L.E., Andrade, J.A.F., and Glass, K. 2019. Chemical characterization of pterosaur melanin challenges color inferences in extinct animals. *Scientific Reports*, 9: 1–8.

Reznikov, N., Bilton, M., Lari, L., Stevens, M.M., and Kröger, R. 2018. Fractal-like hierarchical organization of bone begins at the nanoscale. *Science*, 360: eaao2189.

Rogers, C., Astrop, T., Webb, S., Ito, S., Wakamatsu, K., and McNamara, M. 2019. Synchrotron X-ray absorption spectroscopy of melanosomes in vertebrates and cephalopods: Implications for the affinity of *Tullimonstrum*. *Proceedings of the Royal Society B*, 286: 20191649.

Roy, A., Pittman, M., Saitta, E.T., Kaye, T.G., and Xu, X. 2019. Recent advances in amniote palaeocolour reconstruction and a framework for future research. *Biological Reviews*, 95: 22–50.

Rust, J., Singh, H., Rana, R.S., McCann, T., Singh, L., Anderson, K., Sarkar, N., Nascimbene, P.C., Stebner, F., Thomas, J.C., Solórzano Kraemer, M., Williams, C.J., Engel, M.S., Sahni, A., and Grimaldi, D. 2010. Biogeographic and evolutionary implications of a diverse paleobiota in amber from the early Eocene of India. *Proceedings of the National Academy of Sciences of the United States of America*, 107: 18360.

Schoenhut, K., Vann, D.R., and LePage, B.A. 2004. Cytological and ultrastructural preservation in Eocene *Metasequoia* leaves from the Canadian High Arctic. *American Journal of Botany*, 91: 816–824.

Schwarcz, H., Abueidda, D., and Jasiuk, I. 2017. The ultrastructure of bone and its relevance to mechanical properties. *Frontiers in Physics*, 5: 1–13.

Schweitzer, M.H. 2011. Soft tissue preservation in terrestrial Mesozoic vertebrates. *Annual Reviews in Earth and Planetary Science*, 39: 187–216.

Schweitzer, M.H., Wittmeyer, J.L., and Horner, J.R. 2007. Soft tissue and cellular preservation in vertebrate skeletal elements from the Cretaceous to the present. *Proceedings of the Royal Society B*, 274: 183–197.

Schweitzer, M.H., Zheng, W., Cleland, T.P., and Bern, M. 2013. Molecular analyses of dinosaur osteocytes support the presence of endogenous molecules. *Bone*, 52: 414–423.

Schweitzer, M.H., Zheng, W., Cleland, T.P., Goodwin, M.B., Boatmand, E., Theil, E., Marcus, M.A., and Fakra, S.C. 2014. A role for iron and oxygen chemistry in preserving soft tissues, cells and molecules from deep time. *Proceedings of the Royal Society B*, 281: 20132741.

Schweitzer, M.H., Zheng, W., Organ, C., Avci, R., Suo, Z., Freimark, L.M., Lebleu, V.S., Duncan, M.B., Heiden, M.G.V., Neveu, J.M., Lane, W.S., Cottrell, J.S., Horner, J.R., Cantley, L.C., Kalluri, R., and Asara, J.M. 2009. Biomolecular characterization and protein sequences of the Campanian hadrosaur *B. canadensis*. *Science*, 324: 626–631.

Schweitzer, M.H., Zheng, W., Zanno, L., Werning, S., and Sugiyama, T. 2016. Chemistry supports the identification of gender-specific reproductive tissue in *Tyrannosaurus rex*. *Scientific Reports*, 6: 23099.

Service, R.F. 2017. Researchers close in on ancient dinosaur proteins. "Milestone" paper opens door to molecular approach. *Science*, 355: 441–442.

Taylor, D.W. 1988. Eocene floral evidence of Lauraceae: Corroboration of the North American megafossil record. *American Journal of Botany*, 75: 948–957.

Taylor, T.N., Taylor, E.L., and Krings, M. 2009. *Paleobotany: The Biology and Evolution of Fossil Plants*, 2nd ed. Academic Press, San Diego.

Tütken, T., Pfretzschner, H.-U., Vennemann, T.W., Sun, G., and Wang, Y.D. 2004. Paleobiology and skeletochronology of Jurassic dinosaurs: Implications from the histology and oxygen isotope compositions of bones. *Palaeogeography, Palaeoclimatology, Palaeoecology* 206: 217–238.

Tütken, T., Vennemann, T.W., and Pfretzschner, H.-U. 2011. Nd and Sr isotope compositions in modern and fossil bones—Proxies for vertebrate provenance and taphonomy. *Geochimica and Cosmochimica Acta*, 75: 5951–5970.

Vinther, J. 2015. A guide to the field of palaeo colour: Melanin and other pigments can fossilise: Reconstructing colour patterns from ancient organisms can give new insights to ecology and behaviour. *BioEssays*, 37: 643–656.

Vinther, J., Briggs, D.E.G., Clarke, J., Mayr, G., and Prum, R.O. 2010. Structural coloration in a fossil feather. *Biology Letters*, 6: 128–131.

Vinther, J., Briggs, D.E.G., Prum, R.O., and Saranathan, V. 2008. The colour of fossil feathers. *Biology Letters*, 4: 522–525.

Voigt, E. 1935. Die Erhaltung von Epithelzellen mit Zellkernen, von Chromatophoren und Corium in fossiler Froschhaut aus der mitteleozänen Braunkohle des Geiseltales. *Nova Acta Leopoldina Neue Folge*, 3(14).

Voigt, E. 1937. Weichteile aus Fischen, Amphibien und Reptilien aus der eozänen Braunkohle des Geiseltales. *Nova Acta Leopoldina, Neue Folge*, 5: 116–142.

Weigelt, J. 1927. *Rezente Wirbeltierleichen und ihre paläobiologische Bedeutung*. Verlag von Max Weg, Leipzig.

Wiemann, J., Fabbri, M., Yang, T.-R., Stein, K., Sander, P.M., Norell, M.A., and Briggs, D.E.G. 2018. Fossilization transforms vertebrate hard tissue proteins into *N*-heterocyclic polymers. *Nature Communications*, 9: 4741.

Wolkenstein, K. 2015. Persistent and widespread occurrence of bioactive quinone pigments during post-Paleozoic crinoid diversification. *Proceedings of the National Academy of Sciences of the United States of America*, 112: 2794–2799.

Wolkenstein, K., Gross, J.H., Falk, H., and Schöler, H.F. 2006. Preservation of hypericin and related polycyclic quinone pigments in fossil crinoids. *Proceedings of the Royal Society B*, 273: 451–456.

Wolkenstein, K., Sun, H., Falk, H., and Griesinger, C. 2015. Structure and absolute configuration of Jurassic polyketide-derived spiroborate pigments obtained from microgram quantities. *Journal of the American Chemical Society*, 137: 13460–13463.

Wuttke, M. 1983. Aktuopaläontologische Studien über den Zerfall von Wirbeltieren. Teil 1: Anura. *Senckenbergiana lethaea*, 64: 529–560.

Wuttke, M., and Reisdorf, A.G., eds. 2012. Taphonomic processes in terrestrial and marine environments. *Palaeobiodiversity and Palaeoenvironments, Special Volume*, 92: 1–168.

Yang, H., Huang, Y., Leng, Q., LePage, B.A., and Williams, C.J. 2005. Biomolecular preservation of Tertiary *Metasequoia* fossil *Lagerstätten* revealed by comparative pyrolysis analysis. *Review of Palaeobotany and Palynology*, 134: 237–256.

CHAPTER **2**

Organic Phase Preservation in Fossil Dinosaur and Other Tetrapod Bone from Deep Time

Extending the Probable Osteocyte Record to the Early Permian

KAYLEIGH WIERSMA, SASHIMA LÄBE, AND P. MARTIN SANDER

A B S T R A C T | Since the 1960s, soft tissues and bone proteins have been detected in fossil dinosaur bone, eroding the dogma that the organic phase in bone is completely destroyed during fossilization. Indeed, fossil bone from the Mesozoic frequently appears to retain soft tissues such as cells, blood vessels, and extracellular matrix, which can be liberated by dissolution in weak organic acids. However, the tissue remnants—putative osteocytes, blood vessels, extracellular matrix—found in the residue have led to a controversy over whether these organic remains represent original soft tissue preservation, or if they are instead biofilms produced by bone-degrading bacteria, either during fossilization or in the shallow subsurface. Here we provide an overview of previous research on soft tissue preservation in fossil dinosaur bone to determine the biological origin of these chemically liberated tissues. Although the weight of the evidence tips in favor of the interpretation of the remains as original tissue or its degradation products, the intense discussion on this controversy has led to the realization that the material nature of pre-Quaternary fossil bone remains poorly characterized and that the mode of preservation of the organic phase is not well understood. Our own preliminary observations of osteocyte-like structures liberated from tetrapod bones from the early Permian of Texas extend the previous oldest record of such remains by 40 million years, from the Mesozoic into the Paleozoic. |

Introduction

Among vertebrates in the fossil record, dinosaurs are beyond doubt some of the most spectacular animals. Ever since the first dinosaur skeleton was mounted for exhibition at the Philadelphia Academy of Natural Sciences in 1868 (Prieto-Márquez et al. 2006), these extinct animals have fascinated the public. Exceptional dinosaur fossils with preserved soft tissue, such as skin and other parts of the integument, have been known since the late 1800s, including the London and Berlin specimens of *Archaeopteryx*. Because dinosaurs have been intensively studied scientifically since the early 19th century, scientists have a good understanding of their morphology and paleobiology today (e.g., Curry-Rogers and Wilson 2005; Klein et al. 2011; Brusatte 2012). This is not true, however, for the material nature of dinosaur and other fossil bone, which remains less well-known.

Bone as a living tissue is strikingly complex (Francillon-Vieillot et al. 1990) compared to other mineralized animal tissues, such as tooth enamel, eggshell, mollusk shell, and arthropod cuticle (Halstead 1974; Carter et al. 1990). Vertebrates owe much of their evolutionary success to the adaptability of this tissue in form and function (Halstead 1974; Hall 2015), specifically, its ability to change shape during growth (Francillon-Vieillot et al. 1990). This ability is linked to the fact that bone is a living tissue, containing living cells (osteocytes), vasculature, and nerve tissue set in a biomineral matrix, unlike all of the other tissues listed above, which are acellular biomineral secretions of an epithelium vasculature (Carter et al. 1990).

At the subcellular level, the bone matrix is a complex biocomposite consisting of an extracellular matrix (ECM) of collagen fibrils with embedded bioapatite crystallites (e.g., Dumont et al. 2011). Bone apatite crystallites are nanometer-sized and thus an order of magnitude smaller than tooth enamel crystallites, for example (Edmund 1960, 1969; Glimcher et al. 1990; Sander 1999, 2000). Vascular canals and cells are incorporated during growth into this matrix, the latter differentiating from bone-producing cells (osteoblasts) to bone cells (osteocytes), which reside in a cavity in the bone matrix called an osteocyte lacuna. This incorporation introduces a purely organic component into bone, namely, cell membranes and cell content, in addition to the ECM (Francillon-Vieillot et al. 1990; Hall 2015). On the molecular level, the organics in bone are thus the collagen and other proteins of the ECM and the lipids and proteins of the cell membranes, cell contents, and vessel walls. In addition, red blood cells (erythrocytes) in blood vessels contain other compounds such as heme, a protoporphyrin. A general, often cited proportion is that fresh bone is 30% organics and 70% mineral by weight (Francillon-Vieillot et al. 1990; Currey 2012; Padian and Lamm 2013). Pre-Quaternary fossil bone is generally cited as consisting of more than 99% mineral matter.

Although insight into the biology of dinosaurs is based on fossil remains, little is known about the processes that have turned bone, and in particular the organic phase, of the once-living animals into mineralized fossils. Our review addresses the question of what happens to all the histological, subcellular, and molecular components of bone, collectively called "the organic phase," during fossilization.

It has only been in recent years that the organic phase could be liberated in a systematic fashion from the fossil bone by means of digestion through weak acids, first shown in a seminal paper by Schweitzer et al. (2005). Other studies (Schweitzer et al. 2007b, 2009, 2014; Kaye et al. 2008; Cadena and Schweitzer 2012; Cleland et al. 2015; Surmik et al. 2016, 2019; Schroeter et al. 2017; Wiemann et al. 2018), which are reviewed in detail below, confirmed these results and extended the sample base. Models for the preservation of the organic phase, in which iron from heme triggers the crosslinking of the organic components, making them more stable, were also proposed (Schweitzer et al. 2007b and references therein; Schweitzer 2011; Schweitzer et al. 2014; Briggs and McMahon 2016).

However, doubts were raised soon after the first reports made by Schweitzer and colleagues as to the origin of the organic material. An alternative hypothesis regarding the source of the liberated organic phase suggests that the organics isolated from fossil bone represent biofilms produced by bone-degrading bacteria (Kaye et al. 2008; Saitta et al. 2019). This hypothesis posits that the microbes entered the bone tissue, degraded the original tissue in situ, and ultimately became fossilized within the bone. However, Kaye et al. (2008) did not explain how the fossil biofilm could have been preserved and what its material nature is. In addition, immunohistochemistry and other lines of evidence indicate that the original tissue hypothesis is more likely (Schweitzer et al. 2014, 2016; Surmik et al. 2016, 2019; Lee et al. 2017).

Briggs and McMahon (2016) show that microbial mats do play an important role in soft part preservation in general. Recently, the discussion has been opened again by Saitta et al. (2019), who provide strong evidence for the presence of a modern microbiome in freshly exposed bone fossils several meters below the original land surface. They note that this microbiome makes the preservation of original proteins unlikely (because the bacteria metabolize them) and that the soft tissues liberated from bone may represent recent biofilms. The following review of the evidence will address these hypotheses. At the outset, however, we note that osteocytes are too deeply isolated in the bone to be reached by bacteria. Osteocytes in their lacunae are only accessible through the canaliculi that contain their filopodia. Canaliculi have a diameter an order of magnitude smaller (on the nanometer scale) than bacteria, acting like a micropore filter to remove bacteria from liquids. The conflicting hypotheses on the origin of soft tissue preservation, together with the studies at the different hierarchical organizational levels of fossil bone, indicate that, in any case, fossil bone and its fossilization process are poorly characterized and insufficiently understood. Here we present a largely chronological review of the preservation of the organic phase within fossil bone of previously published work that was accrued primarily in the second decade of the new millennium.

Background on Bone Fossilization and Composition

Fossilization of Bone

Fossilization is the process in which the animal or plant "turns to stone." A simplistic model that is often used for teaching is that of an animal carcass that is rapidly buried and thus protected from scavenging, erosion, and transport. The organics decompose, leaving voids in the hard tissue through which pore waters flow, solubilizing minerals and depositing them in the voids. This model is inadequate, however, because there are more processes that go into the fossilization of a bone.

After the death of an organism, the first process taking place is decay. Degradation starts with autolysis, which begins almost immediately after death and releases intercellular enzymes from lysosomes that break down cellular compounds (Child 1995; Schweitzer 2011). Following autolysis, microbial invasion, either from the resident gut fauna or environmental sources, begins and generally leads to complete skeletonization within as little as two weeks (Cambra-Moo and Buscalioni 2008; Schweitzer 2011). This indicates that for fossilization to occur and to produce a fossil preserved over geological time, little or no microbial degradation must have taken place (Trueman and Martill 2002). The final stage of degradation includes much slower chemical (nonbiological) processes such as hydrolytic bond breakage, oxidation, photooxidation, and other processes that break down molecular structures (Schweitzer 2011). The mineral phase of bone shows a close association with the degradation and may offset the enzymatic and microbial degradation (Child 1995; Kharalkar et al. 2009 and references therein). This phase provides protection against most microbial collagenolytic enzymes that are too large to enter the pores in bone, and the protein coating of the mineral phase forms a barrier to the dissolution of the apatite crystallites (Trueman and Martill 2002; Turner-Walker 2008; Schweitzer 2011; Keenan and Engel 2017). Alternatively, enzymatic degradation may be reduced due to the smaller size and reactivity of bone apatite crystallites (Butterfield 1990, 2003; Trueman et al. 2008; Schweitzer 2011). A final hypothesis puts forth that the constraints of the mineral phase prevent molecular swelling and therefore prevent access to more reactive sites on molecules (Schweitzer 2011).

The next stage in fossilization is recrystallization, permineralization, and/ or authigenic mineralization. Due to sensitivity of bioapatite to pH, these structures are generally thermodynamically unstable (Berna et al. 2004; Keenan and Engel 2017). Recrystallization of hydroxyapatite (HAp) is required for preservation over geological time (Hubert et al. 1996; Nielsen-Marsh and Hedges 1999; Keenan and Engel 2017). During the fossilization of bone, most of the original bioapatite is replaced by fluorapatite (FAp; $Ca_5(PO_4)_3(F)$), in which the size of the exposed apatite crystals increases, reducing solubility and porosity (Dwivedi et al. 1997; Kohn and Cerling 2002; Trueman and Tuross 2002; Wopenka and Pasteris 2005). To preserve the bone, the recrystallization must occur rapidly, within only days to weeks, before the bone interacts with the geochemistry and microbiology of the depositional environment (Trueman and Tuross 2002; Keenan and Engel 2017). Recrystallization of apatite crystallites must hold pace with collagen hydrolysis, since exposed bone crystallites consisting of HAp are susceptible to dissolution (Collins et al. 1995; Bocherens et al. 1997; Trueman and Martill 2002). The changes in crystal size can be observed optically (Pfretzschner 2000, 2001, 2004; Dumont et al. 2011), and several mechanisms for the observed increase in mineral density have been proposed. One was already mentioned above and involves the dissolution of bioapatite and reprecipitation of FAp. Remarkable, however, is that the biogenic orientation of apatite crystallite c axes along collagen fibrils is not affected by fossilization (Padian and Lamm 2013). Hubert et al. (1996) proposed another mechanism, which suggests that FAp grows on biogenic seed crystals, filling in the pore spaces left by collagen, thus preserving the original HAp crystallite alignment (Keenan and Engel 2017).

Recrystallization happens mainly in the inorganic phase in bone, but the soft tissue morphology is mainly preserved via two major processes: permineralization and authigenic mineralization. Permineralization describes the infilling of voids (produced by the degradation of soft tissues) by exogenous minerals that have been solubilized in pore waters (Hedges 2002; Trueman and Tuross 2002; Schweitzer 2011). These voids are osteocyte lacunae, vascular canals that contained blood vessels, or possibly submicrometer spaces once occupied by collagen fibrils (Collins et al. 2002; Schweitzer 2011). Common authigenic mineral infillings are calcite, FAp (francolite), pyrite and marcasite, hematite, sphalerite, barite and witherite, clay minerals, quartz, and gypsum (Trueman and Tuross 2002; Wings 2004). Permineralization can be extremely rapid and is determined by mineral form (e.g., phosphatization or carbonization; Martill 1989; Wilby and Briggs 1997), chemical environment (Hedges and Millard 1995; Bell et al. 1996; Hedges 2002), pH of pore waters, local geochemistry, and the possible presence of microbes known to participate in the precipitation of minerals (Briggs et al. 1993; Schweitzer 2011; Briggs and McMahon 2016; Iniesto et al. 2016).

Traditionally, fossil bone thus has been viewed as "bone turned to stone," that is, the fossil was believed to be fully mineralized without any organic components and compounds remaining. However, classical observations of the perfect histological preservation of the microstructure in most vertebrate hard tissues in petrographic thin sections (Owen 1845–1856; Kiprijanoff 1881–1883) may have served as a challenge to this dogma early on. Nevertheless, the first indications of preservation of the organic phase in fossil dinosaur bone were discovered and described as cells, collagen fibrils, blood vessels, and proteins in the 1960s (Pawlicki et al. 1966; Miller and Wyckoff 1969). These early studies were not followed by in-depth work and are sometimes difficult to reproduce. In the early work of Pawlicki et al. (1966), for example, it is not clear if the observed features such as osteocytes were actually degraded cells or casts of the osteocyte lacunae produced by an embedding medium during sample preparation.

Permineralization may be responsible for original organic soft tissue preservation as well, in which soluble mineral crystallites stabilize the organic tissues and retain their micromorphology (Schweitzer 2011). To preserve original tissue, however, the minerals must be deposited on these labile tissues before decay progresses to the point of loss of integrity, and stabilization must occur within hours to days after death (Martill 1988, 1989; Briggs et al. 1993; Briggs 2003; Carpenter 2007; Daniel and Chin 2010; Schweitzer 2011; Briggs and McMahon 2016). Another explanation for the preservation of endogenous soft tissue is possible crosslinking of original organic molecules to other organic components initiated by unstable metal ions such as heme, thus forming polymers (Schäfer et al. 2000; Stankiewicz et al. 2000; Briggs 2003; Schweitzer et al. 2007a, 2007b, 2014; Schweitzer 2011). The microenvironment of bone provides enough protection from initial degradation so that molecular crosslinks can form, thus providing further resistance against degradation and allowing the original components to act as templates for rapid, authigenic mineralization (Schweitzer et al. 2007a, 2007b, 2014).

Inorganic Phase in Fresh and Fossil Bone

The major inorganic component in bone tissue is bioapatite, a nanocrystalline, nonstoichiometric, highly carbonate-substituted (typical 5–8 wt%), hydroxyl-deficient form of HAp ($Ca_{10-x}[(PO_4)_{6-x}(CO_3)_x]$ $(OH)_{2-x} \cdot nH_2O$) (Elliott et al. 2002; Pasteris et al. 2004). In the bioapatite crystal, lattice carbonate ions can substitute predominantly for the phosphate, as well as for hydroxyl groups (Morgan et al. 2008; Fujisawa and Tamura 2012; Farbod et al. 2014). Furthermore, bone bioapatite also contains some water (Li and Pasteris 2014). Other substitutions for calcium in the inorganic bioapatite phase are magnesium, potassium, strontium, and sodium, as well as chloride and fluoride for the hydroxyl groups (Morgan et al. 2008; Farbod et al. 2014). Bioapatite in bone occurs as small, nanometer-sized crystallites. Fluoride mainly increases the crystallinity of apatite, while other impurities in the inorganic phase of bone reduce the crystallinity, leading to a more soluble bone and optimizing mineral homeostasis and bone remodeling (Farbod et al. 2014).

Organic Phase in Fresh Bone

The organic phase in bone consists mainly of type I collagen fibers, which make up 90% to 95%, and noncollagenous proteins such as osteocalcin, osteonectin, and osteopontin (Collins et al. 2002; Morgan et al. 2008; Palmer et al. 2008; Farbod et al. 2014; Hall 2015). The extracellular matrix, the main component of the organic phase in bone, is a dynamic network composed of organic and inorganic constituents, such as collagen, elastin, polysaccharides, and calcium phosphate (Labat-Robert et al. 1990; Schönherr and Hausser 2000; Morgan et al. 2008; Fujisawa and Tamura 2012; Farbod et al. 2014). Cellular components that make up the organic phase in bone are vascular canals, composed mainly of collagenous and elastic fibers, and bone cells such as osteocytes and osteoclasts (bone destroying cells). Of the bone cells, osteocytes are most abundant, making up 90% to 95% of the total (Florencio-Silva et al. 2015).

Abbreviations of Analytical Methods

Abbreviations of analytical methods used in this chapter. Note that methods only mentioned in table 2.1 are marked with an asterisk (*). 16S rRNA = gene amplicon sequencing;* AFM = atomic force microscopy;* AMS = accelerator mass spectrometry;* ATR FTIR = attenuated total reflectance Fourier transform infrared spectroscopy;* DIBA = dot-immunobinding assay;* EDS = energy-dispersive X-ray spectroscopy; EELS = electron energy-loss spectroscopy; ELISA = enzyme-linked immunosorbent assay; EPMA = electron probe microanalysis; FIB = focused ion beam;* FM = fluorescence light microscopy;* FPLC = fast protein liquid chromatography;* FTIR = Fourier transform infrared spectroscopy; HPLC = high-performance liquid chromatography; ICP-MS = inductively coupled plasma mass spectrometry; IHC = immunohistochemistry;* LC-MS-MS = reversed-phase microcapillary liquid chromatography tandem mass spectrometry; LM = transmitted light microscopy; MS = mass spectrometry; μXANES = micro-X-ray absorption near edge structure; μXRF = micro-focused X-ray fluorescence;* NMR = nuclear magnetic resonance;* Py-GC-MS = pyrolysis–gas chromatography–mass spectrometry; RS = Raman spectroscopy; SAED = selected-area electron diffraction;* SAXS = small angle X-ray scattering;* SDS-PAGE = sodium dodecyl sulfate–polyacrylamide gel electrophoresis; SEM = scanning electron microscopy; SR-FTIR = synchrotron radiation-based Fourier transform infrared spectroscopy; TEM = transmission electron microscopy; ToF-SIMS = time-of-flight secondary ion mass spectrometry; VPSEM = variable pressure scanning electron microscopy;* XPS = X-ray photoelectron spectroscopy; XRD = X-ray diffraction.

A Review of the Literature

First Evidence of the Organic Phase in Fossil Dinosaur Bone

Soft part preservation of collagen fibrils, muscle fibers, epithelial cells, and melanophores have been found in fossil animals from sediments as old as 25 million years from the Oligocene Allier locality in France (Little et al. 1962) and even 47 million years from the Eocene Geiseltal in Germany (Müller 1957), which were identified in peels and thin sections using LM (Voigt 1934, 1988). However, the first documentation of soft part preservation in dinosaurs was made by Pawlicki et al. (1966) on a phalange of an 80-million-year-old dinosaur from the Upper Cretaceous of the Gobi Desert in southern Mongolia. Histological sections of the dinosaur phalange showed the presence of osteocyte lacunae and connecting canaliculi (Pawlicki et al. 1966), as would be expected. After demineralizing the bone, Pawlicki et al. (1966) washed the suspension with Sorensen's phosphate buffer and water, embedded it in methacrylate, then identified an isolated blood vessel and isolated osteocytes in 1 μm thick microtome slides using LM. Images of these soft parts were also studied with an SEM, which suggested the presence of collagenous material in the wall of a vascular canal based on the spiral structure. This structure shows a clear crossbanding of fibrils and a plaited texture. Pawlicki et al. (1966) suggested this preservation was due to a specific mummification process in which the soft parts were separated by a "border barrier."

In 1969, Miller and Wyckoff found more evidence for soft part preservation in dinosaur bone. In their study, six specimens in total were sampled from three Jurassic and two Cretaceous localities, including the famous Howe Quarry, Como Bluff site, and Bone Cabin Quarry in Wyoming, USA (Miller and Wyckoff 1969). The methods of these authors differ from those of Pawlicki et al. (1966) in that 20 mg of each bone sample was powdered and treated with hydrochloric acid (HCl) to dissolve carbonates. After centrifugation, the residues were retreated with HCl until the suspension was colorless, then hydrolyzed with boiling HCl for 24 hours. After evaporation of the HCl, minerals such as quartz were removed by filtering and centrifuging. The dried deposit was then dissolved in water to which ninhydrin was added to detect primary and secondary amines, then analyzed with an automatic amino acid analyzer. The results show a very small amount of more or less intact proteins, containing up to 20 amino acids. Interestingly, only one of six samples showed hydroxyproline characteristic for collagen. However, the authors also state the dissolving method should be improved. In summary, several important amino acids were present in the samples, suggesting that either the original proteins or their degradation products were present.

Pioneer of "Molecular Paleontology": Mary H. Schweitzer and Colleagues

After the studies of Pawlicki et al. (1966) and Miller and Wyckoff (1969), the search for soft part preservation in dinosaur bone in the form of proteins was carried on by several labs. Gurley et al. (1991) reported the discovery of amino acids in a *Seismosaurus*, using HCl, guanidine extracts, and HPLC. Osteocalcin was identified in dinosaur bone using immunological assays (Muyzer et al. 1992; Collins et al. 2000). In 1997, Schweitzer et al. showed the presence of heme proteins in *Tyrannosaurus rex* (specimen MOR 555) using HPLC and RS. Embery et al. (2000) used SDS-PAGE to define proteins such as glycine, aspartate, and serine, which tentatively indicated the presence of phosphoproteins isolated from dinosaur bone.

In 2005, when Mary H. Schweitzer and co-authors published their findings in *Science* (Schweitzer et al. 2005), molecular paleontology entered a new era and is now an area of intense study. This time, their work focused on a *Tyrannosaurus rex* (specimen MOR 1125) from the Upper Cretaceous Hell Creek Formation of Montana, USA. In this study, Schweitzer et al. (2005) demonstrated that pliable, three-dimensional soft tissue blood vessels and osteocytes with cellular contents could be liberated from the bone matrix. They began by demineralizing cortical and endosteal bone using the weak organic acid of ethylenediaminetetraacetic acid (EDTA), during which the mineral phase was removed, leaving behind flexible, elastic, and resilient vascular tissue (Schweitzer et al. 2005). After the cortical bone was completely dissolved, soft tissue vessels were separated from the matrix; the vessels showed a high morphological resemblance to vessels that were liberated from extant ostrich bone samples. In particular, the branching of the vessels is similar in both the *Tyrannosaurus* and ostrich specimens (Schweitzer et al. 2005). One difference, though, is that many of the dinosaur vessels contain small, round, red to dark brown structures, possibly pertaining to the content of the blood vessels preserved in *Tyrannosaurus rex*. Along the blood vessels, elongated three-dimensional structures could be identified. The morphology and size of these structures correlates with those of osteocytes recovered from the ostrich bone sample. After isolation of the *Tyrannosaurus* osteocytes, internal contents, such as possible endothelial cell nuclei, were reported. SEM imaging was used to further verify the presence of the external surface features described above and confirmed the hypothesis of soft tissue preservation in these samples (Schweitzer et al. 2005). The dense mineralization of bone and geochemical and environmental factors were cited as presumably contributing to this organic preservation, but Schweitzer et al. (2005) state that the origin of this preservation is still uncertain. Overall, this pivotal study by Schweitzer and colleagues initiated the start of research on micro- and molecular taphonomic investigations.

Soon afterwards, Schweitzer et al. (2007a) carried out another study on the same *Tyrannosaurus rex* material. Here their main focus was to test for the presence of proteins in the femora of the *T. rex*. Molecular and chemical analyses were conducted after partial demineralization of bone that would be dominated by collagen I in extant animals (Schweitzer et al. 2007a). Cortical and medullary bone extracts were tested by in situ immunohistochemistry, in which thin sections of demineralized cortical and medullary bone were exposed to antibodies raised against chicken collagen I, and the reactivity was measured and compared to the reactivity data measured by ELISA. The reactivity measured with both in situ immunohistochemistry and ELISA is reduced in dinosaur extracts when compared to extant samples, as would be expected, yet it is still higher than in the control samples of sediment and buffer (Schweitzer et al. 2007a).

Another method used to test for the presence of collagen was ToF-SIMS (Schweitzer et al. 2007a). To find amino acid residues indicating the presence of protein, in situ ToF-SIMS analyses were performed, and ratios of glycine and alanine (the two main amino acids found in collagen) were detected. Smaller peaks of proline, lysine, and leucine or isoleucine were also reported.

The published glycine and alanine ratio for collagen I of an extant chicken bone sample is 2.5:1, while the ratio from medullary bone of *Tyrannosaurus* is 2.6:1 (Schweitzer et al. 2007a). This study thus concluded that since the dinosaur protein sequence should be most similar to that of extant birds, the molecular fragments found in the bone tissue of *Tyrannosaurus* are original proteins. Note that a minimum of three repetitions of extraction showing similar results were conducted before reporting an assay as positive, since variations of microenvironments within a single bone are common and bone degradation in modern environments occurs (Schweitzer et al. 2007a).

A possible explanation for the preservation of these proteins is the crosslinking of the originally unstable protein molecules to similar molecules and to other organic components, which could have been initiated by unstable metal ions reacting with organic molecules to form polymers (Schäfer et al. 2000; Stankiewicz et al. 2000; Briggs 2003; Schweitzer et al. 2007a, 2007b). Once crosslinked, the molecules could no longer serve as substrates for further degradative reactions and were therefore preserved.

Broadening the Sample Base in Time and Space

So far, both studies had focused on tyrannosaur bones from the Upper Cretaceous (68 Ma) Hell Creek Formation of Montana, USA. To test for a more general preservation of soft tissue in bone, another study was then carried out by Mary Schweitzer and colleagues using specimens from multiple geological time periods, varied depositional environments, and different taxa (Schweitzer et al. 2007b). The geological age of the specimens used in this study ranged from 90 Ma to the present day. The dinosaurs studied were several *Tyrannosaurus* specimens, a *Triceratops horridus*, again from the Hell Creek Formation of Montana, USA, and a *Brachylophosaurus canadensis* from the Lower Judith River Formation, Canada. All samples showed the preservation of flexible and fibrous bone matrix; transparent, hollow, and pliable blood vessels; intravascular material, including in some cases, possible remains of red blood cells; and osteocytes with filopodia (Schweitzer et al. 2007b). Variations in preservation are evident, though, such as the quantity of osteocytes and the preservation and morphology of the filopodia (e.g., short and stubby in *Triceratops* compared to long in *Brachylophosaurus*).

Apart from the question of whether the same soft tissue structures are preserved in other taxa, three additional major questions were addressed in this paper. (1) How extensive is the preservation of soft tissue and cells is in vertebrate remains? (2) Can patterns in tissue or cell degradation over time be detected? (3) Is there a connection between taxonomic, temporal, geographical, or geochemical factors with the preservation of soft tissue? Schweitzer et al. (2007b) showed that the preservation of soft tissues such as blood vessels and osteocytes is more common than originally thought, as they were found in all 21 samples used in the study. Also, no single environmental component could be identified as favorable for this kind of preservation. However, it should be pointed out that these specimens were mainly derived from fluvial sandstones and no older than middle Late Cretaceous in age.

The last open question, if any patterns in degradation over time could be identified, was framed as a two-stage hypothesis. First, the microenvironment of the cortical bone might work against degradation, providing enough time before degradation reaches these areas and for the previously discussed molecular crosslinks to form. These links might provide resistance against further degradation and allow for the original components to act as templates for rapid, authigenic mineralization, which is the second stage of the degradation pattern (Schweitzer et al. 2007b). Also, a final aspect discussed by Schweitzer and colleagues is possible alternative sources of the soft tissue structures, such as pyrite framboids, remnants of mineralized biofilms, extant fungal hyphae, or contamination from collection or preparation. Yet, most of these hypotheses were discarded after careful consideration, suggesting that the soft tissue remains do indeed consist of original material or their degradational products (Schweitzer et al. 2007b).

Revisiting Previous Samples with New Techniques

The preservation of *Brachylophosaurus canadensis* soft tissue was studied in greater detail by Schweitzer et al. (2009). Since the hypothesis of endogenous proteins in *Tyrannosaurus* was met with criticism, additional protein sequence data were collected from *Brachylophosaurus* to strengthen their hypothesis. Both the chemical extracts and the demineralized bone of *Brachylophosaurus* showed a positive reactivity to antibodies raised against avian collagen I and/or osteocalcin, whereas controls of buffers and sediments were negative (Schweitzer et al. 2009). In situ immunohistochemistry showed the same positive reactivity to avian collagen I. Other antibodies, raised against the proteins elastin, laminin, and hemoglobin, were used to test for epitopes preserved in *Brachylophosaurus* blood vessels. Antibodies to both laminin and elastin showed specific bindings, and antibodies raised against ostrich hemoglobin showed bindings in in situ tests, yet not in extracted blood vessels (Schweitzer et al. 2009). Posttranslational hydroxylation of proline was detected by ToF-SIMS and is a feature typical of collagen that cannot be produced by microbes (Ebert and Prockop 1962; Rasmussen et al. 2003; Schweitzer et al. 2009).

LC-MS-MS was used to test whole bone extracts of *Brachylophosaurus,* and a total of eight collagen peptide sequences, six from collagen α1 type I and two from collagen α2 type I, could be identified. These peptides totaled 149 amino acids, which was nearly double the number found in *Tyrannosaurus rex* (Asara et al. 2007). The four lowest-scoring mass spectra that were acquired with collision-induced dissociation (CID) were used for tests with the MS Search 2.0 spectral comparison algorithm from the National Institute of Standards and Technology (NIST) against over 200,000 random peptide fragmentation spectra and were top matches to high-confidence versions of these spectra, further validating the data. Lastly, the collagen sequences from *Brachylophosaurus* were aligned with the sequences of 21 extant taxa and those of *Mammut americanum* and *Tyrannosaurus rex.* Unfortunately, the amount of data missing from the sequences of *Brachylophosaurus* compared to extant sequences led to low resolution within the Dinosauria. However, the phylogenetic relationship of this specimen was still placed within the Archosauria, closer to extant birds than extant alligators, and some phylogenetic signals still reside within the recovered collagen. All the evidence provided in the study of Schweitzer et al. (2009) thus supports the endogenous origin of the soft tissues found in *Brachylophosaurus.*

The study of Schweitzer et al. (2009) was expanded in 2017 by Schroeter et al. (see also Service 2017), in which additional evidence for preserved protein was found. In general, an updated extraction method was used in which the dinosaur bone and sediment were dissolved in EDTA, followed by resuspension in ammonium bicarbonate (ABC), and a final resuspension in guanidine hydrochloride (GuHCl). In between both resuspension steps, the sample was centrifuged, and the supernatants were collected. ABC and GuHCl were used for silver stain, and gel band excision and in-gel digestion were performed for GuHCl (for detailed methods, see Schroeter et al. 2017). Silver staining results showed distinct bands on the gel from the GuHCl extracts, whereas the controls and the ABC extractions showed none.

Silver stains performed in previous studies show a higher amount of smear than in the *Brachylophosaurus* samples (Tuross 2002; Schweitzer et al. 2007a, 2009; Cleland et al. 2012; Schroeter et al. 2017). However, this reduction of smearing may be caused by the reduction of residual EDTA in these samples, as it has been demonstrated that EDTA causes smearing in gels of extant bone extractions (Cleland et al. 2012). The results show that even though staining is only present in the fossil sample, silver staining alone is not sufficiently specific.

In the study using tandem MS by Schroeter et al. (2017), eight peptides of collagen I could be identified, which included five sequences of collagen α1 type I and three of collagen α2 type I, while no collagen sequences could be identified in the negative controls. Both the bone and sediment samples show environmental signals such as fungal and bacterial proteins that were not observed in the buffer-only control. This indicated that these protein sequences derived from the burial environment and not from collection or preparation contamination. Two of the five collagen α1 type I peptides were also identified by Schweitzer et al. (2009) and are the most abundant peptides in the *Brachylophosaurus* extracts in both studies. This suggests that these peptides may have a particularly high preservation potential due to their location within the collagen fibril and are therefore are more commonly found in the fossil record (Schroeter et al. 2017). Three new peptides of collagen α1 type I (with a total length of 38 amino acids) and three new collagen α2 type I (with a total length of 42 amino acids) were detected, indicating an increase of 33.6% in collagen α1 type I sequence length and 116.7% in collagen α2 type I (Schweitzer et al. 2009; Schroeter et al. 2017). These measurements indicate endogenously derived peptide sequences. Finally, phylogenetic analyses were conducted to further confirm this hypothesis. The newly obtained collagen α1 type I and collagen α2 type I sequences and previous sequences from Schweitzer et al. (2009) were compared with all sequences known for extant archosaurs (21 species) and eight outgroups of snakes, lizards, and turtles. When compared separately, the new collagen sequences from *Brachylophosaurus* indicate a closer relationship to crocodilians, whereas when both sequences are combined, *Brachylophosaurus* is grouped into the basalmost clade of birds. This ambiguous placement of *Brachylophosaurus* indicates that these sequences are not contaminants derived from specific taxa since they are not homologous with either species, and that the sequence from *Brachylophosaurus* shares similarities with both crocodilians and birds (Schroeter et al. 2017). These results therefore strongly support the endogenous origin of the soft tissues preserved in this specimen (Service 2017).

Boatman et al. (2019) hypothesized that early diagenesis iron-mediated Fenton and glycation pathways contributed to the longevity of *Tyrannosaurus* (specimen MOR 555) blood vessels by producing crosslinks. To study this hypothesis, SR-FTIR analyses on crosslink-induced extant chicken type I collagen and untreated chicken type I collagen using either the Fenton reagent or iron-catalyzed glycation were conducted and compared to the SR-FTIR spectrum of *Tyrannosaurus* (Boatman et al. 2019). Three significant changes were recorded. First, the treated chicken samples and *Tyrannosaurus* sample show an apparent blue shift in the amide I band, due to increased intramolecular crosslinking (Barth 2007; Nguyen and Lee 2010), whereas the untreated chicken sample does not show this shift. Second, nonpeptide carbonyl bands are observed in treated chicken samples as well as in the *Tyrannosaurus* sample and may represent a result from Fenton-type reactions leading to peptide crosslinking. Third, a carbohydrate band developed in chicken tissues after glycation could also be observed in the *Tyrannosaurus* sample; however, the shape of this band is highly dependent on the nature and amount of different sugar molecules present during crosslinking (Leopold et al. 2011; Boatman et al. 2019). Boatman et al. (2019) propose that during hemolysis, Fe-containing hemoglobin was liberated and degraded within the vessel lumen, and would bind to collagen, producing hydroxyl radicals. These radicals would instantaneously react with amine groups (Schweitzer et al. 2013), leading to aldol condensation and crosslinking of adjacent protein molecules (Boatman et al. 2019). Alternatively, or simultaneously, glycation would lead to crosslinking of proteins, and would thus enhance the longevity of *Tyrannosaurus* blood vessels (Boatman et al. 2019).

Further studies by the Schweitzer lab (Schweitzer et al. 2008; San Antonio et al. 2011; Cleland et al. 2012, 2015; Zheng and Schweitzer 2012; Boatman et al. 2014; Cadena and Schweitzer 2014; Cadena 2016) are not reviewed in detail here, since they do not use any new techniques for studying soft tissue preservation.

The Biofilm Hypothesis

In 2008, Kaye et al. carried out a scanning electron microscope study to determine if the findings of dinosaurian soft tissue from Schweitzer et al. (2005) could be identified in situ within the bones. In addition to SEM, they used EDS and FTIR to investigate three categories of soft tissues. The first category of tissue was the clusters of spheres with an iron-oxygen signature. Multiple specimens from the Lance, Hell Creek, Chadron, and Pierre Shale Formations all showed these spheres, as well as an ammonite suture sample, which indicated that these spheres are not related to the iron derived from blood. Instead, Kaye et al. (2008) proposed that the spheres were framboids, which are commonly found in algal mats and sediments. The second category of tissue was the soft, pliable, branching tubules resembling blood vessels. SEM images of a fractured bone show a coating peeling away from the blood vessel wall (Kaye et al. 2008). The surface of the coating, as well as the surface of the framboids, have bubble-like pores that are inconsistent with a mineral origin and suggest the release of gases from anaerobic bacteria.

Apart from the pores, cracks (or troughs) cover the surface of a vascular canal, which Kaye et al. interpreted as being formed by free-swimming microbes or bacteria in a viscous medium. The final category of tissue was free floating osteocytes with filopodia. SEM imaging before acid dissolution showed a variety of forms and structures within the original lacunae, consisting of submicrometer spheres and rods that are morphologically similar to bacterial structures (Kaye et al. 2008). Carbon dating indicated a modern origin for the material. Kaye et al. (2008) thus suggested that these structures were the remnants of biofilms, which would have coated the blood vessels and osteocyte lacunae, thus producing an endocast of the structure.

The biofilm hypothesis of Kaye et al. (2008) was further tested by Schweitzer et al. (2013), specifically in regard to preserved osteocytes. For this study, osteocytes from *Tyrannosaurus* and *Brachylophosaurus* were liberated from the bone by EDTA, then isolated and imaged by LM, SEM, TEM, and FLM. The osteocytes varied in shape from broad and roundish to narrow and elongated bodies, as seen in extant bone. Antibodies, both polyclonal (pAb) and monoclonal (mAb), were used to test the presence of specific proteins, such as actin and tubulin, in cells of both dinosaurs. These proteins are components of the cytoskeleton of extant osteocytes (Tanaka-Kamioka et al. 1998; Murshid et al. 2007; Schweitzer et al. 2013). The antibodies to actin bind mostly to linear, filamentous structures, whereas the tubulin antibodies show a more diffuse binding pattern, which is consistent with the patterns seen in extant ostrich and alligator osteocytes (Schweitzer et al. 2013). Neither of the proteins actin and tubulin can be found in prokaryotic cells, and tests with one of the most common biofilm-forming microbes found in soil were negative, indicating that a biofilm origin for the osteocyte remains is unlikely.

Mass spectrometry also indicated the strong presence of the proteins actin and tubulin, providing more proof against a biofilm origin. Histochemical staining was conducted to further test biofilm versus original material using two stains: propidium iodide (PI) and 4′,6-diamidino-2-phenylindole dihydrochloride (DAPI), which both bind to DNA. Both stains showed a reactivity to the internal components of both dinosaur samples, yet appeared greatly reduced when compared to the reactivity of the stains to extant ostrich osteocytes (Schweitzer et al. 2013). These patterns of staining are inconsistent with the patterns of biofilm exposed to the same stains. Both antibodies and histochemical staining indicate the presence of material in the osteocytes that is chemically and structurally similar to DNA. Clear evidence of eukaryotic DNA is provided by the binding of antibodies to histone H4, which is not found in microbes. Contaminant DNA is unlikely since DNA from exogenous sources would not concentrate into one single point. These findings, however, are insufficient to support that the DNA present in these samples is of dinosaurian origin, and further sequence data and antibody research are needed to test this hypothesis (Schweitzer et al. 2013).

A study potentially extending the record of osteocyte preservation as far back as the Jurassic was published by Cadena and Schweitzer (2012). It looked at the morphology of osteocytes in turtle shells, focusing on two different morphotypes (one flat-oblate, the other stellate). One taxon sampled was *Mongolemys elegans* from the Late Cretaceous of Mongolia. Another study by Sukhanov (2000) also reported osteocytes from the Middle Jurassic turtle *Annemys* from the Berezovsk locality in Siberia, Russia (Sukhanov 2000), which is dated as ca. 166 Ma. If verified, this would be the second-oldest published record of osteocyte preservation.

At the start of the 2010s, the preservation of soft tissue structures such as osteocytes and blood vessels remained an intensely studied field. In 2014, Schweitzer et al. introduced the role iron and oxygen might play in this preservation. TEM revealed a close association between iron and the vessels after the demineralization of dinosaur bone, and iron-rich nanoparticles were located on the vessel walls as an amorphous layer (Schweitzer et al. 2014). EELS with the SEM indicated that iron was also localized in the cells and in the intracellular content of both the *Tyrannosaurus* and *Brachylophosaurus* samples. After chelation, the iron signal was greatly decreased and diffuse, which points out a close spatial association between iron minerals and still-soft structures. μXANES spectroscopy showed a chemical speciation of oxy-hemoglobin and Fe oxyhydroxide in hemoglobin (HB) incubated ostrich tissue, as well as goethite and biogenic iron oxyhydroxide (Toner et al. 2009) in both dinosaur samples. High concentrations of crystalline goethite were located in the intravascular structures in dinosaur vessels, combined with iron oxyhydroxides best matching a biogenic iron oxyhydroxide standard (Schweitzer et al. 2014). These observations are consistent with EPMA mapping of bone and the diagenetic infill of bone porosity by Wings (2004) and Dumont et al. (2011), as well as with personal observations that the first generation of void fills in thin sections are opaque and contain dark red minerals containing iron, such as goethite, hematite, and pyrite.

Immunoreactivity of proteins in osteocytes and vessels of dinosaurs increased after iron chelation (Schweitzer et al. 2007a, 2009, 2013, 2014), with osteocytes showing an elevated response to anti-actin antibodies, and blood vessels a raised response to elastin antibodies after chelation. A model using ostrich blood vessels was used to determine postmortem conditions for preserving these structures. During life, iron-containing HB in red blood cells flows through vessels, accessing osteocytes along the way through the lacuna–canalicular network (Aarden et al. 1994; Cowin 2002; Schweitzer et al. 2014). This HB could cause localized, heme-based crosslinking in tissues, which was observed in the ostrich model when vessel stability was increased more than 240-fold after HB incubation. Without HB incubation and in the absence of oxygen, the blood vessels still showed stability, but the most rapid degradation occurred in the presence of oxygen and absence of HB (Schweitzer et al. 2014). Possible explanations for this stabilizing effect are enhanced tissue fixation and the inhibition of microbial growth. These data support a naturally occurring mechanism that stabilizes soft tissues by means of heme-based crosslinking over geological time, thus preserving these soft tissues.

In 2016, Schweitzer et al. tested the biofilm hypotheses experimentally by investigating whether biofilms will colonize and grow within bone from which the organic phase has been removed, and if biofilms are capable of producing structures similar to blood vessels. Schweitzer et al. (2016) set up a model using organic-free bovine bone that was subjected to two different types of biofilm-producing microorganisms, *Bacillus cereus* and *Staphylococcus epidermis*. After approximately two weeks of incubation, the bone fragments were demineralized, the biofilm was studied with SEM, TEM, and antibodies, and the reactivity was compared to that of dinosaur bone exposed to the same antibodies (Schweitzer et al. 2016). Organic-free bone fragments were placed in both sterile water and a nutrient medium, and only the latter showed signs of a biofilm using the bone as a substrate after the incubation period. SEM and TEM showed that biofilm products superficially resemble blood vessels. However, these structures never possessed a lumen and completely lost integrity when manipulated and disintegrated, in comparison to dinosaur blood vessels. Higher magnification also showed a patchy distribution of microbial bodies, which is unlike the blood vessels from dinosaurs under higher magnification.

Antibodies raised against actin bound to dinosaur vessel walls, but showed no reaction to biofilms prepared in the same manner (Schweitzer et al. 2016). The same signal could be observed when the reactivity of antibodies raised against elastin was studied; they bound only to dinosaur vessels and did not react to the biofilm samples. Finally, antibodies raised against peptidoglycan, a peptide exclusively produced by bacteria, bound only to biofilms grown on bone, but showed no reactivity to the dinosaur vessels treated under identical conditions (Schweitzer et al. 2016). When all results are taken into account, an endogenous source of these soft tissues is supported. However, microbes do indeed play an important role during fossilization of the organic phase. Experimental taphonomic studies showed that microbial action promotes a chemical environment favorable for mineralization (Briggs and Kear 1994; Wilby and Briggs 1997; Briggs 2003; Raff et al. 2008, 2013; Daniel and Chin 2010; Darroch et al. 2012; Iniesto et al. 2016). Biofilm formation is indicated as assisting in the preservation of soft tissues, whereby the biofilm acts as a sarcophagus to protect the carcass from external environmental conditions, delaying bacterial decomposition (Gehling 1999; Riding 2000; Kaye et al. 2008; O'Brien et al. 2008; Daniel and Chin 2010; Peterson et al. 2010; Iniesto et al. 2016).

The Past Decade: A Diversity of Labs Evaluate Organic Phase Preservation in Fossil Bone

Although much work on organic phase preservation in fossil bone has been done by the Schweitzer lab, other studies also offer interesting results, inspired by the results of Schweitzer et al. and the criticism offered by Kaye et al. (2008). In 2011, Lindgren et al. proposed a spectroscopic characterization of isolated fibrous bone tissues from an early Maastrichtian mosasaur, *Prognathodon* from Belgium, to provide further proof of the endogenous nature of these soft tissues (fig. 2.1) (Lindgren et al. 2011). Samples from the humerus of *Prognathodon* were digested in EDTA, then analyzed with SEM, TEM, histochemical staining, immunohistochemistry and immunofluorescence, amino acid analyses, infrared microspectroscopy, and MS (for detailed methods, see Lindgren et al. 2011). In situ immunofluorescence showed reactivity to antibodies raised against collagen I. Proteins were extracted from the bone, and amino acids were detected that indicate structural proteins or their breakdown products. These amino acids were then studied using SR-FTIR, which indicated that the amino-acid-containing material is located within bone matrix fibrils that possess typical collagen characteristics (67 nm D-periodicity) and that the spectral signature of the fibrils differs greatly from modern bacterial contaminants (Lindgren et al. 2011). These data thus suggest that the preservation of organic matter is, in fact, endogenous and is not only restricted to fluvial sediments, but occurs in marine sediments as well, as was stated by Schweitzer et al. (2007b).

Eight Cretaceous dinosaur bones were sampled by Bertazzo et al. (2015) and studied using nano-analytical methods. TEM showed one sample with a clear banding at ca. 67 nm, which is diagnostic for collagen, for the length of the preserved fiber in this sample (Bertazzo et al. 2015 and references therein). However, TEM does not allow testing for chemical composition, so ToF-SIMS analyses were conducted. Mass spectra for this sample showed

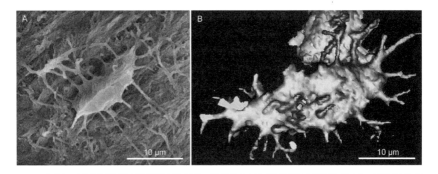

Figure 2.1. (A) SEM image of the fibrous bone tissue of the mosasaur *Prognathodon* showing osteocytes in lacunae. (B) Topographic image of osteocyte indicating three-dimensional arrangement of filopodia. Images from Lindgren et al. (2011), Creative Commons Attribution License.

peaks for the amino acids glycine, alanine, proline, and others (Canavan et al. 2006; Schweitzer et al. 2009; Henss et al. 2013; Bertazzo et al. 2015), which corresponds to the TEM results that the original quaternary structure of the protein is preserved. Also, the spectra obtained from different regions of this sample are strikingly similar to those of emu blood, and partial least square–discriminant analysis (PLS-DA) of these spectra are enclosed within the 95% confidence ellipse, whereas diagenetic cement spectra (negative control) falls outside the ellipse, providing more evidence for the endogenous nature of these soft tissues.

Surmik et al. (2016) performed a spectroscopic study of the organic phase in Middle Triassic marine reptile bones from Poland (plate 2.1A–E). These bones are of earliest Anisian age and, at close to 247.2 Ma (Walker et al. 2018), represent the oldest published record of organic phase preservation in fossil bone. Multiple analytical techniques (XRD, FTIR, ToF-SIMS, XPS) were applied to preserved blood vessels liberated from the bone. XRD, FTIR, and XPS showed typical diagenetically altered fossil bone, in which mainly FAp was found, and goethite in the blood vessels. ToF-SIMS detected the amino acids glycine (plate 2.1B), alanine (plate 2.1C), proline (plate 2.1D), hydroxyproline, leucine, lysine, and hydroxylysine, which are characteristic amino acids for collagen I (Sanni et al. 2002; Schweitzer et al. 2007b, 2009; Henss et al. 2013; Bertazzo et al. 2015). The total ion map outlines a blood vessel in cross section, based on amino acids (plate 2.1E). FTIR analyses show a significant amplification of the organic signal after EDTA incubation, due to a strong connection with ferruginous mineralization of the blood vessels. This indicates that the fixation of organic residues must have taken place at a very early stage of diagenesis, for example, during collagen gelatinization, since the later physiochemical alteration of bone apatite does not seem to affect the organic preservation significantly (Surmik et al. 2016).

In 2017, ribs from the 195 Ma basal sauropodomorph *Lufengosaurus* from China were investigated by Lee et al. The *Lufengosaurus* sample was studied using SR-FTIR and RS, the former analysis showing preserved collagen. Infrared absorption bands characteristic for collagen I and elastin (Cooper and Knutson 1995; Jackson et al. 1995; Vedantham et al. 2000; Wetzel et al. 2005; Karima et al. 2009) were also detected in the *Lufengosaurus* sample. Round, dark red, micrometer-sized particles within the blood vessels were observed as well. RS indicated that these dark particles consist of hematite, which is thought to have been derived from hemoglobin and are localized within the blood vessels and lacunae. Lee et al. (2017) proposed that small chambers were formed within the vascular canals that preserved isolated collagen and protein remains, and thus hematite cementation may have played an important role in organic preservation, echoing the work of Schweitzer et al. (2014). However, this hypothesis proves to be tricky since bone normally creates initially low Eh and pH conditions due to organic matter decay (Tareen and Krishnamurthy 1981; Schwertmann and Murad 1983; pers. comm. Fabian Gäb). Thus, the stability field of Fe_2O_3 might not be initially reached in early bone diagenesis.

Plet et al. (2017) studied an ichthyosaur vertebra that was encapsulated in a 182.7-million-year-old carbonate concretion from the Early Jurassic of Germany. To test what effect a concretion has on the preservation of soft tissue, SEM, TEM, and ToF-SIMS analyses were conducted. Trabecular and cortical bone samples were tested, and SEM imaging of acid-treated surfaces shows elongated fiber bundles with curved geometries that resemble those of modern crocodile collagen. Furthermore, clusters of concave disks were found in close proximity to the fibers that show a striking similarity to red blood cells, and platelet-like structures were found that are similar to modern white blood cells (Plet et al. 2017). It is interesting that all of these blood cell-like structures are four to five times smaller than those of modern mammals (Gregory 2001; Plet et al. 2017); however, since a great size variety in red blood cells in modern mammals has been reported (Undritz et al. 1960; Robert and Pierre 2002), it is hypothesized that the small size of these blood cell-like structures is an evolutionary adaptation to environmental conditions. TEM and ToF-SIMS analyses were carried out on the red blood cell-like structures and showed a composition of carbon (^{12}C) and oxygen (^{16}O) isotopes, indicating an organic origin. Additionally, Me,Et maleimide (3-ethyl, 4-methyl-pyrrole-2,5-dione) was detected in these structures, which is a

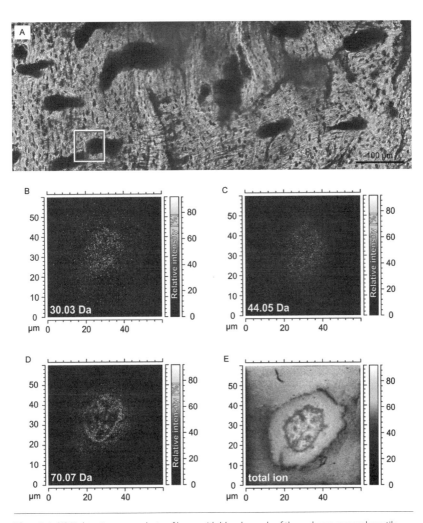

Plate 2.1. (A) Light microscope photo of bone with blood vessels of the archosauromorph reptile *Protanystropheus* sp. Rectangle marks area analyzed with ToF-SIMS. B–E: ToF-SIMS ion distribution maps (fast imaging mode) generated for selected masses and amino acids. (B) Glycine or proline. (C) Alanine. (D) Proline. (E) A blood vessel in cross section outlined by the total ion map based on amino acids. Modified from Surmik et al. (2016), Creative Commons Attribution License.

known oxidative degradation product of heme. Besides fossilized red and white blood cells, an elevated concentration of cholesterol was detected, which gives insight into the diet of this ichthyosaur. These combined findings demonstrate the exceptional preservation potential in this case of the isolating environments within carbonate concretions (Plet et al. 2017).

A fossilized coracoid from the Middle Triassic marine reptiles *Notho-saurus* from Poland provides the focus for the study by Surmik et al. (2017). Enigmatic ring-shaped structures, which occur as a stack or rouleau, can be seen in this sample. These had never before been recognized in fossil nor modern bone samples and, apart from red blood cell stacking, were previously undescribed in the literature (Surmik et al. 2017). SEM, EDS, and ToF-SIMS analyses were performed on eight complete and several dozen fragmentary vessel-like tubes in cortical bone. Mass spectra indicate the presence of the amino acids glycine, alanine, proline, lysine, hydroxyproline, and hydroxylysine, which are typical residues of animal collagen I (Sanni et al. 2002; Schweitzer et al. 2007b, 2009; Henss et al. 2013; Bertazzo et al. 2015). The analyses failed to detect porphyrin, which is a typical residue for blood components (Greenwalt et al. 2013; Bertazzo et al. 2015), making a red blood cell origin less likely and indicating that these structures are most likely blood vessels. The specific morphology of these vessels may represent an evolutionary adaptation for protection against decompression syndrome (Surmik et al. 2017).

Wiemann et al. (2018) also liberated soft tissue (osteocytes, blood vessels, extracellular matrix) from dinosaur and other fossil bones from sediments as old as the Late Jurassic, whereby bones from oxidizing environments, particularly fluvial deposits, produced the best results. They then analyzed the isolates using RS to test the hypothesis that their material nature consisted of altered original organic compounds, primarily protein. Raman spectroscopy suggests that the material represents advanced glycoxidation and lipoxidation end products. These are a class of *N*-heterocyclic polymers that originate from oxidative crosslinking of proteinaceous scaffolds (Wiemann et al. 2018). Crosslinking had previously been hypothesized to be an important aspect of organic phase preservation (Schäfer et al. 2000; Stankiewicz et al. 2000; Briggs 2003; Schweitzer et al. 2007a, 2007b; Boatman et al. 2014).

In 2019, Surmik et al. studied another *Nothosaurus* bone from the Middle Triassic of Poland that contained many brownish to rust-colored osteocytes. XPS and RS was performed on this sample and indicated two to three iron mineral phases: goethite and hematite being the major two, in addition to magnetite. Magnetite or feroxyhyte may indicate the primary stage of iron-mediated diagenesis, which may be transformed into the more stable goethite or hematite (Cornell and Schwertmann 2003; Surmik et al. 2019). In situ RS shows that hematite-like phases mainly correspond to bone areas of greater lacunar density and goethite-like phases to lesser lacunar density, indicating local alterations in bone microstructure (Surmik et al. 2019). ICP-MS showed an amount nearly 13 times higher of iron oxide within the bone compared to quantity in the sediment, suggesting that the first source of iron at least was endogenous in origin (Surmik et al. 2019). However, diagenesis and pore fluids may also enter the bone in form of pyrite and secondary iron mineral infillings (Pfretzschner 1998, 2000), so this must be considered.

In regard to a different aspect of the issue of soft tissue preservation in fossil pre-Quaternary bone, Saitta et al. (2019) tested the hypothesis that fossil bone in a position close to the surface, but before exposure, is inhabited by a microbiome that potentially could lead to erroneous records of preserved organics in bone. SEM and EDS showed different structures in dinosaur bones compared to recent bone when the bone apatite was removed, as well as a relative inorganic composition, which is consistent with a mineralized biofilm (Schultze-Lam et al. 1996; Decho 2010). Results from Py-GC-MS indicate that dinosaur bone lacked clear pyrolysis products indicative of high protein preservation, and its chemical composition more closely resembles that of the environment (Saitta et al. 2019). Using RNA gene sequencing and HPLC of the organics in the fossil dinosaur bone excavated under sterile conditions revealed that a diverse microbiome is indeed present in the bone and to a much lesser extent in the surrounding sediment (Saitta et al. 2019). Saitta et al. (2019) note that the microbiome in near-surface fossil bone presumably thrives on the fossil organic remains in bone and that the biogeochemical signature of the recent microbiome must indeed be taken into account in future analyses of organics in fossil bone. We note, however, that the existence of a shallow-subsurface microbiome in fossil bone does not automatically falsify hypotheses of endogeneity of organics in this fossil bone nor in any other fossil bone.

Two other studies by Saitta and colleagues looked at the preservation of fossil keratin. Saitta et al. (2017) used Py-GC-MS and experimental taphonomy to study the fossilization potential of keratin. Their analysis showed that keratin degrades during diagenesis and catagenesis through microbial action and the diagenetic hydrolysis of peptide bonds. In 2018, Saitta et al. studied the *Shuvuuia* fibers that were previously reported to preserve keratin based on immunohistochemistry (Schweitzer et al. 1999). ToF-SIMS indicated that the fibers were composed of calcium phosphate, which is usually an indication of hardened keratin (Saitta et al. 2018 and references therein), possibly a calcified feather rachis. However, endogenous organic material or amino acid signatures could not be confirmed in the fiber. Saitta et al. (2018) suggest that the immunohistochemistry results from Schweitzer et al. (1999) are false positives due to the cyanoacrylate covering of the fiber and matrix. Cyanoacrylate and potentially calcium phosphate can accumulate antibodies, leading to a false positive for keratin preservation. However, unlike in bone, where the proteins are encapsulated within a mineralized matrix and thus protected, keratin in feathers is not stabilized or shielded from the environment, resulting in microbial decay, hydrolysis, and dissolution (Saitta et al. 2017).

Some Original Observations: Permian Osteocytes

The preceding review raises the question of the maximum length of time that osteocytes and other organic phase remains can survive in fossils. Using the rationale that oxidizing, and in particular fluvial, environments appear to maximize preservation potential, we demineralized amphibian and reptile bones from the oldest fluvial deposits with abundant fossil terrestrial tetrapod remains. These are the classical early Permian redbed deposits of North Texas (e.g., Romer and Price 1940; Olson 1966; Sander 1987, 1989; Reisz 1997), which were extensively studied by leading American vertebrate paleontologists of the time from about 1870 to 1970, such as E.D. Cope, S.W. Williston, E.C. Case, A.S. Romer, and E.C. Olson. The Texas Permian redbeds yielded iconic taxa such as the sail-backed synapsid reptile *Dimetrodon* and the large temnospondyl amphibian *Eryops*. We demineralized bone samples of roughly one to two cubic centimeters in volume in 0.1M EDTA. We changed the EDTA every seven days and exposed the sample to this treatment for four to eight weeks. We report here on the preservation of osteocyte-like structures (fig. 2.2) in bones of these two taxa. These liberated structures show all the same morphological and physical features as those from much younger bone fossils, making the structures the geologically oldest record of putative organic phase preservation in fossil bone. The North Texas redbeds are formed by several formations, with material studied herein primarily originating from the Artinskian-age (286.5–290.1 Ma) Nocona Formation (Shelton et al. 2013; Reisz and Fröbisch 2014; Lucas 2018).

This discovery surpasses what was previously the oldest report of osteocyte-like structures from the Mesozoic (Triassic; Surmik et al. 2019) by ca. 40 million years, extending it back into the Paleozoic (Permian; this study). Based on our results, we suggest that (1) these structures are preserved organic phase and not taphonomic artifacts, and (2) the early Permian record in Texas is far from the actual maximum age for organic phase preservation. Although we did not perform any specific tests of endogeneity beyond direct observation, there is no a priori reason to assume that the preservation potential of such structures in Permian bone would be any different than that

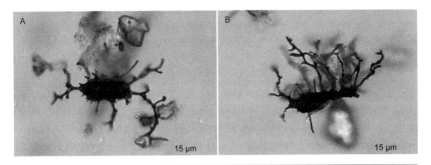

Figure 2.2. (A, B) Two examples of putative osteocytes of the temnospondyl amphibian *Eryops* from the early Permian (Artinskian) of Texas, showing a well-defined osteocyte body and branching filopodia. These specimens represent the geologically oldest liberated osteocytes to date and exceed the previously oldest record from the Middle Triassic (Anisian) by ca. 40 million years.

in Early Triassic and younger bone. We predict that the very earliest fossil bones will eventually yield osteocytes, vessels, and ECM.

Conclusions

The fossil record provides numerous vertebrate fossils of exceptional macroscopic preservation. Only recently has it been confirmed that this exceptional preservation can extend to the cellular and subcellular level of the organic phase in bone, in which delicate tissues such as blood vessels, osteocytes, and organic matter such as proteins are preserved. The preservation of the organic phase was always considered to be the exception to the rule. However, since 2005, researchers have demonstrated that this preservation is much more common than originally thought. The number of pre-Quaternary fossil bone specimens containing organic phase remains and even informative biomolecules have increased drastically over the last decade, and more are yet to be discovered. Organic phase remains have been reported from a wide variety of ages and facies, and the record has been pushed back to 290 Ma. However, there is no reason to assume that the early Permian represents the limit, and we predict that organic phase preservation will eventually be found even in the oldest vertebrate bones. Organic phase remains have been reported in a wide variety of tetrapod taxa: basal tetrapods (*Eryops*), basal synapsids, turtles, the marine reptile clades Sauropterygia and Ichthyosauria, non-avian dinosaurs, and mammals (table 2.1). Thus, organic phase and soft part preservation does not appear to be restricted to any specific clade of tetrapod but is potentially possible in fossilized tetrapod bone. To obtain further insights into the early stages of organic matter survival, we have sampled and will study Quaternary bones next. Clearly, it is crucial to find methods to prove unambiguously the endogenous origin of the organic phase from the fossil bones.

Table 2.1.
Compilation of reports of organic tissue/matter preservation in bone, listed in the same order as they appear in the text; — indicates not applicable or a lack of information

Study	Material	Bone	Age	Locality	Facies
First Evidence of Organic Remains in Fossil Dinosaur Bone					
Pawlicki et al. 1966	Unspecified dinosaur bone	Phalange	Upper Cretaceous	Gobi Desert, Mongolia	—
Miller and Wyckoff 1969	Unspecified dinosaur bone	Phalange	Lance Formation	Gilbert Creek, USA	—
	Unspecified mosasaur bone	Piece of scrap bone	Niobrara Formation	Kansas, USA	—
	Sauropod	Limb bone, sacrum	Morrison Formation	Howe Quarry; Como Bluff; Bone Cabin Quarry, Wyoming, USA	—
Pioneer of Molecular Paleontology: Mary Schweitzer and Colleagues					
Gurley et al. 1991	*Seismosaurus*	Vertebra	Morrison Formation	New Mexico, USA	Fluvial
Muyzer et al. 1992	Bovid	Metatarsal	Pleistocene	Java	—
	Dinosaur	Limb bones	Upper Cretaceous	Alberta, Canada	—
Schweitzer et al. 1997	*Tyrannosaurus rex*	Hind limb	Hell Creek Formation	Montana, USA	Fluvial
Embery et al. 2000	*Iguanodon*	Dorsal rib	Upper Weald Clay	Smokejacks Brickworks, Ockley, Surrey, UK	—
Schweitzer et al. 2005	*Tyrannosaurus rex*	Several elements	Hell Creek Formation	Montana, USA	Estuarine
Schweitzer et al. 2007a	*Tyrannosaurus rex*	Left and right femora	Hell Creek Formation	Montana, USA	Estuarine
Schweitzer et al. 2007b	*Mammuthus columbi*	—	Doeden gravel beds	Montana, USA	Fluvial
	Tyrannosaurus rex	—	Hell Creek Formation	Montana, USA	Fluvial
	Brachylophosaurus canadensis	—	Lower Judith River Formation	Montana, USA	Fluvial
Revisiting Previous Samples with New Techniques					
Schweitzer et al. 2009	*Brachylophosaurus canadensis*	Femur	Lower Judith River Formation	Montana, USA	Fluvial
Schroeter et al. 2017	*Brachylophosaurus canadensis*	Femur	Lower Judith River Formation	Montana, USA	Fluvial
Boatman et al. 2019	*Tyrannosaurus rex*	—	Hell Creek Formation	Montana, USA	Fluvial

As analytical techniques improve in resolution and sensitivity, more organic phase remains, such as other proteins, may be recovered from fossils, and the existing reports will solidify. This could possibly lead to the promise of molecular paleontology becoming a reality, which would allow for a better understanding of the phylogeny of extinct taxa using molecular evi-

Extraction	Descriptive	Analytical	Soft tissues and/or organic matter
Not specified	LM, SEM	—	Osteocyte lacunae with canaliculi and spiral collagen fibrils in vascular canal
HCl	—	Ninhydrin test, Phoenix automatic amino acid analyzer	Amino acids
HCl	—	HPLC	Proteins and amino acids
EDTA	—	Immunology (DIBA)	Proteins and amino acids
EDTA	Spectroscopy	HPLC, NMR, RS, ELISA	Proteins and amino acids
EDTA, GuHCl extraction	—	FPLC, SDS-PAGE, Immunology	Proteins and amino acids
EDTA	LM, SEM	—	Blood vessels with possible cell nuclei and osteocytes
EDTA	AFM, TEM	ELISA, SAED, IHC, ToF-SIMS	Blood vessels with possible cell nuclei, osteocytes, and proteins
EDTA	LM, SEM	—	Collagen fibers, blood vessels, and osteocytes
			Blood vessels, osteocytes, and ECM
			Blood vessels, osteocytes, and ECM
EDTA	SEM	IHC, ToF-SIMS	Bone matrix, blood vessels, possible remains of red blood cells, osteocytes with filopodia, and proteins
EDTA, GuHCl extraction	—	SDS-PAGE, MS	Proteins and collagen
EDTA	SEM, SAXS	SR-FTIR, ICH, µXRF, µXANES	Blood vessels, proteins, and fibrillar collagen

(*continued*)

Table 2.1. (continued)

Study	Material	Bone	Age	Locality	Facies
The Biofilm Hypothesis					
Kaye et al. 2008	Several dinosaurs	—	Hell Creek Formation, Lance Formation	—	—
Schweitzer et al. 2013	*Tyrannosaurus rex*	—	Hell Creek Formation	Montana, USA	Fluvial
	Brachylophosaurus canadensis	—	Lower Judith River Formation	Montana, USA	Fluvial
Schweitzer et al. 2014	Tyrannosaurus rex	—	Hell Creek Formation	Montana, USA	Fluvial
	Brachylophosaurus canadensis	—	Lower Judith River Formation	Montana, USA	Fluvial
The Past Decade: A Diversity of Labs Evaluate Organic Phase Preservation in Fossil Bone					
Bertazzo et al. 2015	Several dinosaurs	—	Dinosaur Park Formation and Lance Formation, North America	—	—
Surmik et al. 2016	*Nothosaurus* sp.	Humerus	Gogolin Formation	Silesia, Poland	Marine
	Protanystropheus sp.	Vertebral centrum	Gogolin Formation	Silesia, Poland	Marine
Lee et al. 2017	*Lufengosaurus*	Long bones and ribs	Early Jurassic	Yunnan Province, China	—
Plet et al. 2017	*Stenopterygius*	Vertebra	Toarcian Posidonia Shale Formation	Dotternhausen, Germany	Marine
Surmik et al. 2017	*Nothosaurus* sp.	Coracoid	Gogolin Formation	Silesia, Poland	Marine
Wiemann et al. 2018	*Allosaurus, Apatosaurus*	—	Morrison Formation	Wyoming, USA	Fluvial
	Oviraptorid	—	Late Cretaceous	Jiangxi Province, China	—
Surmik et al. 2019	*Nothosaurus* cf. *marchicus*	Humerus	Lower Muschelkalk	Miasteczko Śląskie, Poland	Marine
Saitta et al. 2019	*Centrosaurus apertus*	—	Dinosaur Park Formation	Alberta, Canada	Alluvial

dence, physiological strategies, rate and direction of molecular evolution, and the fossilization process at the molecular level.

ACKNOWLEDGMENTS

The authors thank the two anonymous reviewers who greatly improved this chapter, editor Carole Gee for her helpful comments, and Marie Koschowitz, who started the first round of experiments in 2013. This work was funded by the Deutsche Forschungsgemeinschaft (DFG, German Research Foundation), Project number 396703500 to PMS and Christa Müller (both at the University of Bonn). This is contribution num-

Extraction	Descriptive	Analytical	Soft tissues and/or organic matter
EDTA	SEM	EDS	Tube-like structures with iron oxide framboids (previously interpreted by others as "blood vessels with possible cell nuclei") and bacteria (previously interpreted by others as "osteocytes")
EDTA	LM, SEM, TEM, FM	ICH, MS	Osteocytes with filopodia, proteins, and possible DNA
EDTA	TEM	EELS, µXRF, µXANES	Blood vessels and osteocytes
—	SEM, TEM, EDS, FIB	ToF-SIMS, EELS, PLS-DA	Collagen fiber, amino acids, and protein
EDTA	SEM	XRD, FTIR, ToF-SIMS, XPS	Blood vessels, proteins, and amino acids
—	—	SR-FTIR, RS	Proteins and collagen
—	SEM, TEM, FIB	XRD, µXRF, ToF-SIMS	Red and white blood cells, platelets, and collagen fibers
Nitric acid	SEM, EDS	ToF-SIMS	Tube-shaped, vessel-like structures
HCl	Chemospace	RS	Altered proteins from crosslinked structures
EDTA	ESEM	RS, XPS, ICP-MS	Osteocyte-like microbodies
HCl	LM, VPSEM, FM	EDS, ATR FTIR, Py-GC-MS, HPLC, radiocarbon AMS, 16S rRNA amplicon sequencing	Proteins and amino acids

ber 16 of the DFG Research Unit 2685, "The Limits of the Fossil Record: Analytical and Experimental Approaches to Fossilization."

WORKS CITED

Aarden, E.M., Nijweide, P.J., and Burger, E.H. 1994. Function of osteocytes in bone. *Journal of Cellular Biochemistry*, 55: 287–299.

Asara, J.M., Schweitzer, M.H., Freimark, L.M., Phillips, M., and Cantley, L.C. 2007. Protein sequences from mastodon and *Tyrannosaurus rex* revealed by mass spectrometry. *Science*, 316: 280–285.

Barth, A. 2007. Infrared spectroscopy of proteins. *Biochimica et Biophysica Acta*, 1767: 1073–1101.

Bell, L.S., Skinner, S.M.F., and Jones, J. 1996. The speed of post mortem change to the human skeleton and its taphonomic significance. *Forensic Science International*, 82: 129–140.

Berna, F., Matthews, A., and Weiner, S. 2004. Solubilities of bone mineral from archaeological sites: The recrystallization window. *Journal of Archaeological Science*, 31: 867–882.

Bertazzo, S., Maidment, S.C.R., Kallepitis, C., Fearn, S., Stevens, M.M., and Xie, H.N. 2015. Fibres and cellular structures preserved in 75-million-year-old dinosaur specimens. *Nature Communications*, 6: 7352.

Boatman, E.M., Goodwin, M.B., Holman, H.Y., Fakra, S., Schweitzer, M.H., Gronsky, R., and Horner, J.R. 2014. Synchrotron chemical and structural analysis of *Tyrannosaurus rex* blood vessels: The contribution of collagen hypercrosslinking to tissue longevity. *Microscopy and Microanalysis*, 20: 1430–1431.

Boatman, E.M., Goodwin, M.B., Holman, H.-Y.N., Fakra, S., Zheng, W., Gronsky, R., and Schweitzer, M.H. 2019. Mechanisms of soft tissue and protein preservation in *Tyrannosaurus rex*. *Scientific Reports*, 9: 15678.

Bocherens, H., Tresset, A., Bidlack, F.B., Giligny, F., Lafage, F., Lanchon, Y., and Mariotti, A. 1997. Diagenetic evolution of mammal bones in two French Neolithic sites. *Bulletin de la Société Géologique de France*, 168: 555–564.

Briggs, D.E.G. 2003. The role of decay and mineralization in the preservation of soft-bodies fossils. *Annual Review of Earth and Planetary Sciences*, 31: 275–301.

Briggs, D.E.G., and Kear, A.J. 1994. Decay and mineralization of shrimps. *Palaios*, 9: 431–456.

Briggs, D.E.G., Kear, A.J., Martill, D.M., and Wilby, P.R. 1993. Phosphatization of soft-tissue in experiments and fossils. *Journal of the Geological Society*, 150: 1035–1038.

Briggs, D.E.G., and McMahon, S. 2016. The role of experiments in investigating the taphonomy of exceptional preservation. *Palaeontology*, 59: 1–11.

Brusatte, S.L. 2012. *Dinosaur Paleobiology*. Wiley-Blackwell, Oxford.

Butterfield, N.J. 1990. Organic preservation of non-mineralizing organisms and the taphonomy of the Burgess Shale. *Paleobiology*, 16: 272–286.

Butterfield, N.J. 2003. Exceptional fossil preservation and the Cambrian explosion. *Integrative and Comparative Biology*, 43: 166–177.

Cadena, E.A. 2016. Microscopical and elemental FESEM and Phenom ProX-SEM-EDS analysis of osteocyte and blood vessel-like microstructures obtained from fossil vertebrates of the Eocene Messel Pit, Germany. *PeerJ*, 4: e1618.

Cadena, E.A., and Schweitzer, M.H. 2012. Variation in osteocytes morphology vs bone type in turtle shell and their exceptional preservation from the Jurassic to the present. *Bone*, 51: 614–620.

Cadena, E.A., and Schweitzer, M.H. 2014. Preservation of blood vessels and osteocytes in a pelomedusoid turtle from the Paleocene of Columbia. *Journal of Herpetology*, 48: 125–129.

Cambra-Moo, O., and Buscalioni, A.D. 2008. An approach to the study of variations in early stages of *Gallus gallus* decomposition. *Journal of Taphonomy*, 6: 21–40.

Canavan, H.E., Graham, D.J., Cheng, X., Ratner, B.D., and Castner, D.G. 2006. Comparison of native extracellular matrix with adsorbed protein films using secondary ion mass spectrometry. *Langmuir*, 23: 50–56.

Carpenter, K. 2007. How to make a fossil: Part 2—Dinosaur mummies and other soft tissue. *The Journal of Paleontological Science*, JPS.C.07.0002.

Carter, W.G., Kaur, P., Gil, S.G., Gahr, P.J., and Wayner, E.A. 1990. Distinct functions for integrins α3β1 in focal adhesions and α6β4/bullous pemphigoid antigen in a new stable anchoring contact (SAC) of keratinocytes: Relationship to hemidesmosomes. *Journal of Cellular Biology*, 11: 3141–3154.

Child, A.M. 1995. Towards an understanding of the microbial decomposition of archaeological bone in the burial environment. *Journal of Archaeological Science*, 22: 165–174.

Cleland, T.P., Schroeter, E.R., and Schweitzer, M.H. 2015. Biologically and diagenetically derived peptide modifications in moa collagens. *Proceedings of the National Academy of Sciences of the United States of America*, 282: 20150015.

Cleland, T.P., Voegele, K., and Schweitzer, M.H. 2012. Empirical evaluation of bone extraction protocols. *PLOS ONE*, 7: e31443.

Collins, M.J., Gernaey, A.M., Nielsen-Marsh, C.M., Vermeer, C., and Westbroek, P. 2000. Slow rates of degradation of osteocalcin: Green light for fossil bone protein? *Geology*, 28: 1139–1142.

Collins, M.J., Nielsen-Marsh, C.M., Hiller, J., Smith, C.I., and Roberts, J.P. 2002. The survival of organic matter in bone: A review. *Archaeometry*, 44: 383–394.

Collins, M.J., Riley, M.S., Child, A.M.T., and Turner-Walker, G. 1995. A basic mathematical model for the chemical degradation of ancient collagen. *Journal of Archaeological Science*, 22: 175–183.

Cooper, E.A., and Knutson, K. 1995. Fourier transform infrared spectroscopy investigations of protein structure. *Pharmaceutical Biotechnology*, 7: 101–143.

Cornell, R.M., and Schwertmann, U. 2003. *The Iron Oxides: Structure, Properties, Reactions, Occurrences, and Uses.* Wiley, Weinheim, Germany.

Cowin, S.C. 2002. Mechanosensation and fluid transport in living bone. *Journal of Musculoskeletal and Neuronal Interactions*, 2: 256–260.

Currey, J.D. 2012. The structure and mechanics of bone. *Journal of Materials Science*, 47: 41–54.

Curry-Rogers, K., and Wilson, J.A. 2005. *The Sauropods: Evolution and Paleobiology.* University of California Press, Berkeley.

Daniel, J.C., and Chin, K. 2010. The role of bacterially mediated precipitation in the permineralization of bone. *Palaios*, 25: 507–516.

Darroch, S.A.F., Laflamme, M., Schiffbauer, J.D., and Briggs, D.E.G. 2012. Experimental formation of a microbial death mask. *Palaios*, 27: 293–303.

Decho, A.W. 2010. Overview of biopolymer-induced mineralization: What goes on in biofilms? *Ecological Engineering*, 36: 137–144.

Dumont, M., Borbély, A., Kostka, A., Sander, P.M., and Kaysser-Pyzalla, A. 2011. Characterization of sauropod bone structure, pp. 149–170. In: Klein, N., Remes, K., Gee, C.T., and Sander, P.M., eds. *Biology of the Sauropod Dinosaurs: Understanding the Life of Giants.* Indiana University Press, Bloomington.

Dwivedi, S.K., Dey, S., and Swarup, D. 1997. Hydrofluorosis in water buffalo (*Bubalus bubalis*) in India. *Science of the Total Environment*, 207: 105–109.

Ebert, P.S., and Prockop, D.J. 1962. The hydroxylation of proline to hydroxyproline during the synthesis of collagen in chick embryos. *Biochemical and Biophysical Research Communications*, 8: 305–309.

Edmund, A.G. 1960. Tooth replacement phenomena in lower vertebrates. *Royal Ontario Museum, Life Sciences Division Contributions*, 52: 1–190.

Edmund, A.G. 1969. Dentition, pp. 117–200. In: Gans, C., Bellairs, A.d'A., and Parsons, T.S., eds. *Biology of the Reptilia Vol.1: Morphology A*. Academic Press, New York.

Elliott, J.C., Wilson, R.M., and Dowker, S.E.P. 2002. Apatite structures. *Advances in X-Ray Analysis*, 45: 172–181.

Embery, G., Milner, A., Waddington, R.J., Hall, R.C., Langley, M.S., and Milan, A.M. 2000. The isolation and detection of non-collagenous proteins from the compact bone of the dinosaur *Iguanodon*. *Connective Tissue Research*, 41: 249–259.

Farbod, K., Nejadnik, M.R., Jansen, J.A., and Leeuwenburgh, S.C.G. 2014. Interactions between inorganic and organic phases in bone tissue as a source of inspiration for design of novel nanocomposites. *Tissue Engineering Part B: Reviews*, 20: doi:10.1089/ten.teb.2013.0221.

Florencio-Silva, R., Sasso, G.R., Sasso-Cerri, E., Simões, M.J., and Cerri, P.S. 2015. Biology of bone tissue: Structure, function, and factors that influence bone cells. *BioMed Research International*, 2015: doi:10.1155/2015/421746.

Francillon-Vieillot, H., de Buffrénil, V., Castanet, J.D., Géraudie, J., Meunier, F.J., Sire, J.Y., Zylberberg, L., and de Ricqlès, A. 1990. Microstructure and mineralization of vertebrate skeletal tissues, pp. 473–530. In: Carter, J.G., ed., *Skeletal Biomineralization: Patterns, Processes and Evolutionary Trends, Vol. 1*. Van Nostrand Reinhold, New York.

Fujisawa, R., and Tamura, M. 2012. Acidic bone matrix proteins and their roles in calcification. *Frontiers in Bioscience*, 17: 1891–1903.

Gehling, J.G. 1999. Microbial mats in terminal Proterozoic siliciclastics: Ediacaran death masks. *Palaios*, 14: 40–57.

Glimcher, M.J., Cohen-Solal, L., Kossiva, D., and de Ricqles, A. 1990. Biochemical analyses of fossil enamel and dentin. *Paleobiology*, 16: 219–232.

Greenwalt, D.E., Goreva, Y.S., Siljeström, S.M., Rose, T., and Harbach, R.E. 2013. Hemoglobin-derived porphyrins preserved in a Middle Eocene blood-engorged mosquito. *Proceedings of the National Academy of Sciences of the United States of America*, 110: 18496–18500.

Gregory, T.R. 2001. The bigger the C-value, the larger the cell: Genome size and red blood cell size in vertebrates. *Blood Cells, Molecules and Diseases*, 27: 830–843.

Gurley, L.R., Valdez, J.G., Spall, W.D., Smith, B.F., and Gillette, D.D. 1991. Proteins in the fossil bone of the dinosaur, *Seismosaurus*. *Journal of Protein Chemistry*, 10: 75–90.

Hall, B.K. 2015. *Bones and Cartilage: Developmental and Evolutionary Skeletal Biology*. Elsevier Academic Press, San Diego, California.

Halstead, L.B. 1974. *Vertebrate Hard Tissues*. Wykeham Publication, London.

Hedges, R.E.M. 2002. Bone diagenesis: An overview of processes. *Archaeometry*, 44: 319–328.

Hedges, R.E.M., and Millard, A.R. 1995. Bones and groundwater towards the modelling of diagenetic processes. *Journal of Archaeological Science*, 22: 155–165.

Henss, A., Rohnke, M., Khassawna, T.E., Govindaraian, P., Schlewitz, G., Heiss, C., and Janek, J. 2013. Applicability of ToF-SIMS for monitoring compositional changes in bone in a long-term animal model. *Journal of the Royal Society Interface*, 10: 20130332.

Hubert, J.F., Panish, P.T., Chure, D.J., and Prostak, K.S. 1996. Chemistry, microstructure, petrology, and diagenetic model of Jurassic dinosaur bones, Dinosaur National Monument Utah. *Journal of Sedimentary Research*, 66: 531–547.

Iniesto, M., Buscalioni, Á.D., Carmen Guerrero, M., Benzerara, K., Moreira, D., and López-Archilla, A.I. 2016. Involvement of microbial mats in early fossilization by decay delay and formation of impressions and replicas of vertebrates and invertebrates. *Scientific Reports*, 6: 25716.

Jackson, M., Choo, L.P., Watson, P.H., Halliday, W.C., and Mantasch, H.H. 1995. Beware of connective tissue proteins: Assignment and implications of collagen absorptions in infrared spectra of human tissues. *Biochimica et Biophysica Acta*, 1270: 1–6.

Karima, B., Razia, N., Gilles, G., and Cyril, P. 2009. Collagen types analysis and differentiation by FTIR spectroscopy. *Analytical and Bioanalytical Chemistry*, 395: 829–837.

Kaye, T.G., Gaugler, G., and Sawlowicz, Z. 2008. Dinosaurian soft tissues interpreted as bacterial biofilms. *PLOS ONE*, 3: e2808.

Keenan, S.W., and Engel, A.S. 2017. Early diagenesis and recrystallization of bone. *Geochimica et Cosmochimica Acta*, 196: 209–223.

Kharalkar, N.M., Bauserman, S.C., and Valvano, J.W. 2009. Effect of formalin fixation on thermal conductivity of the biological tissue. *Journal of Biomechanical Engineering*, 131: 074508.

Kiprijanoff, A.V. 1881–1883. Studien über die fossilen Reptilien Russlands. *Mémoires de l'Académie Impériale des Sciences de Saint-Pétersbourg*, 7: 1–144.

Klein, N., Remes, K., Gee, C.T., and Sander, P.M., eds. 2011. *Biology of the Sauropod Dinosaurs: Understanding the Life of Giants*. Indiana University Press, Bloomington, Indiana.

Kohn, M.J., and Cerling, T.E. 2002. Stable isotope compositions of biological apatite, pp. 455–488. In: Kohn, M.J., Rakovan, J., and Hughes, J.M., eds., *Phosphates: Geochemical, Geobiological and Material Importance, Reviews in Mineralogy and Geochemistry*. Mineralogical Society of America, Washington, DC.

Labat-Robert, J., Bihari-Varga, M., and Robert, L. 1990. Extra-cellular matrix. *FEBS Letters*, 268: 386.

Lee, Y.-C., Chiang, C.-C., Huang, P.-Y., Chung, C.-Y., Huang, T.D., Wang, C.-C., Chen, C.-I., Chang, R.-S., Liao, C.-H., and Reisz, R.R. 2017. Evidence of preserved collagen in an Early Jurassic sauropodomorph dinosaur revealed by synchrotron FTIR microspectroscopy. *Nature Communications*, 8: 14220.

Leopold, L.F., Leopold, N., Diehl, H.-A., and Socaciu, C. 2011. Quantification of carbohydrates in fruit juices using FTIR spectroscopy and multivariate analysis. *Journal of Spectroscopy*, 26: 93–104.

Lindgren, J., Uvdal, P., Engdahl, A., Lee, A.H., Alwmark, C., Bergquist, K.E., Nilsson, E., Ekström, P., Rasmussen, M., Douglas, D.A., Polcyn, M.J., and Jacobs, L.L. 2011. Microspectroscopic evidence of Cretaceous bone proteins. *PLOS ONE*, 6: e19445.

Li, Z., and Pasteris, J.D. 2014. Tracing the pathway of compositional changes in bone mineral with age: Preliminary study of bioapatite aging in hypermineralized dolphin's bulla. *Biochimica et Biophysica Acta*, 1840: 2331–2339.

Little, K., Kelly, M., and Courts, A. 1962. Studies on bone matrix in normal and osteoporotic bone. *The Journal of Bone and Joint Surgery*, 44: 503–519.

Lucas, S.G. 2018. Permian tetrapod biochronology, correlation and evolutionary events, pp. 405–444. In: Lucas, S.G., and Shen, S.Z., eds., *The Permian Timescale*. Geological Society of London.

Martill, D.M. 1988. Preservation of fish in the Cretaceous of Brazil. *Palaeontology*, 1–18.

Martill, D.M. 1989. The Medusa effect: Instantaneous fossilization. *Geology Today*, 5: 201–205.

Miller, M.F., and Wyckoff, R.W.G. 1969. Proteins in dinosaur bones. *Proceedings of the National Academy of Sciences of the United States of America*, 60: 176–178.

Morgan, E.F., Barnes, G.L., and Einhorn, T.A. 2008. The bone organ system: Form and function, pp. 1–25. In: Marcus, R., Feldman, D., Nelson, D., and Rosen, C.J., eds., *Fundamentals of Osteoporosis*. Elsevier Academic Press, San Diego.

Müller, A.H. 1957. *Lehrbuch der Paläozoologie*. Fischer Verlag, Jena.

Murshid, S.A., Kamioka, H., Ishihara, Y., Ando, R., Sugawara, Y., and Takano-Yamamoto, T. 2007. Actin and microtubule cytoskeletons of the processes of 3D-cultured MC3T3-E1 cells and osteocytes. *Journal of Bone and Mineral Metabolism*, 25: 151–158.

Muyzer, G., Sandberg, P., Knapen, M.H.J., Vermeer, C., Collins, M., and Westbroek, P. 1992. Preservation of the bone protein osteocalcin in dinosaurs. *Geology*, 20: 871–874.

Nguyen, T.-H., and Lee, B.-T. 2010. Fabrication and characterization of cross-linked gelatin electro-spun nano-fibers. *Journal of Biomedical Science and Engineering*, 3: 1117–1124.

Nielsen-Marsh, C.M., and Hedges, R.E.M. 1999. Bone porosity and the use of mercury intrusion porosimetry in bone diagenesis studies. *Archaeometry*, 41: 165–174.

O'Brien, N.R., Meyer, H.W., and Harding, I.C. 2008. The role of biofilms in fossil preservation, Florissant Formation, Colorado. *The Geological Society of America Special Paper*, 435: 19–31.

Olson, E.C. 1966. Community evolution and the origin of mammals. *Ecology*, 47: 291–302.

Owen, R., ed.1845–1856. *Descriptive and Illustrated Catalogue of the Fossil Organic Remains of Mammalia and Aves Contained in the Museum of the Royal College of Surgeons of England*. Royal College of Physicians of London.

Padian, K., and Lamm, E. 2013. *Bone Histology of Fossil Tetrapods: Advancing Methods, Analysis, and Interpretation*. University of California Press, Berkeley.

Palmer, L.C., Newcomb, C.J., Kaltz, S.R., Spoerke, E.D., and Stupp, S.I. 2008. Biomimetic systems for hydroxyapatite mineralization inspired by bone and enamel. *Chemical Reviews*, 108: 4754.

Pasteris, J.D., Wopenka, B., Freeman, J.J., Rogers, K., Valsami-Jones, E., van der Houwen, J., and Silva, M.J. 2004. Lack of OH in nanocrystalline apatite as a function of degree of atomic order: Implications for bone and biomaterials. *Biomaterials*, 25: 229–238.

Pawlicki, R., Korbel, A., and Kubiak, H. 1966. Cells, collagen fibrils and vessels in dinosaur bone. *Nature*, 6: 655–657.

Peterson, J.E., Lenczewski, M.E., and Scherer, R.P. 2010. Influence of microbial biofilms on the preservation of primary soft tissue in fossil and extant archosaurs. *PLOS ONE*, 5: e13334.

Pfretzschner, H.-U. 1998. Frühdiagenetische Prozesse bei der Fossilisation von Knochen. *Neues Jahrbuch für Geologie und Paläontologie*, 210: 369–397.

Pfretzschner, H.-U. 2000. Collagen gelatinization: The key to understand early bone-diagenesis. *Palaeontographica Abteilung A*, 278: 135–148.

Pfretzschner, H.-U. 2001. Pyrite in fossil bone. *Neues Jahrbuch für Geologie und Paläontologie Abhandlung*, 220: 1–23.

Pfretzschner, H.-U. 2004. Fossilization of Haversian bone in aquatic environments. *Comptes Rendus Palevol*, 3: 605–616.

Plet, C., Grice, K., Pagès, A., Verrall, M., Coolen, M.J.L., Ruebsam, W., Rickard, W.D.A., and Schwark, L. 2017. Palaeobiology of red and white blood cell-like structures, collagen and cholesterol in an ichthyosaur bone. *Scientific Reports*, 7: 13776.

Prieto-Márquez, A., Weishampel, D.B., and Horner, J.R. 2006. The dinosaur *Hadrosaurus foulkii*, from the Campanian of the East Coast of North America, with a reevaluation of the genus. *Acta Palaeontologica Polonica*, 51: 77–98.

Raff, E.C., Andrews, M.E., Turner, F.R., Toh, E., Nelson, D.E., and Raff, R.A. 2013. Contingent interactions among biofilm-forming bacteria determine preservation or decay in the first steps toward fossilization of marine embryos. *Evolution and Development*, 15: 243–256.

Raff, E.C., Schollaert, K.L., Nelson, D.E., Donoghue, P.C.J., Thomas, C.-W., Turner, F.R., Stein, B.D., Dong, X., Bengtson, S., Huldtgren, T., Stampanoni, M., Chongyu, Y., and Raff, R.A. 2008. Embryo fossilization is a biological process mediated by microbial biofilms. *Proceedings of the National Academy of Sciences of the United States of America*, 105: 19360–19365.

Rasmussen, M., Jacobsson, M., and Björck, L. 2003. Genome-based identification and analysis of collagen-related structural motifs in bacterial and viral proteins. *Journal of Biological Chemistry*, 278: 32313–32316.

Reisz, R.R. 1997. The origin and early evolutionary history of amniotes. *Trends in Ecology and Evolution*, 12: 218–222.

Reisz, R.R., and Fröbisch, J. 2014. The oldest caseid synapsid from the Late Pennsylvanian of Kansas, and the evolution of herbivory in terrestrial vertebrates. *PLOS ONE*, 9: e94518.

Riding, R. 2000. Microbial carbonates: The geological record of calcified bacterial-algal mats and biofilms. *Sedimentology*, 47: 179–214.

Robert, V., and Pierre, M.D. 2002. Red cell morphology and peripheral blood film. *Clinics in Laboratory Medicine*, 22: 25–61.

Romer, A.S., and Price, L.I. 1940. Review of the Pelycosauria. *Geological Society of America Special Paper*, 28: 1–538.

Saitta, E.T., Fletcher, I., Martin, P., Pittman, M., Kaye, T.G., True, L.D., Norell, M., Abbott, G.D., Summons, R.E., Penkman, K.E., and Vinther, J. 2018. Preservation of feather fibers from the Late Cretaceous dinosaur *Shuvuuia deserti* raises concern about immunohistochemical analyses on fossils. *Organic Geochemistry*, 125: 142–151.

Saitta, E.T., Liang, R., Lau, M.C., Brown, C.M., Longrich, N.R., Kaye, T.G., Novak, B.J., Salzberg, S.L., Norell, M.A., Abbott, G.D., Dickinson, M.R., Vinther, J., Bull, I.D., Brooker, R.A., Martin, P., Donohoe, P., Knowles, T.D., Penkman, K.E., and Onstott, T. 2019. Cretaceous dinosaur bone contains recent organic material and provides an environment conducive to microbial communities. *eLife*, 8: e46205.

Saitta, E.T., Rogers, C., Brooker, R.A., Abbott, G.D., Kumar, S., O'Reilly, S.S., Donohoe, P., Dutta, S., Summons, R.E., and Vinther, J. 2017. Low fossilization potential of keratin protein revealed by experimental taphonomy. *Palaeontology*, 60: 547–556.

San Antonio, J.D., Schweitzer, M.H., Jensen, S.T., Kalluri, R., Buckley, M., and Orgel, J.P.R.O. 2011. Dinosaur peptides suggest mechanisms of protein survival. *PLOS ONE*, 6: e20381.

Sander, P.M. 1987. Taphonomy of the Lower Permian Geraldine Bonebed in Archer County, Texas. *Palaeogeography, Palaeoclimatology, Palaeoecology*, 61: 221–236.

Sander, P.M. 1989. Early Permian depositional environments and pond bonebeds in central Archer County, Texas. *Palaeogeography, Palaeoclimatology, Palaeoecology*, 69: 1–21.

Sander, P.M. 1999. The microstructure of reptilian tooth enamel: Terminology, function, and phylogeny. *Münchener Geowissenschaftliche Abhandlungen*, 38: 1–102.

Sander, P.M. 2000. Longbone histology of the Tendaguru sauropods: Implications for growth and biology. *Paleobiology*, 26: 466–488.

Sanni, O.D., Wagner, M.S., Briggs, D.E.G., Castner, D.G., and Vickerman, J.C. 2002. Classification of adsorbed protein static ToF-SIMS spectra by principal component analysis and neural networks. *Surface and Interface Analysis*, 33: 715–728.

Schäfer, F.Q., Yue Qian, S., and Buettner, G.R. 2000. Iron and free radical oxidations in cell membranes. *Cellular and Molecular Biology*, 46: 657–662.

Schönherr, E., and Hausser, H.J. 2000. Extracellular matrix and cytokines: A functional unit. *Developmental and Comparative Immunology*, 7: 89–101.

Schroeter, E.R., DeHart, C.J., Cleland, T.P., Zheng, W., Thomas, P.M., Kelleher, N.L., Bern, M., and Schweitzer, M.H. 2017. Expansion for the *Brachylophosaurus canadensis* collagen I sequence and additional evidence of the preservation of Cretaceous protein. *Journal of Proteome Research*, 16: 920–932.

Schultze-Lam, S., Fortin, D., Davis, B.S., and Beveridge, T.J. 1996. Mineralization of bacterial surfaces. *Chemical Geology*, 132: 171–181.

Schweitzer, M.H. 2011. Soft tissue preservation in terrestrial Mesozoic vertebrates. *Annual Review of Earth and Planetary Sciences*, 39: 187–216.

Schweitzer, M.H., Avci, R., Collier, T., and Goodwin, M.B. 2008. Microscopic, chemical and molecular methods for examining fossil preservation. *Comptes Rendus Palevol*, 7: 159–184.

Schweitzer, M.H., Marshall, M., Carron, K., Bohle, D.S., Busse, S.C., Arnold, E.V., Barnard, D., Horner, J.R., and Starkey, J.R. 1997. Heme compounds in dinosaur trabecular bone. *Proceedings of the National Academy of Sciences of the United States of America*, 94: 6291–6296.

Schweitzer, M.H., Suo, Z., Avci, R., Asara, J.M., Allen, M.A., Teran Arce, F., and Horner, J.R. 2007a. Analyses of soft tissue from *Tyrannosaurus rex* suggest the presence of protein. *Science*, 316: 277–280.

Schweitzer, M.H., Watt, J.A., Avci, R., Knapp, L., Chiappe, L., Norell, M., and Marshall, M. 1999. Beta-keratin specific immunological reactivity in feather-like structures of the Cretaceous alvarezsaurid, *Shuvuuia deserti*. *Journal of Experimental Zoology*, 285: 146–157.

Schweitzer, M.H., Wittmeyer, J.L., and Horner, J.R. 2007b. Soft tissue and cellular preservation in vertebrate skeletal elements from the Cretaceous to the present. *Proceedings of the Royal Society B*, 274: 183–197.

Schweitzer, M.H., Wittmeyer, J.L., Horner, J.R., and Toporski, J. K. 2005. Soft-tissue vessels and cellular preservation in *Tyrannosaurus rex*. *Science*, 307: 1952–1955.

Schweitzer, M.H., Zheng, W., Cleland, T.P., and Bern, M. 2013. Molecular analyses of dinosaur osteocytes support the presence of endogenous molecules. *Bone*, 52: 414–423.

Schweitzer, M.H., Zheng, W., Cleland, T.P., Goodwin, M.B., Boatman, E., Theil, E., Marcus, M.A., and Fakra, S.C. 2014. A role of iron and oxygen chemistry in preserving soft tissues, cells and molecules from deep time. *Proceedings of the Royal Society B*, 281: 20132741.

Schweitzer, M.H., Zheng, W., Organ, C.L., Avci, R., Suo, Z., Freimark, L.M., Lebleu, V.S., Duncan, M.B., Vander Heiden, M.G., Neveu, J.M., Lane, W.S., Cottrell, J.S., Horner, J.R., Cantley, L.C., Kalluri, R., and Asara, J.M. 2009. Biomolecular characterization and protein sequences of the Campanian hadrosaur *B. canadensis*. *Science*, 324: 626–631.

Schweitzer, M.H., Moyer, A.E., and Zheng, W. 2016. Testing the hypothesis of biofilm as a source for soft tissue and cell-like structures preserved in dinosaur bone. *PLOS ONE*, 11: e0150238.

Schwertmann, U., and Murad, E. 1983. The effect of pH on the formation of goethite and hematite from ferrihydrite. *Clays and Clay Minerals*, 31: 277–284.

Service, R.F. 2017. Researchers close in on ancient dinosaur proteins. "Milestone" paper opens door to molecular approach. *Science*, 355: 441–442.

Shelton, C., Sander, P.M., Stein, K., and Winkelhorst, H. 2013. Long bone histology indicates sympatric species of *Dimetrodon* (Lower Permian, Sphenacodontidae). *Earth and Environmental Science Transactions of the Royal Society of Edinburgh*, 103: 217–236.

Stankiewicz, B.A., Briggs, D.E.G., Michels, R., Collinson, M.E., Flannery, M.B., and Evershed, R.P. 2000. Alternative origin of aliphatic polymer in kerogen. *Geology*, 28: 559–562.

Sukhanov, V.B. 2000. Mesozoic turtles of Middle and Central Asia, pp. 309–367. In: Benton, M.J., Shishkin, M.A., Unwin, D.M., and Kurichkin, E.N., eds., *The Age of Dinosaurs in Russia and Mongolia*. Cambridge University Press, UK.

Surmik, D., Boczarowski, A., Balin, K., Dulski, M., Szade, J., Kremer, B., and Pawlicki, R. 2016. Spectroscopic studies on organic matter from Triassic reptile bones, Upper Silesia, Poland. *PLOS ONE*, 11: e0151143.

Surmik, D., Dulski, M., Kremer, B., Szade, J., and Pawlicki, R. 2019. Iron-mediated deep-time preservation of osteocytes in a Middle Triassic reptile bone. *Historical Biology*, doi:10.1080/08912963.2019.1599884.

Surmik, D., Rothschild, B.M., and Pawlicki, R. 2017. Unusual intraosseous fossilized soft tissues from the Middle Triassic *Nothosaurus* bone. *Naturwissenschaften*, 104: 25. doi:10.1007/s00114-017-1451-y.

Tanaka-Kamioka, K., Kamioka, H., Ris, H., and Sim, S.S. 1998. Osteocyte shape is dependent on actin filaments and osteocyte processes are unique actin-rich projections. *Journal of Bone and Mineral Research*, 13: 1555–1568.

Tareen, J.A.K., and Krishnamurthy, K.V. 1981. Hydrothermal stability of hematite and magnetite. *Bulletin of Materials Science*, 3: 9–13.

Toner, B., Santelli, C.M., Marcus, M.A., Wirth, R., Chan, C.S., McCollum, T., Bach, W., and Edwards, K.J. 2009. Biogenic iron oxyhydroxide formation at mid-ocean ridge hydrothermal vents: Juan de Fuca Ridge. *Geochimica et Cosmochimica Acta*, 73: 388–403.

Trueman, C.N., and Martill, D.M. 2002. The long-term survival of bone: The role of bioerosion. *Archaeometry*, 44: 371–382.

Trueman, C.N., Privat, K., and Field, J. 2008. Why do crystallinity values fail to predict the extent of diagenetic alteration of bone mineral? *Palaeogeography, Palaeoclimatology, Palaeoecology*, 266: 160–167.

Trueman, C.N., and Tuross, N. 2002. Trace elements in recent and fossil bone apatite, pp. 489–522. In: Kohn, M.J., Rakovan, J., and Hughes, J.M., eds., *Phosphates: Geochemical, Geobiological and Material Importance, Reviews in Mineralogy and Geochemistry*. Mineralogical Society of America, Washington, DC.

Turner-Walker, G. 2008. The chemical and microbial degradation of bones and teeth, pp. 3–29. In: Pinhasi, R., and Mays, S., eds., *Advances in Human Palaeopathology*. John Wiley & Sons, Hoboken, New Jersey.

Tuross, N. 2002. Alterations in fossil collagen. *Archaeometry*, 44: 427–434.

Undritz, E., Betke, K., and Lehmann, H. 1960. Sickling phenomenon in deer. *Nature*, 187: 333–334.

Vedantham, G., Sparks, H. G., Sane, S. U., Tzannis, S., and Przybycien, T. M. 2000. A holistic approach for protein secondary structure estimation from infrared spectra in H_2O solutions. *Analytical Biochemistry*, 285: 33–49.

Voigt, E. 1934. Weichteile an Säugetieren aus der eozänen Braunkohle des Geiseltales. *Nova Acta Leopoldina*, 4: 301–310.

Voigt, E. 1988. Preservation of soft tissues in the Eocene lignite of the Geiseltal near Halle/S. *Courier Forschungsinstitut Senckenberg*, 107: 325–343.

Walker, J.D., Geissman, J.W., Bowring, S.A., and Babcock, L.E. 2018. Geologic Time Scale v. 5.0. *Geological Society of America*, doi:10.1130/2018.

Wetzel, D.L., Post, G.R., and Lodder, R.A. 2005. Synchrotron infrared microspectroscopic analysis of collagens I, III, and elastin on the shoulders of human thin-cap fibroatheromas. *Vibrational Spectroscopy*, 38: 53–59.

Wiemann, J., Fabbri, M., Yang, T.-R., Stein, K., Sander, P.M., Norell, M.A., and Briggs, D.E.G. 2018. Fossilization transforms vertebrate hard tissue proteins into *N*-heterocyclic polymers. *Nature Communications*, 9: 4741.

Wilby, P.R., and Briggs, D.E.G. 1997. Taxonomic trends in the resolution of detail preserved in fossil phosphatized soft tissues. *Geobios*, 20: 493–502.

Wings, O. 2004. Authigenic minerals in fossil bones from the Mesozoic of England: Poor correlation with depositional environments. *Palaeogeography, Palaeoclimatology, Palaeoecology*, 204: 15–32.

Wopenka, B., and Pasteris, J.D. 2005. A mineralogical perspective on the apatite in bone. *Materials Science and Engineering C*, 25: 131–143.

Zheng, W., and Schweitzer, M.H. 2012. Chemical analyses of fossil bone. *Methods in Molecular Biology*, 915: 153–172.

CHAPTER 3

Fossilization of Reproduction-Related Hard and Soft Tissues and Structures in Non-Avian Dinosaurs and Birds

TZU-RUEI YANG AND AURORE CANOVILLE

ABSTRACT | Growth and reproduction are the two most important goals during an organism's life. The growth history of extinct vertebrates such as dinosaurs has been extensively studied based on commonly preserved mineralized tissues such as bones and teeth. The reproductive biology of extinct species, however, remains poorly understood because reproduction-related tissues (e.g., medullary bone, reproductive tracts, etc.) or traits (e.g., number of broods, type of parental care, etc.) are less likely or not preserved in the fossil record. In recent years, an unprecedented number of fossils that preserve reproduction-related tissues, including a gravid oviraptorosaur with a pair of eggs within the pelvis and Mesozoic birds recovered with purported ovarian follicles, has contributed to a deeper understanding of the evolution of dinosaur–bird reproduction. Even more recently, an emerging field—molecular paleontology—in combination with advances in chemical analytical methods has illuminated the preservation of several reproduction-related tissues and compounds, including the cuticle layer atop eggshell, pigments within eggshell, and medullary bone tissue. While exciting, some of these discoveries require further investigations and validation. How these inorganic and organic reproductive tissues and compounds were preserved in the fossil record is therefore an open key question for future studies. Moreover, modern chemical analytical methods such as Raman spectroscopy and matrix-assisted laser desorption/ionization–time-of-flight mass spectrometry (MALDI-ToF MS) allow molecular paleontologists to explore preservation processes in different sedimentary environments and thus help resolve relevant taphonomic questions. |

Introduction

Natural history museum collections, and more specifically paleontology collections, are well known for their huge abundance of dinosaur bones and eggs. Bones, teeth, and eggshells are the three most common types of biomineralized remains in the fossil record of dinosaurs. Paleontologists commonly utilize these types of remains to interpret dinosaur growth and reproduction (Grellet-Tinner et al. 2006; Erickson 2014), mainly through morphological and microstructural analyses. While dinosaur growth has been relatively well studied based on bones and teeth, dinosaur reproductive biology has been less well-investigated because some reproduction-related tissues and compounds were not mineralized (i.e., they were mainly made of soft organic tissues) and are thus less likely to be preserved in the fossil record.

However, in recent years, studies using high-resolution chemical analyses have uncovered soft part preservation in bones and eggshells and, accordingly, have revealed many of the previously hidden details of dinosaur reproduction (e.g., eggshell structure and color associated with egg burial type and nesting ecology; Wiemann et al. 2017; Yang et al. 2018). Here we review many of the recent exciting discoveries of dinosaur reproductive tissues in the fossil record and discuss their implications for unraveling the reproductive biology of non-avian dinosaurs and for a better understanding of the origin and evolution of the unique reproductive biology of birds. Furthermore, we discuss how such studies should be developed and validated.

Dinosaurs, including birds, are a group of oviparous animals, like their closest living relatives, the crocodilians. A great quantity of fossil eggs attributed to this group have been recovered but rarely include in ovo embryos (e.g., Norell et al. 1994; Varricchio et al. 1997; Chiappe et al. 1998). While it is possible that some dinosaurs laid soft-shelled eggs, all dinosaur eggs discovered thus far have had a mineralized eggshell (Mikhailov 1997; Varricchio and Jackson 2016; Stein et al. 2019), although a thick and rigid eggshell evolved independently in several dinosaur groups from a thin and flexible shell (Stein et al. 2019).

We know from modern birds that the egg of non-avian dinosaurs was encapsulated in a soft tissue sac within the ovary before maturity, at which point the eggs were coated with the calcitic eggshell. Such a sac containing an immature egg, or oocyte, is known as an ovarian follicle. During ovulation, a mature egg is released into the magnum, where albumen (the egg white) is added. Afterwards, the egg with albumen is encapsulated by a layer of shell membrane. Finally, just before oviposition, the egg is coated by the hard eggshell and soft cuticle in the uterus.

Modern avian eggs are composed of the cuticle layer, calcitic shell, and shell membrane, from outside to inside (plate 3.1A, B). Unlike the rigid calcitic eggshells, soft tissues and organic compounds such as the cuticle layer covering the eggshell, the pigments within the eggshell, and the shell membrane (*membrana testacea*) inside the eggshell have been rarely reported in

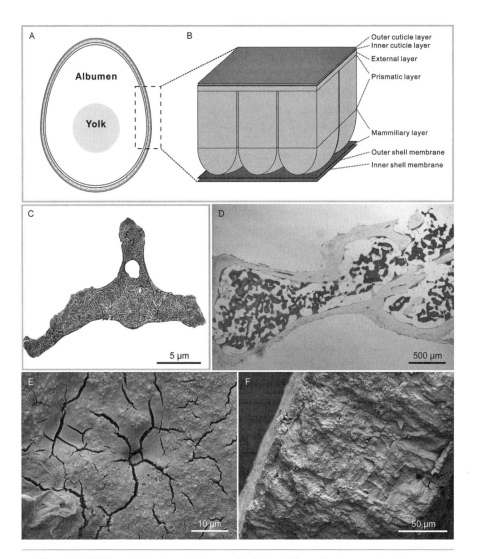

Plate 3.1. A, B: Schematic view through the egg of a chicken (*Gallus gallus domesticus*) and its eggshell. Modified from Yang et al. (2018), Creative Commons Attribution License. (**A**) The anatomy of an egg. (**B**) The eggshell is composed of two crystalline layers, the prismatic layer and mammillary layer. The cuticle layer overlying the calcareous eggshell is further divided into two layers, a proteinaceous outer layer, and a hydroxyapatite inner layer. The shell membrane (*membrana testacea*) is also characterized by two layers. **C, D:** Histological images of medullary bone in a black swan, *Cygnus atratus* (CM-S16508). Modified from Canoville et al. (2019), Creative Commons Attribution License 4.0. (**C**) Medullary bone is evident in the cavities of the ground section of a caudal vertebra. (**D**) An Alcian blue-stained paraffin section shows the medullary bone as a blue area. **E, F:** Scanning electron microscope (SEM) images of a chicken eggshell. From Yang et al. (2018), Creative Commons Attribution License. (**E**) The cracked pattern of the cuticle layer on the surface of the eggshell. (**F**) A radial section through the eggshell.

fossils and have been considered unlikely to fossilize (Mikhailov 1991; Kohring and Hirsch 1996; Varricchio and Jackson 2004; Grellet-Tinner 2005; Igic et al. 2009; Jackson and Varricchio 2010; Oskam et al. 2010; Reisz et al. 2013; Wiemann et al. 2017; Bailleul et al. 2019b).

Dinosaur eggshells are mostly two-layered, except for some derived maniraptoran eggs (maniraptorans are the group that include birds and the dinosaurs most closely related to birds) that show a possible third layer similar to that in extant bird eggs (Varricchio et al. 2002). The biomineralized eggshell can display various structural morphotypes and different numbers of crystallized layers, which are used for assignment to ootaxa (Mikhailov 1997). In addition, there are some amazing fossils that also reveal the intermediate stage in the phylogenetic bracket formed by crocodilians and birds along the non-avian dinosaur lineage. For example, modern crocodilians possess two functional reproductive tracts (two ovaries and two oviducts), each simultaneously producing multiple eggs during a laying cycle. However, most modern birds have only a single functional ovary and oviduct that produces a single egg at any one time. The other oviduct was reduced during maniraptoran evolution. One of the most amazing fossils revealing details of dinosaur reproductive biology is a pair of eggs that were recovered in the pelvic region of an oviraptorosaur (Sato et al. 2005). This fossil indicates an ovulation mode for this group that is intermediate between crocodilians and birds. Oviraptorosaurs retained two functional reproductive tracts but had already evolved sequential egg production, whereby each reproductive tract produced a single egg at any one time (Sato et al. 2005). While this kind of phylogenetic bracketing enables paleontologists to interpret dinosaur reproductive biology based on additional information from the fossil record, the paucity of preserved soft tissues that would have been associated with eggshells or located within bones has hindered a more detailed understanding of dinosaur egg structure and, more broadly, the reproductive biology of dinosaurs.

Information on ontogenetic stages of dinosaurs can be recovered from the study of bone histology (e.g., Klein and Sander 2008), but bone histology also informs us directly about reproduction. This is due to a special bone tissue called medullary bone, which is unique to birds among living amniotes. Medullary bone is a transient tissue that functions as a calcium reservoir for shell formation. Medullary bone can be recognized by its histology and with immunohistochemical analyses (Plate 3.1C, D) (Bonucci and Gherardi 1975; Yamamoto et al. 2001). Crocodilians lack medullary bone (Schweitzer et al. 2007), suggesting that medullary bone evolved in the dinosaur–bird lineage after the divergence from crocodilians. Several studies have indeed reported medullary bone-like tissues in non-avian dinosaurs (Schweitzer et al. 2005, 2016; Lee and Werning 2008; Hübner 2012). However, these reports were simply based on histological observation of fossils without a comprehensive understanding of medullary bone in living birds and have therefore been repeatedly questioned (Prondvai and Stein 2014; Chinsamy et al. 2016; Prondvai 2017).

In this chapter, we will summarize the discoveries of the aforementioned reproductive tissues, review studies on chemical compounds that have been discovered in dinosaur fossils, and discuss the methods that have been used to investigate these tissues. Open questions and concerns will be also reviewed and explored. Finally, we will indicate where future research should be focused.

Institutional Abbreviations

CM, Carnegie Museum of Natural History, Pittsburgh, Pennsylvania, USA; MOR, Museum of the Rockies, Bozeman, Montana, USA; STM, Shandong Tianyu Museum, Linyi, Shandong Province, China.

Inorganic and Organic Structure and Components Associated with Reproductive Tissues in Extant and Extinct Avemetatarsalian Dinosaurs

Ovarian Follicles

Before the onset of eggshell formation in utero in modern birds, the developing egg moves through the oviduct as an ovarian follicle. An ovarian follicle is a fluid-filled sac that encapsulates an immature egg, or oocyte. The nature of the ovarian follicle led to the preconceived idea that it is unlikely to be preserved in the fossil record (Mayr and Manegold 2013). A recent study reported on groups of round structures hypothesized to be ovarian follicles in a specimen of the long bony-tailed basal bird *Jeholornis* (STM 2-51) and in two enantiornithine birds (STM 29-8 and STM10-45) from the Early Cretaceous Jehol Biota of northeastern China (Zheng et al. 2013). The purported ovarian follicles were found aligned to the left side of the body, suggesting one functional reproductive tract (monochronic ovulation) similar to that in modern birds rather than two functional tracts (monoautochronic ovulation) as in oviraptors (Sato et al. 2005), *Troodon* (Varricchio et al. 1997), and extant crocodilians. Further examples of Mesozoic birds with preserved ovarian follicles have also been reported (O'Connor et al. 2013). Although the lacustrine deposits of the Jehol Group have produced numerous fascinating specimens with preserved feathers, the preservation of ovarian follicles is nonetheless controversial. Mayr and Manegold (2013) questioned whether the preservation of ovarian follicles as described by Zheng et al. (2013) is plausible when less degradable organs such as muscle are not preserved in the same specimen. Future work using high-resolution analytical methods (e.g., Bailleul et al. 2019a) will aim at deciphering whether these structures are indeed ovarian follicles or simply food items such as round seeds that had been ingested by the fossil bird, as proposed by Mayr and Manegold (2013).

Cuticle in Fossil Eggs

Cuticle is the thin outer, mainly organic, cover of bird eggshell. The cuticle consists of a proteinaceous outer layer and hydroxyapatite inner layer. A cuticle layer has been reported in several fossil dinosaur eggs, but these reports were not well supported. Chow (1951) and Young (1954) described a thin and transparent layer forming both the innermost and outermost eggshell in eggs from Laiyang, China, as fossil cuticle. However, Chao and Chiang (1974) reexamined the thin sections and suggested that the transparent layer described by Chow (1951) and Young (1954) is simply a calcite layer resulting from recrystallization during diagenesis. Under the polarized light microscope, the cuticle-like layer of dinosaurian eggshells from Laiyang shows columnar cleavage, obvious extinction, and twinkling, which are all characteristic for carbonate minerals with a low refractive index and high birefringence (Chow 1951; Young 1954). Vianey-Liaud et al. (1994) proposed that the outer calcite layer observed in eggshells from France is the result of the decomposition of the egg's organic material and is therefore not preserved cuticle. However, Kohring and Hirsch (1996) studied crocodilian and ornithoid (bird-like) eggshells from the middle Eocene of the Geiseltal open-pit mines, Germany, and noted potential fossilized cuticle and shell membrane preservation from this site.

All of these previous studies were limited to morphological and histological comparisons between the cuticle covering modern avian eggshells and putative cuticle structures in fossilized eggshells. The first ever chemical support for cuticle preservation on oviraptorid eggshells from Upper Cretaceous deposits of southern China was provided by Raman spectroscopy and electron probe microanalysis (EPMA) (Yang et al. 2018). Elemental analysis with EPMA shows a high concentration of phosphorus on the boundary between the eggshell and sediment, probably representing the inner, hydroxyapatite cuticle layer. A lack of phosphorus in the embedding sediment excludes an allochthonous origin for phosphorus in the eggshells (Yang et al. 2018; Bailleul et al. 2019b). The chemometric analysis of Raman spectra derived from fossil and extant eggs provides further supporting evidence for the cuticle preservation on oviraptorid and alvarezsaurid eggshells (Yang et al. 2018).

Shell Membrane (*membrana testacea*) in Dinosaur Eggs

In modern amniote eggshell, the shell unit layer consisting of calcium carbonate (calcite) crystals is underlain by a layer of proteinaceous fibers called the *membrana testacea*. The *membrana testacea* protects the embryo from bacterial and fungal invasion (Palmer and Guillette 1991). Packard and De-Marco (1991) investigated the microstructure of the *membrana testacea* in extant eggs and determined that it has a perplexing fibrous pattern.

Although fossil preservation does not typically favor organic remains, the *membrana testacea* is occasionally preserved in dinosaur eggs as a mineralized layer. Kohring (1999) summarized the discoveries of fossil *membrana testacea* and provided strong evidence for fossilized *membrana testacea* in several non-avian dinosaur and bird eggshells from Spain, Germany, and Czechia. Schweitzer et al. (2013) suggested that endogenous organic material (e.g., proteinaceous fibers, various biomolecules) may be preserved in specific taphonomic environments and can be detected by immunological assays and mass spectrometry. However, this idea has never been tested. *Membrana testacea* preservation in dinosaurs has been only reported for sauropodomorph eggshells from Auca Mahuevo in Argentina, the Haţeg Basin in Romania, the Tremp Basin in Spain, and Dawa in China (Kohring 1999; Peitz 2000; Grellet-Tinner 2005; Grellet-Tinner et al. 2012; Reisz et al. 2013).

Pigmentation

The color of all reptile eggs is white, which reflects the color of the calcium carbonate of the shell and the hydroxyapatite of the cuticle. However, most living birds, especially passerines, produce colored eggs by employing two specific pigments: biliverdin, which is distributed throughout the prismatic layer of the eggshell, and protoporphyrin, which is stored in the cuticle layer (Wiemann et al. 2017). In previous reconstructions, dinosaur eggs were generally depicted as white-colored, similar to modern crocodile eggs.

A recent study using high-resolution electrospray ionization time-of-flight mass spectrometry (HR-ESI-ToF-MS) found evidence for both pigments, biliverdin and protoporphyrin, in *Macroolithus yaotunensis* oviraptorid dinosaur eggs from the redbed basins of southeastern China (Wiemann et al. 2017). The presence of these pigments strongly suggested that oviraptorid dinosaur eggs were blue-green in color. This blue-green color may have functioned as camouflage, similar to that in the eggs of living emus, or it may have mechanically strengthened the eggshell. However, determining the selective advantages of colored oviraptorid eggshells requires further comparisons with modern birds. In general, our understanding of the evolution of coloration in bird eggs from a paleontological perspective is currently emerging. So far, it is not clear if oviraptorids evolved coloration convergently with modern birds or if these dinosaurs were the first clade in which egg coloration appeared. The latter hypothesis is slightly better supported at the moment, because the oviraptorid eggs show the same pigments as modern bird eggs.

Another recent study on eggshell coloration using Raman spectrometry found that eggshell pigments have only been detected in eumaniraptorans such as oviraptors, troodontids, and all extant birds (Wiemann et al. 2018b). While exciting, this study has been criticized by Shawkey and D'Alba (2019). They pointed out that negative controls, such as fossil eggs classified as unpigmented or white eggs with small amounts of pigment or maculation that do not show up in the Raman surface mapping, were not included in Wiemann et al. (2018b), thus making the results questionable pending further analyses.

Medullary Bone

Medullary bone is an estrogen-induced bone tissue unique to modern birds that functions as a calcium reservoir for eggshell formation during the egg-laying period (e.g., Simkiss 1961; Dacke et al. 1993). Since modern birds are derived from non-avian dinosaurs, paleontologists have been looking for medullary bone in the fossil record. So far, endosteal tissues hypothesized to be homologous to avian medullary bone have been described in three Lower Cretaceous birds, including *Confuciusornis sanctus* from the Jehol group of northeastern China (Chinsamy et al. 2013), an unnamed pengornithid enantiornithine from the Jiufotang Formation, China (O'Connor et al. 2018), and another enantiornithine, *Avimaia schweitzerae*, from the Xiagou Formation of China (Bailleul et al. 2019b). In non-avian dinosaurs, purported medullary bone has been described in a femur of *T. rex* (MOR1125; Schweitzer et al. 2005), in a tibia of *Allosaurus* (Lee and Werning 2008), a femur and tibia of *Tenontosaurus* (Lee and Werning 2008), and in *Dysalotosaurus* (Hübner 2012).

These putative occurrences of medullary bone display various textures and thus the identification of medullary bone in the fossil record is complicated and, as of yet, controversial. Chinsamy and Tumarkin-Deratzian (2009) reported on a pathological bone microstructure occurring both periosteally and endosteally in an extant turkey vulture (*Cathartes aura*) and in a non-avian dinosaur from Transylvania. Since medullary bone is only found endosteally (Simkiss 1961; Dacke et al. 1993), the similarity of the pathological bone microstructure to the previously described medullary bone in dinosaurs suggests that caution is necessary to avoid pathologically derived endosteal bone tissue from being mistakenly identified as medullary bone.

The occurrence of medullary bone has also been used as histological evidence for gender recognition because medullary bone forms only in gravid bird females (Dacke et al. 1993). For instance, Varricchio et al. (2008) suggested that the absence of medullary bone is consistent with their hypothesis that the clutch-associated oviraptors were male, implying paternal care in these dinosaurs. Indeed, this study triggered widespread enthusiasm for identifying the mode of parental care in the fossil record. However, the utility of medullary bone for inferring the sex of fossil species has been challenged by follow-up studies. Prondvai and Stein (2014) reported on medullary bone-like tissue in a pterosaur and suggested that it might not be associated with reproduction since pterosaur eggs are thin-shelled and thus contain less calcium than hard-shelled eggs. However, medullary bone also acts as a calcium and phosphorous reservoir for the production of the egg yolk, not only for the eggshell. A later study also found medullary bone in juvenile pterosaurs (Prondvai 2017). Caution is thus warranted when using medullary bone as an indicator of sex in the fossil record.

Understanding the nature of modern medullary bone is crucial for any further studies of this tissue in dinosaurs. For instance, a recent comprehensive paper has surveyed the distribution of medullary bone in the avian skeleton (Canoville et al. 2019) and noted that it is found in most bones of the skeleton. However, they also concluded that skeletal distribution of medullary bone varies interspecifically. The authors also noted that medullary bone is more widely distributed in small-bodied birds than in large-bodied ones. However, medullary bone is also associated with pneumaticity. In birds of considerable size, such as large ratites, the skeleton is well pneumatized, and thus medullary bone is restricted to nonpneumatized regions such as the hindlimb bones. Thus, in *T. rex*, we would expect medullary bone to be restricted to similar parts of the body.

In addition to the aforementioned studies based on bone histology, the nature of medullary bone-like tissues has been tested using different histochemical techniques. One is histochemical staining with different agents (Alcian blue and high-iron diamine) known to preferentially react with medullary bone and not the surrounding cortical bone. Another is immunohistochemistry using antibodies targeting keratan sulfate (a chemical compound found in medullary bone but not in the surrounding cortical bone tissue; see Schweitzer et al. 2016 for further details). A groundbreaking study identified a gravid *T. rex* based on the immunohistochemical detection of medullary bone in her femur (Schweitzer et al. 2016). Although using medullary bone-like tissues for sex identification can be problematic, as reviewed above, Schweitzer et al. (2016) opened up a new avenue in molecular paleontology by showing that some organic molecules indicative of medullary bone persist through geological time. Applying immunohistochemical analyses to the medullary bone that has been reported in various fossils would be beneficial for the validation of their methodology and for the elucidation of how medullary bone is preserved in the fossil record.

Analytical Methods for Detecting Reproductive Tissues in the Fossil Record

Raman Spectroscopy

Raman spectroscopy is a nondestructive analytical method that targets specific molecular bindings (e.g., Sander and Gee, chap. 1; Geisler and Menneken, chap. 4). Commonly used in mineralogical studies, Raman spectroscopy is most often performed on materials such as amber or biominerals (Brody et al. 2001; Thomas et al. 2011). In addition to the detection of biominerals such as those found in the cuticular hydroxyapatite layer of bird eggshell (Plate 3.1E, F) (Yang et al. 2018), paleontologists have attempted to detect organic remains in the fossil record using Raman spectroscopy. For instance, Thomas et al. (2014) mapped the distribution of pigments and found melanin, but no carotenoids, in six feathers preserved in amber ranging in age from mid-Cretaceous to Miocene, as well as in an Eocene feather preserved as a compression fossil. Another recent study also attempted to identify organic remains such as eggshell pigments using Raman spectroscopy (Wiemann et al. 2018b), as discussed above.

Both studies, however, were only based on data from Raman spectroscopy and lacked additional lines of evidence using other chemical methods. Indeed, melanin, carotenoids, and the eggshell pigments protoporphyrin and biliverdin are detectable using high-performance liquid chromatography mass spectrometry (HPLC-MS) (Stradi et al. 1995; Zhao et al. 2006; Gorchein et al. 2009; Igic et al. 2009; Mendes-Pinto et al. 2012; Wiemann et al. 2017). A major advantage of Raman spectroscopy, in contrast to HPLC-MS or other methods, is that it is nondestructive. However, Raman spectroscopic results must be verified through comparisons to other, more precise methods.

Raman spectroscopy has also been used to illuminate how diagenetic and fossilization processes transform tree resins (Winkler et al. 2001) and carotenoids (Marshall and Marshall 2010), as well as vertebrate tissues (Wiemann et al. 2018a; Yang et al. 2018). Most studies have indicated that diagenetic and fossilization processes transform molecular structure through a series of reactions, which can be elucidated using Raman spectroscopy. For instance, a decrease of band intensity at around 1640 cm^{-1} was recognized in diagenetically altered resins and was associated to the loss of ν(C=C) stretching vibrations (the vibrational mode of carbon–carbon double bonds; Winkler et al. 2001). Such a loss of ν(C=C) stretching vibrations was also identified in the diagenetically altered carotenoids found in fossils, indicating that the loss of double bonds is a common reaction during the fossilization process.

On the other hand, a recent study identified an intensity increase of advanced glycoxidation and lipoxidation end products (AGE/ALE) bands between 1550 cm^{-1} and 1600 cm^{-1} with the increasing geological age of fossil tissues such as scale, tooth, bone, and eggshell (Wiemann et al. 2018a). Such an increase of AGE/ALE is associated with the more intense brown stain of fossils recovered from oxidizing depositional environments. Since all aforementioned studies have shown that Raman spectroscopy helps to further understand the chemical reactions during the fossilization process, this calls for further experimental taphonomic studies using Raman spectroscopy to compare fossil and experimentally altered material.

Matrix-Assisted Laser Desorption/Ionization–Time-of-Flight Mass Spectrometry (MALDI-ToF MS)

MALDI-ToF MS is an ionization technique that is commonly applied to the detection of biopolymers (i.e., proteins and peptides; Hillenkamp et al. 1991). Such a nondestructive technique has been embraced by some paleontologists who have used MALDI-ToF MS to detect various large organic molecules in the fossil record (Ostrom et al. 2006; Schweitzer et al. 2008; Lindgren et al. 2012). This method, however, has never been applied to the confirmation of reproduction-related organic compounds (i.e., eggshell pigments). Because reproductive tissues usually contain characteristic biomolecules that are detectable with MALDI-ToF MS, this analysis will certainly play a crucial role in the identification of preserved soft tissues in future research.

Future Research Directions

Some key questions in the study of dinosaur reproductive tissues concern the processes by which they are preserved in the fossil record. Most reports on dinosaur reproductive tissues emphasize the excitement of the discovery and its biological interpretations (fig. 3.1). Only few studies have attempted to elucidate the series of chemical reactions that have occurred during the fossilization process. Here we formulate several questions that should guide future studies. First, are the reproductive tissues preserved in the fossil record unaltered, or have they been diagenetically altered? If they are diagenetically altered, what are the mechanisms and relevant chemical reactions behind the transformation? Second, are these transformation mechanisms and relevant chemical reactions observable in taphonomic experiments? Third, how does the depositional environment affect the preservation of reproduc-

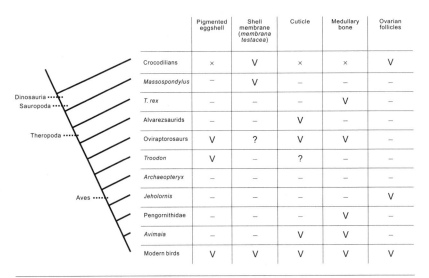

		Pigmented eggshell	Shell membrane (membrana testacea)	Cuticle	Medullary bone	Ovarian follicles
	Crocodilians	×	V	×	×	V
	Massospondylus	−	V	−	−	−
	T. rex	−	−	−	V	−
	Alvarezsaurids	−	−	V	−	−
	Oviraptorosaurs	V	?	V	V	−
	Troodon	V	−	?	−	−
	Archaeopteryx	−	−	−	−	−
	Jeholornis	−	−	−	−	V
	Pengornithidae	−	−	−	V	−
	Avimaia	−	−	V	V	−
	Modern birds	V	V	V	V	V

Figure 3.1. A simplified phylogeny of saurischian dinosaurs with reports of five different reproductive tissues in the fossil record. Symbols: V = presence; x = absence; ? = uncertain; – indicates the lack of a record in fossil saurischians.

tive tissues? All of these open questions require studies using the various high-resolution chemical analytical tools suggested in this chapter.

Thus, we urge the use of high-resolution techniques in future studies to confirm previous results. For organic molecules such as eggshell pigments, MALDI-ToF MS could be carried out to provide potential additional lines of supporting evidence, as has been shown in previous investigations (Wiemann et al. 2017). Second, taphonomic experiments for comparison are urgently needed as they offer a time-lapse view of the fossilization process. Third, an increase in Raman spectroscopic studies of dinosaur reproductive tissues will not only increase our understanding of molecular transformation but also reveal the series of chemical reactions that occur during the fossilization process.

Conclusions

New fossil discoveries and high-resolution chemical analytical tools have offered paleontologists many opportunities for discovering reproduction-related tissues preserved in dinosaur fossils and, accordingly, have helped elucidate dinosaur reproductive biology. In this chapter, we review recent discoveries of reproductive tissues in dinosaurs. It should be remembered, however, that our understanding of the processes behind the preservation of reproductive tissues in the fossil record is still in its infancy because it is based on a small number of specimens. Moreover, all discoveries of soft tissue preservation thus far, while exciting, should only be considered starting points for future investigations applying modern techniques in chemistry, histology, immunohistochemistry, and molecular biology in order to confirm, refute, or refine the original interpretations. We recommend the integration of various chemical analytical methods to provide multiple lines of evidence, for they will surely shed light on how these delicate reproductive tissues have survived the fossilization process in the millions of years since the Mesozoic.

ACKNOWLEDGMENTS

The authors thank two anonymous reviewers, as well as Victoria E. McCoy, P. Martin Sander, and Carole T. Gee for their helpful comments and editorial efforts. This work was supported by the Deutsche Forschungsgemeinschaft (DFG, German Research Foundation), Project number 396768613 to P. Martin Sander and Marianne Engesser (both at the University of Bonn) and the Ministry of Science and Technology (MOST) of Taiwan, Project number 108-2116-M-178-003-MY2 to Tzu-Ruei Yang. This is contribution number 17 of the DFG Research Unit 2685, "The Limits of the Fossil Record: Analytical and Experimental Approaches to Fossilization."

WORKS CITED

Bailleul, A.M., O'Connor, J.K., Li, Z., Wu, Q., and Zhou, Z. 2019a. First histological examination of ovarian follicles in a Cretaceous enantiornithine, p. 23. *Program and Abstracts Book, 5th International Symposium on Paleohistology*. Cape Town, South Africa.

Bailleul, A.M., O'Connor, J., Zhang, S., Li, Z., Wang, Q., Lamanna, M.C., Zhu, X., and Zhou, Z. 2019b. An Early Cretaceous enantiornithine (Aves) preserving an unlaid egg and probable medullary bone. *Nature Communications*, 10: 1275.

Bonucci, E., and Gherardi, G. 1975. Histochemical and electron microscope investigations on medullary bone. *Cell and Tissue Research*, 163: 81–97.

Brody, R.H., Edwards, H.G.M., and Pollard, A.M. 2001. A study of amber and copal samples using FT-Raman spectroscopy. *Spectrochimica Acta Part A: Molecular and Biomolecular Spectroscopy*, 57: 1325–1338.

Canoville, A., Schweitzer, M.H., and Zanno, L.E. 2019. Systemic distribution of medullary bone in the avian skeleton: Ground truthing criteria for the identification of reproductive tissues in extinct Avemetatarsalia. *BMC Evolutionary Biology*, 19: 71.

Chao, T.K., and Chiang, Y.K. 1974. Microscopic studies on the dinosaurian eggshells from Laiyang, Shantung Province. *Scientia Sinica*, 17: 73–90.

Chiappe, L.M., Coria, R.A., Dingus, L., Jackson, F., Chinsamy, A., and Fox, M. 1998. Sauropod dinosaur embryos from the Late Cretaceous of Patagonia. *Nature*, 396: 258–261.

Chinsamy, A., Chiappe, L.M., Marugán-Lobón, J., Chunling, G., and Fengjiao, Z. 2013. Gender identification of the Mesozoic bird *Confuciusornis sanctus*. *Nature Communications*, 4: 1381.

Chinsamy, A., and Tumarkin-Deratzian, A. 2009. Pathologic bone tissues in a turkey vulture and a nonavian dinosaur: Implications for interpreting endosteal bone and radial fibrolamellar bone in fossil dinosaurs. *The Anatomical Record*, 292: 1478–1484.

Chow, M. 1951. Notes on the Late Cretaceous dinosaurian remains and the fossil eggs from Laiyang, Shantung. *Bulletin of the Geological Society of China*, 31: 89–96.

Dacke, C.G., Arkle, S., Cook, D.J., Wormstone, I.M., Jones, S., Zaidi, M., and Bascal, Z.A. 1993. Medullary bone and avian calcium regulation. *Journal of Experimental Biology*, 184: 63–88.

Erickson, G.M. 2014. On dinosaur growth. *Annual Review of Earth and Planetary Sciences*, 42: 675–697.

Gorchein, A., Lim, C.K., and Cassey, P. 2009. Extraction and analysis of colourful eggshell pigments using HPLC and HPLC/electrospray ionization tandem mass spectrometry. *Biomedical Chromatography*, 23: 602–606.

Grellet-Tinner, G. 2005. *Membrana testacea* of titanosaurid dinosaur eggs from Auca Mahuevo (Argentina): Implications for exceptional preservation of soft tissue in Lagerstätten. *Journal of Vertebrate Paleontology*, 25: 99–106.

Grellet-Tinner, G., Chiappe, L., Norell, M., and Bottjer, D. 2006. Dinosaur eggs and nesting behaviors: A paleobiological investigation. *Palaeogeography, Palaeoclimatology, Palaeoecology*, 232: 294–321.

Grellet-Tinner, G., Fiorelli, L.E., and Salvador, R.B. 2012. Water vapor conductance of the Lower Cretaceous dinosaurian eggs from Sanagasta, La Rioja, Argentina: Paleobiological and paleoecological implications for South American faveoloolithid and megaloolithid eggs. *Palaios*, 27: 35–47.

Hillenkamp, F., Karas, M., Beavis, R.C., and Chait, B.T. 1991. Matrix-assisted laser desorption/ionization mass spectrometry of biopolymers. *Analytical Chemistry* 63: 1193A–1203A.

Hübner, T.R. 2012. Bone histology in *Dysalotosaurus lettowvorbecki* (Ornithischia: Iguanodontia)—Variation, growth, and implications. *PLOS ONE*, 7: e29958

Igic, B., Greenwood, D.R., Palmer, D.J., Cassey, P., Gill, B.J., Grim, T., Brennan, P.L.R., Bassett, S.M., Battley, P.F., and Hauber, M.E. 2009. Detecting pigments from colourful eggshells of extinct birds. *Chemoecology*, 20: 43–48.

Jackson, F.D., and Varricchio, D.J. 2010. Fossil eggs and eggshells from the lowermost Two Medicine Formation of western Montana, Sevenmile Hill locality. *Journal of Vertebrate Paleontology*, 30: 1142–1156.

Klein, N., and Sander, P.M. 2008. Ontogenetic stages in the long bone histology of sauropod dinosaurs. *Paleobiology*, 34: 247–263.

Kohring, R., and Hirsch, K.F. 1996. Crocodilian and avian eggshells from the Middle Eocene of the Geiseltal, Eastern Germany. *Journal of Vertebrate Paleontology*, 16: 67–80.

Kohring, R.R. 1999. Calcified shell membranes in fossil vertebrate eggshell: Evidence for preburial diagenesis. *Journal of Vertebrate Paleontology*, 19: 723–727.

Lee, A.H., and Werning, S. 2008. Sexual maturity in growing dinosaurs does not fit reptilian growth models. *Proceedings of the National Academy of Sciences of the United States of America*, 105: 582–587.

Lindgren, J., Uvdal, P., Sjövall, P., Nilsson, D.E., Engdahl, A., Schultz, B.P., and Thiel, V. 2012. Molecular preservation of the pigment melanin in fossil melanosomes. *Nature Communications*, 3: 824.

Marshall, C.P., and Marshall, A.O. 2010. The potential of Raman spectroscopy for the analysis of diagenetically transformed carotenoids. *Philosophical Transactions of the Royal Society A*, 368: 3137–3144.

Mayr, G., and Manegold, A. 2013. Can ovarian follicles fossilize? *Nature*, 499: E1.

Mendes-Pinto, M.M., LaFountain, A.M., Stoddard, M.C., O. Prum, R.O., Harry, A.F., and Robert, B. 2012. Variation in carotenoid–protein interaction in bird feathers produces novel plumage coloration. *Journal of the Royal Society Interface*, 9: 3338–3350.

Mikhailov, K.E. 1991. Classification of fossil eggshells of amniotic vertebrates. *Acta Palaeontologica Polonica*, 36: 193–238.

Mikhailov, K.E. 1997. Fossil and recent eggshell in amniotic vertebrates: Fine structure, comparative morphology and classification. *Special Papers in Palaeontology*, 56: 1–80.

Norell, M.A., Clark, J.M., Dashzeveg, D., Barsbold, R., Chiappe, L.M., Davidson, A.R., McKenna, M.C., Altangerel, P., and Novacek, M.J. 1994. A theropod dinosaur embryo and the affinities of the Flaming Cliffs dinosaur eggs. *Nature*, 266: 779–782.

O'Connor, J.K., Zheng, X., Wang, X., Wang, Y., and Zhou, Z. 2013. Ovarian follicles shed new light on dinoꞏ·ꞏar reproduction during the transition towards birds. *National Science Review*, 1: 15–17.

Oskam, C.L., Haile, J., McLay, E., Rigby, P., Allentoft, M.E., Olsen, M.E., Bengtsson, C., Miller, G.H., Schwenninger, J., Jacomb, C., Walter, R., Baynes, A., Dortch, J.,

Parker-Pearson, M., Gilbert, M.T.P., Holdaway, R.N., Willerslev, E., and Bunce, M. 2010. Fossil avian eggshell preserves ancient DNA. *Proceedings of the Royal Society of London B*, 277: 1991–2000.

Ostrom, P.H., Gandhi, H., Strahler, J.R., Walker, A.K., Andrews, P.C., Leykam, J., Stafford, T.W., Kelly, R.L., Walker, D.N., Buckley, M., and Humpula, J. 2006. Unraveling the sequence and structure of the protein osteocalcin from a 42ka fossil horse. *Geochimica et Cosmochimica Acta*, 70: 2034–2044.

Packard, M.J., and DeMarco, V.G. 1991. Eggshell structure and formation in eggs of oviparous reptiles, pp. 53–69. In: Deeming, C., and Ferguson, M.W.J., eds., *Egg Incubation: Its Effects on Embryonic Development in Birds and Reptiles*. Cambridge University Press, Cambridge, UK.

Palmer, B.D., and Guillette, L. 1991. Oviductal proteins and their influence on embryonic development in birds and reptiles, pp. 29–46. In: Deeming, C., and Ferguson, M.W.J., eds., *Egg Incubation: Its Effects on Embryonic Development in Birds and Reptiles*. Cambridge University Press, Cambridge, UK.

Peitz, C. 2000. Megaloolithid dinosaur eggs from the Maastrichtian of Catalunya (NE-Spain): Parataxonomic implications and stratigraphic utility. *1st International Symposium on Dinosaur Eggs and Babies*. Isona i Conca Dellà, Catalonia, Spain, pp. 155–159.

Prondvai, E. 2017. Medullary bone in fossils: Function, evolution and significance in growth curve reconstructions of extinct vertebrates. *Journal of Evolutionary Biology*, 30: 440–460.

Prondvai, E., and Stein, K.H.W. 2014. Medullary bone-like tissue in the mandibular symphyses of a pterosaur suggests non-reproductive significance. *Scientific Reports*, 4: 6253.

Reisz, R.R., Huang, T.D., Roberts, E.M., Peng, S., Sullivan, C., Stein, K., LeBlanc, A.R., Shieh, D., Chang, R., and Chiang, C. 2013. Embryology of Early Jurassic dinosaur from China with evidence of preserved organic remains. *Nature*, 496: 210–214.

Sato, T., Cheng, Y.-N., Wu, X., Zelenitsky, D.K., and Hsiao, Y.-F. 2005. A pair of shelled eggs inside a female dinosaur. *Science*, 308: 375.

Schweitzer, M.H., Avci, R., Collier, T., and Goodwin, M.B. 2008. Microscopic, chemical and molecular methods for examining fossil preservation. *Comptes Rendus Palevol*, 7: 159–184.

Schweitzer, M.H., Elsey, R.M., Dacke, C.G., Horner, J.R., and Lamm, E.T. 2007. Do egg-laying crocodilian (*Alligator mississippiensis*) archosaurs form medullary bone? *Bone*, 40: 1152–1158.

Schweitzer, M.H., Wittmeyer, J.L., and Horner, J.R. 2005. Gender-specific reproductive tissue in ratites and *Tyrannosaurus rex*. *Science*, 308: 1456–1460.

Schweitzer, M.H., Zheng, W., Cleland, T.P., and Bern, M. 2013. Molecular analyses of dinosaur osteocytes support the presence of endogenous molecules. *Bone*, 52: 414–423.

Schweitzer, M.H., Zheng, W., Zanno, L., Werning, S., and Sugiyama, T. 2016. Chemistry supports the identification of gender-specific reproductive tissue in *Tyrannosaurus rex*. *Scientific Reports*, 6: 23099.

Shawkey, M.D., and D'Alba, L. 2019. Egg pigmentation probably has an early archosaurian origin. *Nature*, 570: E43–E45.

Simkiss, K. 1961. Calcium metabolism and avian reproduction. *Biological Reviews*, 36: 321–359.

Stradi, R., Celentano, G., and Nava, D. 1995. Separation and identification of carotenoids in bird's plumage by high-performance liquid chromatography-diode-array detection. *Journal of Chromatography B*, 670: 337–348.

Thomas, D.B., McGoverin, C.M., Fordyce, R.E., Frew, R.D., and Gordon, K.C. 2011. Raman spectroscopy of fossil bioapatite—A proxy for diagenetic alteration of the oxygen isotope composition. *Palaeogeography, Palaeoclimatology, Palaeoecology*, 310: 62–70.

Thomas, D.B., Nascimbene, P.C., Dove, C.J., Grimaldi, D.A., and James, H.F. 2014. Seeking carotenoid pigments in amber-preserved fossil feathers. *Scientific Reports*, 4: 5226.

Varricchio, D.J., Horner, J.R., and Jackson, F.D. 2002. Embryos and eggs for the Cretaceous theropod dinosaur *Troodon formosus*. *Journal of Vertebrate Paleontology*, 22: 564–576.

Varricchio, D.J., and Jackson, F.D. 2004. A phylogenetic assessment of prismatic dinosaur eggs from the Cretaceous Two Medicine Formation of Montana. *Journal of Vertebrate Paleontology*, 24: 931–937.

Varricchio, D.J., and Jackson, F.D. 2016. Reproduction in Mesozoic birds and evolution of the modern avian reproductive mode. *Auk*, 133: 654–684.

Varricchio, D.J., Jackson, F.D., Borkowski, J.J., and Horner, J.R. 1997. Nest and egg clutches of the dinosaur *Troodon formosus* and the evolution of avian reproductive traits. *Nature*, 385: 247–250.

Varricchio, D.J., Moore, J.R., Erickson, G.M., Norell, M.A., Jackson, F.D., and Borkowski, J.J. 2008. Avian paternal care had dinosaur origin. *Science*, 322: 1826–1828.

Vianey-Liaud, M., Mallan, P., Buscail, O., Montgelard, C., Carpenter, K., Hirsch, K.F., and Horner, J.R. 1994. Review of French dinosaur eggshells: Morphology, structure, mineral and organic composition, pp. 151–183. In: Carpenter, K., Hirsch, K.F., and Horner, J.R., eds., *Dinosaur Eggs and Babies*. Cambridge University Press, UK.

Wiemann, J., Fabbri, M., Yang, T.-R., Stein, K., Sander, P.M., Norell, M.A., and Briggs, D.E.G. 2018a. Fossilization transforms vertebrate hard tissue proteins into *N*-heterocyclic polymers. *Nature Communications*, 9: 4741.

Wiemann, J., Yang, T.-R., and Norell, M.A. 2018b. Dinosaur egg colour had a single evolutionary origin. *Nature*, 563: 555–558.

Wiemann, J., Yang, T.-R., Sander, P.N., Schneider, M., Engeser, M., Kath-Schorr, S., Müller, C.E., and Sander, P.M. 2017. Dinosaur origin of egg color: Oviraptors laid blue-green eggs. *PeerJ*, 5: e3706.

Winkler, W., Kirchner, E.C., Asenbaum, A., and Musso, M. 2001. A Raman spectroscopic approach to the maturation process of fossil resins. *Journal of Raman Spectroscopy*, 32: 59–63.

Yamamoto, T., Nakamura, H., Takehito, T., and Hirata, A. 2001. Ultracytochemical study of medullary bone calcification in estrogen injected male Japanese quail. *Anatomical Record*, 264: 25–31.

Yang, T.-R., Chen, Y.-H., Wiemann, J., and Spiering, B. 2018. Fossil eggshell cuticle elucidates dinosaur nesting ecology. *PeerJ*, 6: e5144.

Young, C.C. 1954. Fossil reptilian eggs from Laiyang, Shantung, China. *Scientia Sinica*, 3: 505–522.

Zhao, R., G.-Y. Xu, Z.-Z. Liu, J.-Y. Li, and N. Yang. 2006. A study on eggshell pigmentation: Biliverdin in blue-shelled chickens. *Poultry Science*, 85(3): 546–549.

Zheng, X., O'Connor, J., Huchzermeyer, F., Wang, X., Wang, Y., Wang, M., and Zhou, Z. 2013. Preservation of ovarian follicles reveals early evolution of avian reproductive behaviour. *Nature*, 495: 507–511.

CHAPTER 4

Raman Spectroscopy in Fossilization Research

Basic Principles, Applications in Paleontology,
and a Case Study on an Acanthodian Fish Spine

THORSTEN GEISLER AND MARTINA MENNEKEN

A B S T R A C T | Confocal Raman spectroscopy has been an essential
analytical method in materials science and the geosciences for several
decades. With improvements of the technique in recent years with respect
to spatial resolution and overall sensitivity, it has also become increasingly
important in paleontology and paleobotany to chemically and mineralog-
ically characterize fossil materials. Even though the possibilities of Raman
spectroscopy may appear to be limitless, it is important to know how to
proceed when selecting, preparing, and measuring samples, as well as how
to interpret resulting data. Here we introduce the reader to some possibilities, difficulties, and potential pitfalls when using Raman spectroscopy
in fossilization studies. This is mainly done by guiding the reader through
a Raman spectroscopic case study of a ca. 405-million-year-old acanthodian fish spine from the Devonian Hunsrück Slate in Germany. This study
reveals, for instance, the occurrence of two populations of carbon with
different structural disorder that are interpreted as reflecting two different
carbon sources, one being the altered extracellular bone matrix and the
other a metamorphic fluid that has infiltrated the bone. This fluid has also
mineralized the bone with quartz, ankerite, muscovite, and secondary
hydroxyapatite. |

Introduction

Over the last decades, Raman spectroscopy has become an increasingly important tool in materials science and geosciences because it offers a wealth
of possibilities in investigating synthetic and earth materials (McCreery 2000;
Nasdala et al. 2004; Fries and Steel 2012; Vandenabeele et al. 2014; Chou and
Wang 2017; King and Geisler 2018) and their behavior at high temperatures
and pressures (e.g., Salje 1992; Gillet 1996). Two-dimensional Raman spectro-

scopic imaging has even been used very recently to study solid–water reactions (Geisler et al. 2019; Lönartz et al. 2019) and high-temperature solid–solid reactions in situ and in real time at the micrometer scale (Stange et al. 2018; Böhme et al. 2019; Hauke et al. 2019). Raman spectroscopy is based on the interaction between electromagnetic radiation and matter, and it offers the possibility of nondestructively analyzing even very small crystals or particles ranging in size down to a few tenths of nanometers, or even less when also considering surface- and tip-enhanced Raman spectroscopy (SERS and TERS). However, in this chapter, we focus only on confocal Raman spectroscopy, where the resolution is limited by the diffraction of light (Dieing et al. 2012), since this technique is comparably easy to use, requires little sample preparation, and is available in many geoscientific research institutions. Measurements can be performed on loose grains and microparticles or on routinely prepared rock and fossil thin sections.

Confocal Raman spectroscopy is not limited to solid-state analysis, but can also be applied to liquids and gases that are, for example, entrapped as fluid inclusions in minerals (Burke 2001; Menneken et al. 2017). It can be used to identify unknown phases (Frost et al. 1999; Socrates 2004), to structurally and chemically characterize all kinds of organic and inorganic materials, and to quantify them in polyphase materials (Pelletier 2003). In solid-state research, Raman spectroscopy has been applied, for instance, to characterize and quantify (1) materials' defects (Kitajima 1997; Gouadec and Colomban 2007), (2) internal and external stress imposed on solids (Atkinson and Jain 1999; Nasdala et al. 2003; Gouadec and Colomban 2007), used as a geobarometer (e.g., Izraeli et al. 1999; Enami et al. 2007), (3) self-irradiation damage, for instance, in U- and Th-bearing minerals (Zhang 2017) and defect annealing (Geisler 2002), (4) elastic strain in solid solutions (Geisler et al. 2005, 2016), (5) orientational disorder in plastic crystals (Guinet et al. 1988), (6) structural ordering (Sidorov et al. 2014), (7) the crystallite size in nanocrystalline materials (Kitajima 1997; Gouadec and Colomban 2007), and (8) structural phase transitions (Salje 1992; Gillet 1996). It has also been used to visualize the relative crystallographic orientation of anisotropic organic or inorganic crystallites in composite or polyphase materials, such as bones, teeth, and shells (Kazanci et al. 2007; Schrof et al. 2014; He and Bismayer 2019). Although Raman spectroscopy provides a very valuable means for obtaining information about the nature of bonding in condensed matter, it cannot be considered as a method of structure determination. However, Raman spectroscopy combined with group theoretical analysis can be used to check a given crystal structure obtained by X-ray diffraction analysis (see, e.g., Geisler et al. 2006).

Because of this plethora of possibilities to identify and characterize inorganic and organic materials, Raman spectroscopy has also increasingly been applied to study all kinds of fossil materials. For example, Raman spectroscopy has been used to infer a biological origin of presumed carbonaceous microfossils found in early Archean rocks (Brasier et al. 2002; Schopf et al. 2002). However, the poor state of sample preservation and similarities of the spectral signatures with poorly ordered carbonaceous materials generated by abiotic processes has resulted in this interpretation remaining a point of debate (Marshall et al. 2010). More successfully, Raman spectroscopy enabled the detection and localization of organic remnants of lignin and cellulose in fossil wood (Kuczumow et al. 2019) and of lipids, proteins, and possibly cholesterol in fossil bone fragments (Kiseleva et al. 2019). It has also been used to characterize the silica phases of fossil wood (Liesegang et al., chap. 7). Thomas et al. (2014) were even able to identify carotenoid pigments in fossil bird feathers. Raman spectroscopy has also been used to study fossil bone and tooth material to identify their diagenetic alteration (Thomas et al. 2011; France et al. 2014). It further allows the estimation of the maturity of fossil tree resins (i.e., copal and amber) and is used as provenance technique to locate their origin by comparing the Raman spectrum of a sample of unknown origin with those from reference materials (Brody et al. 2001; Drzewicz et al. 2016). Moreover, inclusions in fossil resins, such as pollen, insects, or lichen, have been investigated by Raman spectroscopy (Hartl et al. 2015). These are just some examples of the variety of potential applications of Raman spectroscopy in fossilization studies. For a more comprehensive review of vibrational spectroscopic studies on fossils, we refer to Marshall and Marshall (2015).

In order to understand how to use Raman spectroscopy in fossilization research, it is necessary to understand the physics behind the Raman effect, the technical principles of Raman spectroscopy, and how the information mentioned above can be deduced from a Raman spectrum. In this chapter, we briefly introduce the reader to the theoretical and technical background of confocal Raman spectroscopy without the use of extensive mathematics. The information summarized in this chapter provides the necessary theoretical background for the measurement and interpretation of Raman spectra of organic and inorganic fossil matter. The interpretation of Raman spectra is illustrated by intriguing results of a Raman spectroscopic investigation of a ca. 405-million-year-old acanthodian fish fin spine from the Hunsrück Slate in Germany. The first Raman data from this fossil fin spine offer a tantalizing look at how Raman spectroscopy can be useful in paleontological research, but they also point out potential pitfalls.

Fundamentals of Raman Spectroscopy

The Raman Effect

The Raman effect was named after C.V. Raman, an Indian physicist who was the first to experimentally observe the effect in 1928. However, it was already predicted based on theoretical considerations by Adolf Smekal in 1923. In Raman spectroscopy, monochromatic laser light is directed at a solid, liquid, or gas sample, and the spectrum of the scattered light is analyzed. Most of the light interacting with matter is elastically scattered, the so-called *Rayleigh scattering* in which energy is neither gained nor lost (thus, the wavelength of the scattered light does not change). However, a small portion of the light is scattered inelastically in that the incoming light photons gain or lose energy during the scattering process. If a molecular vibration is excited from its ground to a virtual vibrational energy state by the interaction with incident photons of frequency and then falls back to a higher-energy vibrational state than before, the scattered photons have less energy (a longer wavelength and lower frequency) than the incident photons. This is called *Stokes scattering*. On the other hand, if a molecular vibration is already in a higher vibrational energy state to begin with (i.e., shows a larger displacement of atoms) and falls back to its ground state after interaction with the incoming photons, the scattered photons have more energy (a shorter wavelength and higher frequency) than the incident photons. This is called *anti-Stokes scattering*. Only about one of 10^8 photons undergoes Stokes Raman scattering and even fewer photons anti-Stokes scattering. This is because the excited vibrational state, namely, vibrations involving larger atomic displacements, will only be thermally populated, that is, the fraction of excited vibrations involving larger atomic displacements increases with temperatures. Hence, at room temperature, the intensity of spectral bands related to Stokes scattering is much higher than for bands related to anti-Stokes scattering.

A common misconception by nonspecialists is that the Raman spectral signals are related to the occurrence of specific elements in a material, like in X-ray fluorescence spectroscopy. However, Raman spectroscopy is a method for direct measurement of specific chemical bonds in a substance. Different Raman bands can thus be related to the same molecular group, but reflect different molecular motions. A nonlinear compound with n atoms exhibits $3n$ fundamental modes of atom displacements in three dimensions from which three modes represent rotational and translational movements, leaving $3n - 6$ fundamental vibrational modes. For linear molecules, this number is reduced by one, that is, $3n - 5$, because a linear molecule has only two axes of rotation. However, not all molecular vibrations are Raman-active in that not all vibrations in a solid, liquid, or gas can be observed using Raman spectroscopy (Raman selection rules). For a vibrational mode to be Raman-active, it must involve a change in the polarizability (α) of the molecule. The electric polarizability is the relative tendency of a charge distribution, such as the electron cloud of an atom or molecule, to be distorted from its normal shape by an external electric field (E). When a molecule is in an electric field, its distribution of charge is displaced, which may or may not induce a dipole. This explains, for instance, why pure metals do not give a Raman spectrum, because there is no change in the polarizability involved in their vibrations.

The polarizability is related to the dipole moment (P) and the electric field by $\alpha = P/E$. Let us consider, for instance, the CO_2 molecule that exhibits $3n - 5 = 4$ different vibrational modes. These are (1) the symmetric stretching mode that can be visualized as ⟨○—●—○⟩ and that does not destroy the symmetry of the molecule, (2) the antisymmetric stretching mode, representable by ○⟩◄●—○⟩, and (3) two bending vibrations, representable by ○—●—○ (in-plane bending) and ⊕—●—⊕ (out-of-plane bending), that are energetically indistinguishable (so-called degenerate modes). In this case, only the symmetric stretching vibration produces a change in the polarizability, that is, a deformation of the electron cloud, and will thus be Raman-active. Although the three other modes do not change the polarizability, they produce a dipole moment. Such modes fulfill infrared (IR) selection rules and are thus IR-active and can be measured by IR spectroscopy that is based on the absorption of light in the infrared region. The number and symmetry of Raman-active vibrations of a given substance can be calculated by group theoretical analysis (e.g., Fateley et al. 1971; Rousseau et al. 1981). However, the number of observable Raman bands may be reduced because the polarizability of a vibrational mode is not equal along and across a bond. Therefore, the intensity of Raman scattering is different when the natural polarization of the laser light is along or orthogonal to a particular bond axis. It should be emphasized that the number of observable Raman modes may be further reduced by (1) similar vibrations at the same but spatially separated molecule fragments, (2) accidental frequency coincidence of different vibrations, and (3) limited sensitivity of the Raman spectrometer. On the other hand, aside from pure fundamental vibrational modes described above, rotational, combination, or overtone modes may also be excited by the incoming laser, so the number of bands in a Raman spectrum may also exceed the theoretically predicted number.

As a rule of thumb, more energy is needed to stretch than to bend a particular set of atomic bonds, so stretching modes commonly vibrate with higher frequency (wavenumber) than bending or deformation modes. Also, antisymmetric modes usually, though not always, vibrate with higher frequencies than corresponding symmetric modes. In general, the lower the bond strength between the atoms involved in a vibration and the higher their masses, the lower the vibrational frequency. In some cases, the mass dependence of the vibrational frequency allows the quantification of the isotope composition of condensed phases by Raman spectroscopy (e.g., Geisler et al. 2012; King and Geisler 2018). The effect of the bond strength on the vibrational frequency of a molecular vibration can be illustrated by the bonding between carbon and nitrogen. A C=N double bond is about twice as strong as a C–N single bond, and a C≡N triple bond is likewise stronger than a double bond. Accordingly, the stretching frequencies of these molecular groups vary in the same order, ranging from about 1100 cm^{-1} for C–N, to 1660 cm^{-1} for C=N, and to 2220 cm^{-1} for C≡N. These frequencies are called group frequencies because the same functional groups in different molecules will typically vibrate with specific frequencies. One of the most comprehensive tables of group frequencies for organic compounds is given by Socrates (2004), while a table with characteristic frequencies for anionic molecular groups in minerals is given by Frost et al. (1999). These tables of known group frequencies are helpful for interpreting the Raman spectrum of organic and inorganic substances. However, although group frequencies generally occur within narrow limits, the influence of the atomic environment may cause a frequency shift of the characteristic bands. In molecular and ionic crystals, collective translational and rotational motions of atoms produce lattice modes (phonons) in the lowest frequency region of the spectrum, that is, usually below about 400 cm^{-1}.

Instrumentation

A typical true confocal Raman spectrometer consists of a few simple components (plate 4.1A). The integral part of a Raman system is a laser that produces monochromatic and polarized light. Laser is an acronym for Light Amplification by Stimulated Emission of Radiation. Laser wavelengths between the ultraviolet (UV) and IR are used in Raman spectroscopy. The most commonly applied lasers are a semiconductor (diode), gas (e.g., He–Ne, Ar), and solid-state lasers (e.g., Neodymium:Y–Al–garnet) with excitation wavelengths (λ_0) between 515 and 1064 nm, whereby the most frequently used wavelength is 532 nm. The laser power can either be controlled

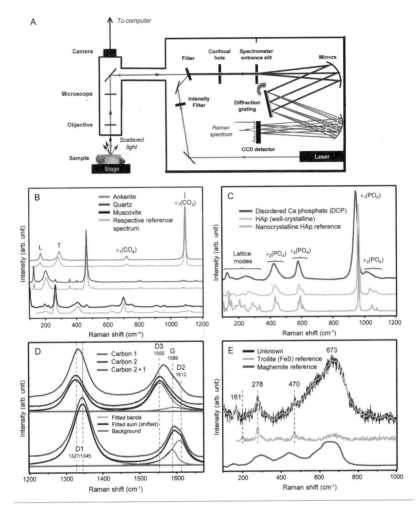

Plate 4.1. (A) Schematic sketch of a typical confocal Raman spectrometer setup (not to scale). See text for explanation. **B–E:** Representative Raman spectra of all phases detected within the mapped area of the acanthodian spine under study. **(B)** Ankerite, quartz, and muscovite, along with reference spectra from the RRUFF database (ankerite ID: R050181; quartz ID: R040031; muscovite ID: R040104). Ankerite bands are labeled according to the Herzberg notation and Rividi et al. (2010). **(C)** Calcium phosphate phases (DCP and HAp) along with a reference spectrum from nanocrystalline HAp (Asjadi et al. 2019). **(D)** Spectra of two different carbon phases (red and dark blue curves) and a mixed spectrum (light blue curve). The carbon spectra were fitted with three Gauss-Lorentz functions (gray curves) and a linear background (green line) for crystallite size and metamorphic temperature determination (see explanation in Chapter 4). The different Raman bands of the carbonaceous matter are labeled according to Kouketsu et al. (2014). **(E)** Raman spectrum of a yet unidentified phase that, however, shows some similarities with a RRUFF spectrum of troilite (RRUFF ID R070242) and maghemite (graphically extracted from Dubois et al. 2008). The yellow line represents a smoothed spectrum of the unknown phase using polynomial smoothing.

by the laser unit itself or by special filters that are included in the beam path (plate 4.1A).

A microscope objective usually collects the scattered light in a 180° back-scattered geometry, as shown in plate 4.1A. It is also used to optically investigate the sample and to restrict the spatial distribution of the laser spot, thus determining the lateral resolution of a Raman measurement. Different objectives can usually be installed, including long-working distance and immersion objectives. After the light has passed through the objective, it reaches the Rayleigh filter, which removes the elastically scattered light that otherwise would overwhelm the detector. In true Raman confocal systems, the scattered and filtered light must subsequently pass through a confocal aperture or hole, the width of which controls the depth (axial) resolution.

After the confocal aperture, the scattered light passes through a spectrometer entrance slit, which resembles a pinhole camera. Smaller slit widths produce a narrower beam that is then dispersed by a diffraction grating, as illustrated in plate 4.1A. A typical diffraction grating used in Raman spectroscopy consists of a substrate, which usually has between 150 and 2400 parallel grooves ruled or replicated in a surface that is coated with a reflecting material such as aluminum. In general, the more grooves per millimeter, the higher its dispersion efficiency and, in turn, the higher the spectral resolution. Spectral resolution is the ability to resolve spectral features and bands into their separate components and usually varies as a function of wavenumber. It can be calculated from the optical and geometrical parameters of the spectrometer, but is best determined empirically by measuring distinct gas emission lines (e.g., Ne or Hg) that have very narrow line widths (e.g., Hauke et al. 2019). Unlike a typical Raman spectrum derived from scattered light, only the uncertainty principle contributes to their broadening, which relates the lifetime of an excited state to the uncertainty of its energy.

After the scattered light has been dispersed by the optical grating, it reaches the detector, usually a charge-coupled device (CCD). A CCD consists of a large number of surface-mounted, photosensitive semiconductor elements, generally consisting of silicon. Each semiconductor element is a single photodetector, representing a pixel at a given position on the detector that converts the incident photons into electrons. During the integration phase (the exposure time), the electrons that are released by the incident photons in the semiconductor material are collected in a potential well, that is, in each pixel. An actual CCD consists of a large number of pixels (i.e., potential wells) arranged horizontally in rows and vertically in columns. The resolution of the CCD is defined by the pixel size and their separation (the pixel pitch). The accumulated amount of charge in each pixel is proportional to the intensity of the incident light and to the exposure time. The amount of charge per pixel is then read out and visualized by a computer as a Raman spectrum (plate 4.1A) according to detector-specific readout techniques. An electron multiplication CCD, for example, uses the latest technology, which allows the signal-to-noise ratio to be increased more than 10-fold in low-intensity areas compared to conventional CCDs. This may be important for applications where extremely low signal levels have to be detected or where fast acquisition times are necessary, such as for Raman imaging.

Two-dimensional Raman imaging (or mapping) is possible if the system is equipped with an automated x-y-z stage. Raman imaging is usually performed either by point-by-point measurements during which the sample stage is automatically moved by programmed steps in x, y, and z directions or by continuous stage movement with step sizes in the nanometer-range (e.g., Lohumi et al. 2017; Hauke et al. 2019). In transparent solids, even three-dimensional Raman maps of a volume within the sample can be taken (e.g., Nasdala et al. 2003; Schrof et al. 2014). The numerous spectra are then analyzed for a correlation between the two-dimensional spectral band distribution to the existence of specific chemical components in the sample. After locating a specific band that corresponds to the presence of a specific molecular group of a substance, a two- or three-dimensional image of the intensity response of that band can be displayed in false-color images by the instrument software. Since a multitude of information is contained in the Raman spectrum of each point (pixel) of an image that can individually be visualized, this technique is called hyperspectral Raman imaging. For instance, the spatial variations of the frequency or width of a particular Raman band within a solid phase can give quantitative information about compositional and/or structural variations within that phase (see, e.g., Nasdala et al. 2004; King and Geisler 2018; Hauke et al. 2019). In this respect, it should be noted that the reliable quantitative analysis of Raman spectra requires well-calibrated and -maintained Raman spectrometer systems. The calibration of the instrument is particularly important when comparing Raman spectra obtained from different instruments or on different days. McCreery (2000), Hutsebaut et al. (2005), Berg and Nørbygaard (2006), Bowie et al. (2006), and Everall (2010) have presented detailed protocols about how to maintain and control the performance of a confocal Raman spectrometer system, including recommendations how to avoid instrumental artifacts and systematic errors.

Sample Preparation

Raman spectroscopy works under atmospheric conditions and best on flat, polished surfaces. Samples must not be coated by any material such as gold or carbon. Such coating, for example, is necessary for electron or ion microprobe analysis. Hence, if Raman spectroscopy data should be correlated with chemical or isotope data obtained by these techniques, Raman measurements must be performed first. This also ensures that potential electron beam or ion beam modifications do not affect the Raman spectrum (see, e.g., Nasdala et al. 2004).

Studies on fossils often have to make compromises in preparation, since sample preservation is of great importance. Fortunately, Raman spectroscopy can also be applied to rough surfaces, glassy materials, and organic materials, as well as to soft tissue in its pristine state. The only limitation is the size of the fossil, as it must fit under the microscope, but open Raman microscope systems are available, in which even several decimeter large samples can be fitted under the microscope. Other options to analyze large samples are to attach a flexible optical fiber with an objective (Raman probe) to the Raman spectrometer or to use portable Raman spectrometer systems that allow the analysis of fossil samples even in the field, though with a worse spectral resolution (e.g., Vandenabeele et al. 2014).

The Raman Spectrum

The Raman spectrum is a plot of the intensity of the scattered light as a function of the Raman shift, which is traditionally given in reciprocal centimeters, that is, in wavenumbers. By convention, the excitation laser wavelength is given the reference wavenumber "0," so the position of the maximum of an intensity profile (band) in the Raman spectrum directly corresponds to the wavenumber (frequency, energy) of the corresponding molecular vibration. For instance, for laser light with a wavelength of 514.308 nm, one obtains $v_0 = 10^7 / (514.308$ nm$) = 19444$ cm^{-1}. It follows that the scattered light with a wavelength of 490.1 nm, corresponding to a frequency of $6.12 \cdot 10^8$ MHz, will appear at a Raman shift (v) of 960 cm^{-1}, as given by $v = 10^7 / (490.1$ nm$) - 19444$ cm^{-1}. Raman shifts are plotted from left (low shift) to right (high shift), in contrast to the usual practice for infrared frequencies. Since modern CCD detectors count photons rather than measure power, the intensity is usually given as photon counts per second (cps, photon flux) or just the number of accumulated counts (cts).

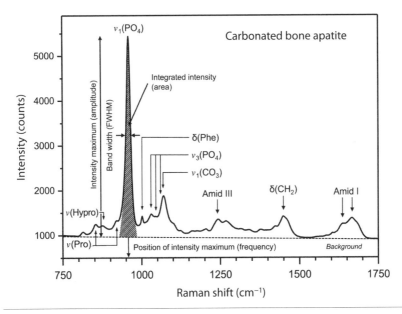

Figure 4.1. Raman spectrum of carbonated bone apatite of a kestrel. The position of the intensity maximum (equals the frequency of the corresponding vibration), the band width measured as full width at half maximum (FWHM), the intensity maximum (amplitude), and the integrated intensity (area) of the $v_1(PO_3)$ Raman band are illustrated. Some bands are labeled according to their origin as explained in the text. Abbreviations: Hypro = hydroxyproline; Phe = phenylalanine; Pro = proline.

Fig. 4.1 shows a Raman spectrum of carbonated bone hydroxyapatite that is characterized by a number of distinct maxima (Raman bands), some of which are labeled in the figure using the so-called Herzberg notation for molecular groups (Herzberg 1945). In this notation, the $v_1(PO_4)$ band, for instance, located near 960 cm^{-1} reflects the fully symmetric stretching vibration of the P–O bond in the PO_4 tetrahedra of the bone apatite structure. This vibration can be considered as a kind of pumping motion in which the local symmetry of the tetrahedron remains unchanged during the vibration. The same holds true for the symmetric stretching mode $v_1(CO_3)$ of the planar carbonate group in the bone apatite structure. The symmetric bending vibrations are labeled v_2, while the antisymmetric stretching and bending modes of the tetrahedra are designated v_3 and v_4, respectively. In organic vibrational spectroscopy, symmetric and antisymmetric stretching and bending modes (or deformation modes) of certain structural units, such as functional groups of organic compounds, are commonly labeled with v_s, v_{as}, δ_s, and δ_{as}, respectively, followed by the condensed chemical formula in parentheses or an abbreviation of the compound. For example, $v_s(CH_2)$ in fig. 4.1 denotes the symmetric stretching mode of a CH_2 group and δ(Phe) the phenyl ring angular C–C–C bending vibrations of phenylalanine (Hernández et al. 2013).

A Raman band profile is characterized by four parameters (fig. 4.1): (1) The position of the intensity maximum (Raman shift, v), (2) the full width at half maximum (band width, $FWHM$), (3) the intensity maximum (amplitude, I), and (4) the integrated intensity (area, A). To extract quantitative information from a Raman band, a combination of a Gauss and Lorentz function (=Voigt function) is usually fitted to the Raman band by least-squares techniques. The Lorentzian part originates from the physical interactions in the substance under consideration, that is, it reflects the intrinsic broadening, whereas the Gaussian part considers broadening due to instrumental effects mainly due to diffraction at the spectrometer entrance slit. Thus, to quantitatively compare the widths of Raman bands obtained with different spectrometer entrance slit widths, namely, spectral resolutions (s), the measured band width ($FWHM_{meas}$) must be corrected for the instrumental broadening to obtain the corrected, intrinsic width ($FWHM_{corr}$). To do so, a so-called apparatus function has been defined (Tanabe and Hiraishi 1980):

$$FWHM_{corr} = FWHM_{meas} [1 - (s / FWHM_{meas})^2] \qquad \text{(Eq. 1)}$$

The fitting procedure may be complicated when there are multiple bands overlapping each other and/or when there is significant background noise that has to be correctly accounted for. For a detailed elucidation and discussion of potential pitfalls and problems of curve fitting spectroscopic data, the reader is referred to Meier (2005) and Saltonstall et al. (2019).

Information from Raman Frequencies

The Raman shift is defined by maximum intensity of the Raman band and corresponds to the vibrational frequency of the respective molecular species (fig. 4.1). Each substance has its own bonding properties and atomic structure (symmetry), so that the band positions along with the number of bands in a Raman spectrum (characteristic frequencies) can be considered as a kind of fingerprint allowing phase identification. This also holds true for polymorphic materials such as calcite and aragonite. However, the vibrational frequencies in solid materials are also influenced by (1) solid solution formation (chemical disorder), (2) permanent stress imposed on the material (i.e., in mineral inclusions or at grain boundaries), (3) crystallite size, and (4) structural defects produced during crystal growth, by self-irradiation, or also by polishing and grinding. While this may pose a problem for mineral identification, band positions may be used, in particular, to characterize and quantify the solids' defect structure and bonding properties or in certain cases the composition of solid solutions. Disordered materials are characterized by larger variations in intermolecular and intramolecular distances. This variation results in a range of different frequencies for the vibrations and therefore in shifted and broadened Raman bands, which gives (quantitative) information about the degree of disorder. A special case of disordered materials is amorphous or noncrystalline solids, such as glasses, gels, or organic polymers, that lack any long-range order but exhibit interconnected structural units with short-range order.

In practice, phase identification is usually performed by computer-based algorithms that compare a measured Raman spectrum of an unknown substance with reference spectra collected in a Raman spectral database, such as the comprehensive RRUFF database, which contains several thousand Raman spectra from natural minerals (Lafuente et al. 2015). This database can be used free of charge and accessed at https://rruff.info. The Raman database is connected to the Windows-based software CrystalSleuth, which is capable of comparing multiple spectra. Several thousand Raman spectra of organic compounds are stored in the spectral database, SDBS, of the National Institute of Advanced Industrial Science and Technology, Japan, which can be freely accessed at https://sdbs.db.aist.go.jp.

The usefulness of a Raman spectral database is defined in part by its accuracy and how much information is included in it (e.g., spectral range, chemical composition). The critical decision to be made by the analyst is whether the measured spectrum can be considered consistent with the reference spectrum. Furthermore, the Raman spectrum of an unknown phase may in fact represent a mixture of different mineral or organic phases that are tightly intergrown with each other, so it may be difficult to identify the component mixture from a Raman spectrum.

Information from Raman Band Widths

The width of a Raman band, which is given as the FWHM (fig. 4.1), is inversely correlated with the lifetime of an excited vibration, that is, the broader the band, the faster the vibrational damping. In organic and inorganic crystals, the decay or damping of an excited molecular or atomic vibration occurs faster at structural and mass defects, but also at surfaces, so the width of the corresponding Raman band is increased in nanocrystalline and defect-rich crystals compared to the defect-free, bulk crystal (Gouadec and Colomban 2007). The width of Raman bands is thus used to characterize and quantify structural disorder as well as crystallite size.

Information from Raman Intensities

There are two Raman band parameters that are related to the intensity, namely, the integrated intensity, reflecting the area under the Raman band, and the maximum intensity, which corresponds to the amplitude of the curve (fig. 4.1). Based on the quantum interpretation of the Raman effect, it can be shown that the power of the scattered light is proportional to the product of the intensity of the incident photons and the so-called Raman scattering cross section. The Raman cross section of an atomic vibration mainly depends on the electric polarizability of the bonds involved in that vibration, and is thus a material property. However, it also depends on the scattering geometry and the inverse of the excitation wavelength to the fourth power, that is, on $1/\lambda_0^4$. It follows that for a given laser power, the lower the excitation wavelength and the greater the Raman scattering cross section of a particular bond, the higher the intensity of the scattered light. However, in Raman spectroscopy of solid materials, the absolute intensity also depends on how well the incident laser is focused on the solid surface. Whereas the optimal laser focus can be well controlled by reliable auto-focusing systems, the analyst has no control on how much light is lost by surface scattering processes (e.g., on rough surfaces) and how well the analyzed material is crystallized or ordered, which affects its Raman cross sections and thus Raman intensities. Therefore, the intensity of a Raman band is mostly used only in relation to the intensity of other Raman bands in a spectrum. However, if a material is anisotropic (i.e., material properties are different in different crystallographic directions), the relative Raman intensities also depend on the crystallographic orientation, which complicates quantitative phase determination, but on the other hand allows the visualization of crystallite orientation (Kazanci et al. 2007; Schrof et al. 2014; He and Bismayer 2019).

Optimizing Raman Spectroscopic Settings for Analysis

The quality of a Raman spectrum and thus the extractable (quantitative) information depends, on one hand, on the sample properties, but it also depends on a range of analytical parameters and settings that the analyst can vary or select to optimize the signal-to-noise ratio and the spatial and spectral resolution of a confocal Raman measurement, for example, (1) the laser excitation wavelength, (2) the laser power, (3) the objective, (4) the confocal hole size, (5) the spectrometer entrance slit width, (6) the optical diffraction grating, and (7) the exposure or acquisition time. The analyst must adjust the analytical parameters and settings with respect to the desired information to be obtained from the sample being studied. For instance, if a high spectral resolution is required, the spectrometer entrance slit width can be decreased, but at the expense of intensity reaching the detector. The intensity loss can in principle be compensated by increasing the laser power, changing the excitation wavelength into the blue region of the visible spectrum, or by increasing the exposure time. Thereby, the analyst has to be aware of the consequences of such measures on various aspects of spectrum quality. However, at the very end, the selected analytical parameters will always be a compromise between the best achievable signal-to-noise ratio and the best spectral and spatial resolution, as well as an optimized acquisition time. In the following, we briefly introduce the reader to the various effects of choosing particular analytical parameters and settings on spectrum quality, spectral, and spatial resolution.

Choice of the Laser Wavelength

UV lasers usually produce the highest Raman intensity at a given laser power because of the $1/\lambda_0^4$ dependence of the scattered light intensity on the excitation wavelength, as described above. Nonetheless, UV lasers are less often used in Raman spectroscopy as they may cause severe fluorescence signals overwhelming the Raman intensities. The fluorescence process is excited by the incident photons, lifting some orbital electrons from certain atoms in the material under investigation to an excited singlet state. When the electrons relax, they emit photons of distinct energy (fluorescence). The fluorescence commonly appears as extremely broad signals in the Raman spectrum that may cover the entire frequency range, but may also occur as discrete, relative sharp bands that may be indistinguishable from Raman bands. However, fluorescence bands will not appear in the same wavenumber region in a Raman spectrum when using different laser excitation wavelengths, so they can be identified when different lasers are available. Even trace amounts of rare earth elements in a material may cause discrete fluorescence bands that in some cases can be used to study the distribution of these elements in a sample (Lenz et al. 2015). As electrons in many organic materials are also excited by light in the visible region, fluorescence may in particular occur during Raman spectroscopic analysis of fossil material, where it may even overwhelm the Raman intensities. If fluorescence is a problem, longer wavelength lasers (e.g., 784 nm or 1064 nm) can be used, but at the price of intensity due to the $1/\lambda_0^4$ relationship and a higher risk of sample heating. Another option is photobleaching by prolonged laser light exposure, but sample damage and increased data acquisition time are strong limitations associated with photobleaching.

Choice of the Laser Power

In general, the higher the laser power at the sample surface, the higher the scattered light intensity, and thus less counting time is necessary to reach a desired signal-to-background ratio. However, when the laser power exceeds a material-specific threshold value, the absorption of light creates a detectable temperature increase at the sample surface around the laser spot, which may result in a frequency shift and broadening of Raman bands. This effect is called anharmonic (or thermal) broadening because the potential energy change associated with an atomic vibration cannot be described anymore by simple harmonic (parabolic) relationships between the potential energy and the deflection of the atoms from their equilibrium positions as derived from laws of classical mechanics. Laser heating may also exceed certain mineral transformation temperatures, indicated by spectrum variations over time, or may even induce evaporation of the sample. Therefore, careful optical examination of the irradiated spot of an unknown sample before and after analysis, in order to check for any modifications of the sample, is recommended. In addition, when quantitative information is to be extracted from a Raman spectrum of a substance, the influence of the laser power on the frequency and width of its Raman bands should first be investigated.

Choice of Objective

If a high spatial resolution is required, microscope objectives with a high magnification and a high numerical aperture (NA) should be chosen. The NA is a measure of the ability of an objective to collect light at a fixed object distance. For instance, with a 532 nm excitation wavelength and a 100x objective having an NA of 0.9, the diffraction-limited theoretical lateral resolution (d_x) at the sample surface is 721 nm, as given by:

$$d_x \approx 1.22\, \lambda_0 / \text{NA} \qquad \text{(Eq. 2)}$$

At the sample surface, the diffraction-limited depth (axial) resolution (d_z) can reasonably be estimated from following equation:

$$d_z \approx 4\, n\, \lambda_0 / \text{NA}^2 \qquad \text{(Eq. 3)}$$

where n is the refractive index of the sample (Everall 2010). Using this equation and $n = 1.5$, we obtain a depth resolution of 3.9 μm for our example. It follows that a high spatial resolution can be achieved with objectives with a high NA, but also with lasers of shorter wavelengths (which, however, may produce fluorescence problems as explained above). For example, the use of an immersive objective, having an NA of 1.4, would reduce d_x and d_z to ca. 460 nm and ca. 1.6 µm, respectively. We emphasize that these calculations of the diffraction-limited resolution are strictly valid only with the laser focus at the sample surface. Based on theoretical considerations, Everall (2004) showed that within a transparent sample the true laser focus is always deeper than the distance below the surface given by the z drive of the stage. The axial laser focus also broadens upon focusing deeper into the sample. In any case, due to possible optical aberrations of the instrument, the calculated values should be considered as lower limits for the spatial resolution (Everall 2010). Therefore, if spatial resolution is of concern, it should be determined empirically (Geisler et al. 2019; Hauke et al. 2019).

Choice of the Confocal Hole Size

If a high-depth resolution is required, as for three-dimensional Raman imaging within a transparent sample, or stray light and/or fluorescence is a problem and needs to be removed or reduced, the confocal hole size can be decreased. This, however, comes at the cost of scattered light intensity reaching the detector, and the analyst has to find the best compromise between spectrum quality and spatial resolution. Finally, we would like to mention here that the penetration depth of the laser beam strongly depends on the optical absorption properties of the investigated material, so that in some cases Raman spectroscopy is also a good method for surface-sensitive analysis (e.g., Tian et al. 1998).

Choice of Optical Grating and Spectrometer Entrance Slit Widths

If a high spectral resolution is required because closely overlapping bands must be resolved, an optical grating with a large number of grooves per millimeter, that is, with a high dispersion efficiency and a small spectrometer entrance slit, should be used. The thinner the slit and the better the dispersion efficiency of the grating, the better the spectral resolution (plate 4.1A). A spectral resolution down to 0.2 cm^{-1} or better is achievable with modern Raman spectrometers. However, such a high spectral resolution can only be achieved with a small entrance slit (e.g., 10 µm) and highly dispersive gratings. These spectrometer settings, however, give rise to a significant decrease in the scattered light intensity reaching the detector, as more light is filtered out by small spectrometer entrance slit widths and less scattered light is dispersed on each individual pixel of the CCD when using highly dispersive gratings (plate 4.1A). A high dispersive efficiency grating also decreases the wavenumber range that is detected by the CCD array, so the grating must be rotated incrementally to measure a larger wavenumber range, which is time consuming and thus increases the overall acquisition time.

Choice of Acquisition/Counting Time

The choice of the acquisition or counting time for a single Raman measurement is always a compromise between the time available at the instrument and the desired quality of the spectrum, especially in two- and three-dimensional Raman imaging, where several thousand to 1 million individual measurements are collected. With modern CCD detectors, exposure or counting times between 1 ms to several minutes can be chosen, whereby the upper time limit is constraint by the signal rate and the maximum number of electrons each pixel of the CCD can store (full well capacity). If a high fluorescence or stray light intensity is a concern, for instance, during automatic mapping of unknown substances, the required total counting time can be divided into multiple smaller time steps to avoid detector saturation. In any case, at least one repetition is recommended, that is, two accumulations, since sharp intensity spikes from cosmic rays striking the detector are then identifiable and can be removed automatically (Bowie et al. 2006). In general, it can be said that under given experimental conditions, the quality of a Raman spectrum, quantified by the standard deviations of signal and background, is better when counting times are maximized instead of performing multiple accumulations with shorter counting times that are summed to the same total acquisition time, that is, 2 times 10 seconds give better results than 20 times 1 second, contradicting the expectation from Poisson statistics.

To find the best compromise between acquisition time and spectrum quality, it is helpful to consider that the photon counting statistics follow a Poisson distribution, which means that the acquisition time must be increased fourfold to reduce the standard deviation of the measured intensity by half. It follows from this square root relationship between the counting error and the counting time that at a certain point a further increase of the counting time only insignificantly reduces the standard deviation, that is, only insignificantly improves the spectrum quality. If increasing the counting time does not improve the spectrum quality, other measures must be taken to improve the counting statistics, such as increasing the laser power if possible. If considering the detection of trace amounts of organic substances in fossil material, it is important to be aware of the limits of detection (LOD). There are several ways of determining the LOD, but the simplest is based on the signal-to-noise (S/N) ratio. The LOD is reached when S/N is at least greater than three. For more details on the statistics of the LOD as well as quantification in analytical chemistry, the reader is referred to a review article from Shrivastava and Gupta (2011).

Case Study: Raman Spectroscopy of a Devonian Acanthodian Fin Spine

The specimen investigated in this study is a ca. 405-million-year-old fin spine of an acanthodian fish from the Lower Devonian Hunsrück Slate in the state of Rhineland-Palatinate, western Germany. It is deposited in the collections of the Goldfuss Museum, Division of Paleontology, Institute of Geosciences at the University of Bonn, Germany. Acanthodians are teleostomes, a group of early jawed fish with a cartilaginous interior skeleton but well-developed bony fin spines. The first acanthodians appeared in the Silurian and the last went extinct at the end of the Permian. In addition to their morphology, the fin spines have been traditionally studied using paleohistological thin sections to elucidate their microstructure and biological hard tissues. Acanthodians possessed large asymmetrically curved fin spines that are composed of dentin-like material with thin outer and inner layers, and a thick middle layer that is composed of trabecular dentin or both bone and dentin forming the bulk of the spine (Denison 1979; Jerve et al. 2017). The Raman data presented in the following are part of a larger Raman spectroscopic study on acanthodian spine samples with the aim of assessing the inorganic and organic composition of the bone material for reconstructing and comparing taphonomic processes in fossils from different locations of the unique conservation lagerstätte in the Central Hunsrück mountains, Germany, as well as from peripheral fossils sites in the southeast Eifel. Here we focus on the information that can be gathered from a single hyperspectral Raman image or map that was taken from a polished section of the thick middle layer of the fossil spine to gain a quick overview of the phase composition of the fossil and chiefly to search for any remnant organic matter. The survival of organic remnants in a fossil bone that has been exposed to greenschist facies conditions (Wagner and Boyce 2006) may not be as unlikely as it would seem at first glance, considering that chitin has recently been found in a fossil demosponge from the Burgess Shale in British Columbia, Canada, which also underwent greenschist facies conditions (Ehrlich et al. 2013).

In the following, we will guide the reader through the analysis and evaluation of the two-dimensional Raman spectroscopic data set, first identifying the phases and their spatial distribution occurring in the mapped area of the acanthodian spine. We then illustrate the possibility to semiquantitatively determine the composition of solid solutions, to identify, characterize, and quantify structural and/or chemical disorder, and to gain an estimate of the metamorphic temperature that the fossil spine has experienced.

Hyperspectral Raman Mapping: Analytical Conditions

The Raman map was acquired in the point-by-point mode with a confocal Horiba Scientific LabRAM HR800 Raman spectrometer, equipped with an Olympus BX41 microscope. The instrument is located at the Institute of Geosciences of the University of Bonn, Germany. Raman scattering was excited with a 2W solid-state Nd:YAG laser (532.18 nm) and ca. 50 mW at the sample surface. A 50x objective with NA = 0.75 was used, yielding a diffraction-limited spatial resolution in the order of 0.9 μm lateral and 5.7 μm axial (n = 1.5, Eq. 2 and 3, respectively), and thus an optical intensity at the sample surface of about 0.065 W/μm^2. The mapped area includes three vascular canals and comprises an area of 850 × 590 μm^2. The pixel spacing was 4 μm in each direction. The hyperspectral data set thus contains of 31,524 individual spectra, each taken 5 times per second. This low counting time was selected to minimize the risk of overloading the detector by fluorescence or stray light during automatic imaging. The laser focus was automatically adjusted at each point during the mapping procedure by the instrument software. For this depth profile, data were acquired automatically at three corners of the mapped area prior to the start of the mapping process; from these data, the plane of focus of the polished sample was determined by the instrument software. The confocal hole was set to 300 μm to maximize the depth resolution, which, for our spectrometer (Hauke et al. 2019), was found to be a good compromise between the depth resolution and the loss of intensity associated with closing the confocal aperture.

Measurements were performed with an optical grating containing 600 grooves/mm and a 100 μm spectrometer entrance slit, because a high spectral resolution was not necessary for the purpose of phase detection and identification. Spectra were recorded in the frequency range between 90 and 1730 cm^{-1} with a spectral resolution of 3.5 cm^{-1}, as given by the full width at half maximum of the Ne line at 1707.06 cm^{-1} from a Ne lamp that was positioned below the thin section. This Ne line was also used as internal standard to correct any spectrometer drift that resulted from small temperature fluctuations of ± 0.5°C in the Raman laboratory during the long-time imaging procedure. With such an internal standard, the propagated precision of the Ne-corrected Raman shift (frequency), obtained from least-squares fitting, was usually between $\pm$ 0.2 cm^{-1} and $\pm$ 0.5 cm^{-1} for narrow (FWHM < 10 cm^{-1}) and broad Raman bands (FWHM > 10 cm^{-1}), respectively (given at the 2-sigma level). The accuracy is thereby coupled to the precision of the Ne line wavelength determination, which is better than $\pm$ 10^{-5}% (Saloman 2006). All spectra were white-light corrected, that is, corrected for the wavelength-dependence of the instrument sensitivity. Reported Raman band widths are corrected for instrumental broadening (Eq. 1) and have an error between $\pm$ 0.2 cm^{-1} for FWHM values < 10 cm^{-1} and $\pm$ 1.0 cm^{-1} for larger widths. Least-square fitting and all spectrum corrections were performed with the instrument software LabSpec 6.4.4.15, while the freeware software Fityk (Wojdyr 2010) was used for error evaluations.

Phase Identification

Within the imaged area of the fossil acanthodian spine, we could unambiguously identify six different phases, based on comparisons with spectra from the RRUFF database using the CrystalSleuth software (plate 4.1B–E). These are ankerite, quartz, muscovite (plate 4.1B), a disordered calcium phosphate phase (DCP), and well crystalline hydroxyapatite (HAp), the latter being unambiguously identified by the frequency and width of their internal $\nu_1(PO_4)$ to $\nu_4(PO_4)$ bands (plate 4.1C). DCP is characterized by significantly broadened and shifted Raman bands, indicating its structural disorder. The Raman spectra of both calcium phosphate phases do not show the fully symmetric $\nu_1(CO_3)$ carbonate band near 1070 cm^{-1} that is typically observed in carbonated bone apatite and amorphous calcium phosphate (ACP) (cf. fig. 4.1).

In the mapped region, no bands could be detected that would indicate the survival of collagen degradation products such as amino acids or higher hydrocarbons. However, two types of carbonaceous matter can clearly be distinguished (plate 4.1D). Carbon 1 is characterized by two broad band profiles with maxima located near 1340 and 1585 cm^{-1}, respectively, and a shoulder near 1620 cm^{-1} (plate 4.1D). Perfectly crystalline graphite, however, only displays one sharp band near 1590 cm^{-1} (Tuinstra and Koenig 1970), reflecting C=C stretching vibrations of sp^2 atom bonds (labeled G in plate 4.1D). In contrast, disordered and nanocrystalline graphite also contains random sp^3 bonds, volatile elements, and broken bonds, leading to new disorder bands near 1292, 1340 (D1 in plate 4.1D), and 1620 (D2 in plate 4.1D) cm^{-1} (e.g., Kouketsu et al. 2014). The second carbon phase (Carbon 2 in plate 4.1D) is characterized by (1) a significantly lower D1 frequency (down to 1326 cm^{-1}), (2) a new broad band located near 1555 cm^{-1} (D3 in plate 4.1D), and (3) a significantly different relative intensity of the D1 and G bands (plate 4.1D). The band near 1555 cm^{-1} has previously been assigned to amorphous, sp^3-bonded forms of carbon (Knight and White 1989; Nikiel et al. 1993; Tamor et al. 1994). If such assignment is correct, the absence of a significant fraction of sp^2 bonds, indicated by the almost complete absence of the G band, suggests that carbon 2 is largely amorphous, whereas carbon 1 has much in common with highly disordered nanocrystalline graphite.

One phase in the mapped area has not yet been identified with certainty. This case, however, illustrates well the problems that can arise in phase identification (plate 4.1E). The Raman spectrum of this phase (or possibly phases) is characterized by a low signal-to-noise ratio when compared to the other identified minerals, making its identification more difficult. Relatively sharp bands near 161, 278, and 470 cm^{-1}, and a relative broad profile with a maximum near 673 cm^{-1}, which is likely composed of two or more bands, are nevertheless clearly detectable, particularly in the smoothed spectrum (plate 4.1E). The absence of any bands at higher wavenumbers implies that the observed bands are likely related to an inorganic phase. An automatic search in the RRUFF mineral database and the CrystalSleuth software did not provide a satisfactory match (> 80%) with a geologically meaningful mineral. The relatively weak Raman shifts (frequencies), however, indicate that the mineral may be a sulfide or oxide, which reduces the number of potential candidates substantially. In fact, the spectrum shows some similarities (55% match) with a spectrum from hexagonal troilite (stoichiometric FeS) from the Union Mine, California, USA (RRUFF ID: R070242), that is also of relatively poor quality, as well as with maghemite (γ-Fe$_2$O$_3$). The latter mineral, however, is not yet included in the RRUFF database, so we visually compared and graphically reproduced the maghemite spectrum published by Dubois et al. (2008) in plate 4.1E, which matches very well with the observed intensity profile between 500 and 800 cm^{-1}. Avril et al. (2013) published Raman spectra from synthetic and natural troilite crystals, which display bands near 160, 310, 360 and 160, 240, 290, and 335 cm^{-1}, respectively. As noted by Avril et al. (2013), troilite has a very low Raman scattering cross section, and not all Raman bands may appear under a given experimental orientation due to the crystal orientation effects on Raman intensities (polarization selection rules). In fact, a group theory analysis using the crystal structure data for troilite from Skála et al. (2006) revealed that 29 Raman-active modes are expected theoretically.[*]

---

[*] For those readers who are familiar with symmetry analysis of vibrational modes, the irreducible representation (Γ) for Raman-active modes of troilite is $\Gamma = 6A'_1 + 12E' + 11E''$. Hence, six modes involving atomic displacements along the c axis and 23 modes vibrating within the a,b plane of the hexagonal structure (double degenerate E modes) are predicted.

Therefore, the absence of Raman modes in this case is not necessarily an exclusion criterion, and we may conclude that both troilite and maghemite are intergrown with each other. However, it is known that high laser power may oxidize troilite (Dubois et al. 2008; Avril et al. 2013). Since in our case the optical intensity at the sample surface was significantly higher than that used by Avril et al. (2013), who did not detect the formation of Fe_2O_3, the maghemite signals may just be an analytical artifact. Clarification of this issue is ongoing.

Visualization of the Spatial Distribution of the Identified Phases

The visualization of the spatial distribution of the various phases within the mapped area of the fossil acanthodian spine in a Raman false-color phase distribution image provides substantial textural and mineralogical information, such as the paragenetic relationship of the different phases or spatial variations in crystallinity. Plate 4.2A shows such a false-color image of the investigated area of the fossil spine. This image was generated using the classical least-squares (CLS) fitting procedure, where pure Raman spectra of the different phases are selected from the map and then their relative contribution to each spectrum is calculated by a least-squares algorithm. The phase with the largest contribution to the Raman spectrum at each pixel is attributed with a color that is assigned to that particular phase. Colors are mixed according to the relative Raman intensities of the different phases recorded at each point (Stange et al. 2018; Hauke et al. 2019). When viewing plate 4.2A, it becomes immediately apparent that ankerite, quartz, muscovite, HAp, and carbon 1 are concentrated along cracks and, in particular, inside the vascular canals, while carbon 2 and DCP are located exclusively in the bone matrix. We also note the tabular intergrowth between ankerite and the unknown phase, which is in agreement with the typical platy or tabular habit of troilite (see "troilite" at http://webmineral.com/Alphabetical _Listing.shtml).

The most notable observation in the false-color Raman image shown in plate 4.2D is that there are distinct halos along the cracks and around the vascular canals, where the $v_1(PO_4)$ frequency is higher than in areas further within the bone matrix. New HAp seems to have replaced the former bone matrix along a reaction front that penetrated into the bone matrix (white arrows in plate 4.2A). Furthermore, there are also subtle frequency variations within the inner bone matrix itself, forming a wavy texture that resembles the lamellae texture of an osteon. Note also that a similar texture is also observable in the spatial distribution of the amorphous carbon and

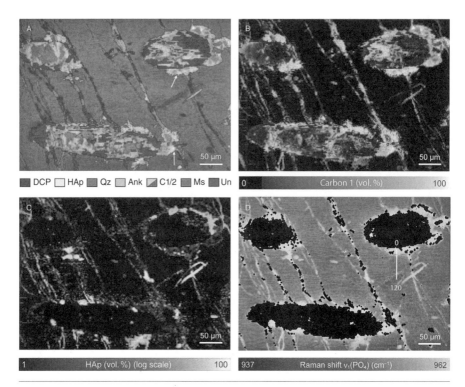

Plate 4.2. False-color hyperspectral Raman images of the fossil acanthodian spine. (A) Raman phase distribution image. In this image, the phase with the highest intensity contribution in a Raman spectrum is displayed in a color assigned to that particular phase. Note that this does not exclude that other phases are also present in a certain area (cf. B and C). Gray (B) and color scale (C) Raman images of the volume fractions of carbon 1 and HAp within the mapped area. Note that in B, the color-coding is in the log scale. (D) False-color Raman image of the $v_1(PO_4)$ frequency of calcium phosphate with location of the transect shown in fig. 4.2. Color-coding is based on the viridis colormap that is perceptually uniform consistently across the entire range of values. Abbreviations: DCP = disordered calcium phosphate; HAp = hydroxyapatite; Qz = quartz; Ank = ankerite; C1/2 = carbon 1 and 2 (cf. plate 4.1D); Ms = muscovite; Un = unidentified phase(s). White arrows in **A** point to replacement fronts that spread out from the vascular canals into the bone matrix.

DCP in plate 4.2A. However, the frequency fluctuations within the bone matrix, which significantly exceed the experimental error, can be observed more clearly in a line transect of the frequency of the $v_1(PO_4)$ band that runs 120 μm away from a vascular canal into the bone matrix (white line in plate 4.2D) and that is reproduced in figure 4.2. Moreover, this transect clearly shows that the $v_1(PO_4)$ frequency decreases nearly exponentially away from the bone replacement front, suggesting gradual chemical and/or structural changes in the DCP phase ahead of the replacement front.

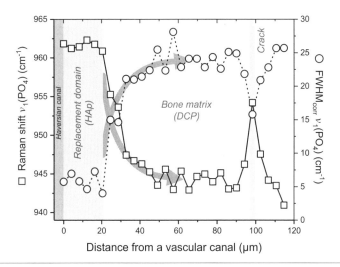

Figure 4.2. Line profile of the frequency (Raman shift) and corrected width (FWHM$_{corr}$) of the v_1(PO$_4$) band of calcium phosphate (HAp and DCP). The line profile extends from a vascular canal ca. 120 μm into the bone matrix (for location, see white line in plate 4.2D). The arrows mark the smooth changes of both parameters away from the bone replacement front and are visual guides only. Note the fluctuations of both Raman parameters inside the bone matrix. These are significantly larger than the experimental errors that are well within the symbol size.

Such a phase distribution image well reflects the surface distribution of the various phases (Hauke et al. 2019). However, due to limitations of the lateral and depth resolution, more than one phase may be excited by the laser beam and analyzed at a given position (pixel), particularly in multiphase, nano-sized materials such as bones and teeth. Here, the actual Raman spectrum represents the phase distribution within the actually excited and analyzed volume (ca. 6.7 μm^3 in our case, assuming a cylindrical focal volume). The real spread of a particular phase within the actually excited and analyzed volume is thus concealed in a multiphase image. However, the volume fraction of each phase obtained from a CLS fit can be visualized individually, as shown for carbon 1 and HAp in plate 4.2B and C, respectively. In such a representation, the real distribution of both phases within the analyzed area or better volume is completely reproduced. From this image, it is immediately evident that carbon 1 is not only located along cracks and within the vascular canals, where it is intertwined with the other mineral phase filling the canals, it also spreads into the bone matrix, along with HAp (plate 4.2C) and sometimes with quartz (plate 4.2A) but not with ankerite. Within these replacement domains, Raman spectra of carboniferous matter are observed that we interpret as mixed spectra rather than as a new discrete carbon phase (plate 4.1D).

Determination of the Solid Solution Composition of the Ankerite Fillings Using Raman Frequencies

Raman shifts of simple binary solid solutions across their compositional space are often linearly correlated with composition, allowing the (semi) quantitative determination of the solid solution composition from Raman data, such as, for example, a forsterite-fayalite solid solution (Breitenfeld et al. 2018). However, Raman spectroscopy also facilitates the quantification of more complex, multicomponent solid solutions, such as amphiboles (Leissner et al. 2015) and garnets (Bersani et al. 2009). To assess the potential of Raman spectroscopy as a rover-based technique for stromatolite characterization during future Mars missions, Rividi et al. (2010) studied carbonate solid solutions in the Ca–Mg–Fe–Mn system. They found that the frequencies of Raman-active vibrations in siderite-magnesite and ankerite-dolomite solid solutions display a high positive correlation ($r^2 > 0.9$) with Mg# = 100 · Mg / (Ca + Mg + Fe + Mn). Using their correlations and the average Raman shifts of four bands obtained from a fit of Gauss-Lorentz functions to 240 Raman spectra of ankerite (see fig. 4.4A) from the imaged area, we obtained a low average Mg# value of 14.6 ± 1.7 (table 4.1), which is even lower than in all ankerite samples analyzed by Rividi et al. (2010). This implies that the ankerite is Fe-dominated, that is, it contains only a minor dolomite component.

Spatial Distribution and Characterization of Structural and/or Chemical Variations in the Calcium Phosphate Phases Using the Frequency and Width of the $v_1(PO_4)$ Band

In disordered materials, the Raman shifts can be used to characterize and quantify the defect structure, including chemical disorder. Here we use the Raman shifts (frequencies) of the $v_1(PO_4)$ band to visualize any structural and/or chemical variations among both calcium phosphate phases in a false-color Raman image shown in plate 4.2D. The frequencies and FWHMs were determined by fitting an asymmetric Gauss-Lorentz function to the $v_1(PO_4)$ band, which accounts for an asymmetry toward the low wavenumber side in some spectra that is also typically observed for amorphous and nanocrystalline HAp (Asjadi et al. 2019). Note that we have chosen the *viridis*

Table 4.1.
Raman data from ankerite fillings in acanthodian fine spine used to obtain the dolomite content, given as $Mg\# = 100 \cdot Mg / (Ca + Mg + Fe + Mn)$

Band	Observed frequency (cm^{-1})		Calibration	Mg#		Assignment
T	167.3	± 1.2	$Mg\# = 3.34 \cdot T - 545$	13.8	± 4.0	Translational lattice mode
L	285.5	± 1.6	$Mg\# = 1.95 \cdot L - 539$	17.7	± 3.1	Librational lattice mode
$v_4(CO_3)$	719.6	± 0.5	$Mg\# = 6.17 \cdot v_4 - 4431$	8.9	± 3.1	In plane bending vibration of CO_3
$v_1(CO_3)$	1091.5	± 0.5	$Mg\# = 4.24 \cdot v_1 - 4612$	16.0	± 2.1	Fully symmetric stretching of CO_3
$v_3(CO_3)$	1436	*Not given*	Antisymmetric stretching of CO_3			
			Weighted average:	14.6	± 1.7	

Notes: The calibrations of Raman shift against *Mg#* for four bands from Rividi et al. (2010) were used. All errors are reported at the 1-sigma level and represent the standard deviation around the mean of 240 data points. The errors of *M#* do not include the errors and covariance of the calibration lines, as these values were not reported by Rividi et al. (2010). To account for the actual scatter of the data rather than the predicted scatter and the fact that the true error is only an estimate that was derived from four data points only, the internal standard deviation of *M#* was multiplied by the square root of the mean square weighted deviates and the Student's t-value for $4 - 1 = 3$ degrees of freedom.

colormap to visualize the Raman shift in a false-color image because this colormap has an even perceptual contrast, in both color and black and white, and is also perceived by viewers with common forms of color blindness (Karim et al. 2019). In contrast, more commonly used colormaps, such as the *jet* colormap, often have perceptual discontinuities that may even induce the appearance of false features in an image (Kovesi 2015; Karim et al. 2019).

To further analyze the structural and/or compositional variations within the DCP phase, we have plotted the Raman shift of the $v_1(PO_4)$ band from 11786 spectra against its FWHM along with the distribution of both parameters (fig. 4.3). The first observation from this diagram is that the data from the acanthodian fish spine define a distinctly negative trend that fans out to larger FWHMs and that connects the two end members HAp and DCP. These are defined by two widely separated distribution maxima at 945.3 (DCP) and 961.9 (HAp) and 21.9 (DCP) and 4.1 cm^{-1} (HAp) in the distribution of the Raman shift and the FWHM values, respectively. Note that the strong correlation between the Raman shift and FWHM is also noticeable in the line transect shown in figure 4.2, where both parameters clearly behave antithetically.

Whereas the average frequency and FWHM values for HAp fit very well with those from synthetic, nanocrystalline HAp with crystallite sizes between 70 and 100 nm (Asjadi et al. 2019), the values for the DCP phase are

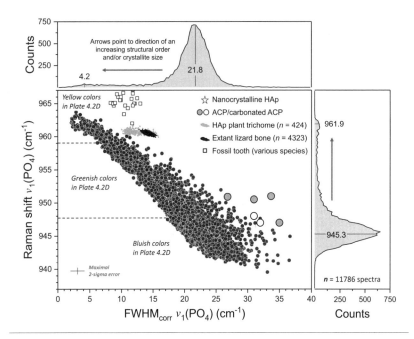

Figure 4.3. The Raman shift of the $v_1(PO_4)$ band of calcium phosphate from the acanthodian spine plotted against the corrected full width at half maximum ($FWHM_{corr}$). Also plotted are data from synthetic nanocrystalline HAp with crystallite sizes between 70 and 100 nm (Asjadi et al. 2019), ACP of various origins (Termine and Lundi 1974; Leung et al. 1995; De Grauw et al.1996; Ou-Yang et al. 2000; Stammeier et al. 2018), an extant lizard bone (Barthel et al. 2020), various fossil teeth (Thomas et al. 2011), and nanocrystalline HAp in tips and hooks of plant trichomes (Ensikat et al. 2016). The distribution of both band parameters is also shown along with their maxima values, which were obtained by fitting two asymmetric Gauss functions and a broad Gaussian background to the distribution data.

about 10 and 5 cm^{-1} lower than typical Raman shift and FWHM values reported for ACP, respectively (Termine and Lundy 1974; Leung et al. 1995; De Grauw et al. 1996; Ou-Yang et al. 2000; Stammeier et al. 2018). This indicates that the DCP phase in the fish spine is structurally and/or chemically different to synthetic ACP. The observed trend also neither overlaps with data from an extant lizard bone (Barthel et al. 2020) and fossil teeth (Thomas et al. 2011) nor with recently discovered nanocrystalline HAp in tips and hooks of plant trichomes (Ensikat et al. 2016). Note that these significant differences cannot be explained by data processing errors, since some of the reference data stem from our laboratory and were also fitted with an asymmetric Gauss-Lorentz function. They must thus represent special properties of the acanthodian spine.

Metamorphic Temperatures from Both Carbon Phases Using Band Widths and Raman Intensities

Several Raman scientists have proposed calibrations to estimate the maximum temperature of a metamorphic rock from a Raman spectrum of carbonaceous material (Beyssac et al. 2002; Rahl et al. 2005; Aoya et al. 2010; Kouketsu et al. 2014). Most Raman carbon geothermometers are based on the integrated intensity (A) ratio R2 to estimate the metamorphic temperature from graphitic material (Beyssac et al. 2002; Rahl et al. 2005; Aoya et al. 2010) that is given by $R2 = A(D1) / [A(D1) + A(D2) + A(G)]$. While the first calibration by Beyssac et al. (2002) uses only the R2 ratio and is calibrated for temperatures between 330°C and 650°C with an error of about ± 50°C, Rahl et al. (2005) extended the temperature range of the geothermometer to temperatures as low as 100°C, while reducing the error to ± 25°C. The latter authors used the R2 ratio and the R1 intensity (I) ratio, given as $R1 = I(G) / I(D1)$, which correlates well with the crystallite size (see "Estimation of the Crystallite Size of Carbon 1 from the R1 Intensity Ratio," below), and fitted a bivariate polynomial function to the observed R1-R2-T relationship.

Recently, Kouketsu et al. (2014) proposed a new geothermometer that is based on highly significant correlations of the width of the D1 and D2 band with metamorphic temperature. Kouketsu et al. (2014) have also developed a flow chart for the band fitting procedure to offer a user-friendly and, in particular, robust geothermometer. This is helpful because the number of Raman bands observed for carbonaceous matter strongly depends on the degree of structural disorder.

We recall that Raman spectra from carbon 1 are characterized by the intense D1 disorder band as well as the G band with the D2 band as a shoulder at its high-wavenumber side (plate 4.1D). Spectra of carbon 2 can also be described by three bands that, however, are shifted significantly to lower wavenumbers compared to those of carbon 1. They are assigned to the D1 disorder band, the D3 band indicative of amorphous carbon (see "Phase Identification," above), and the G band that, however, is hidden below the D3 band (plate 4.1D). We were thus able to use three Gauss-Lorentz functions along with a linear background between 1200 and 1670 cm^{-1} to automatically fit the entire Raman data set ($n = 31,524$), that is, both carbon phases with one set of functions. The obtained distributions of the Raman parameters (fig. 4.4) are characterized by two distinguishable distribution maxima (fig. 4.4A) that sometimes are accompanied by an additional broad background (fig. 4.4B). This background stems from mixed Raman spectra and from erroneous fitting results due to, for example, the absence of car-

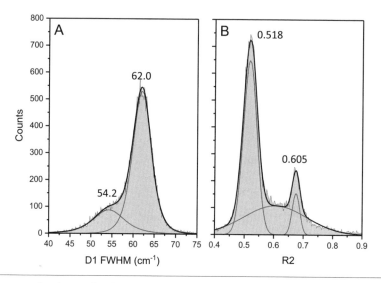

Figure 4.4. Distribution of the full width at half maxima (FWHM) (**A**) and the R2 intensity values (**B**) obtained from more than 25,000 fitted carbon spectra from the mapped area (gray area) along with the results of a least-squares fit of two asymmetric Gauss functions to the distributions and, in the case of **B**, a broad Gaussian function that accounts for the background (see text).

bon at a particular location, where the least-squares fitting procedure finds a local minimum somewhere within the background noise. As with the treatment of the Raman shift and FWHM distributions from the calcium phosphate phases (fig. 4.3), we fitted the bimodal distributions to obtain the average FWHM parameters of both carbon phases (fig. 4.4A). From these, we then calculated the metamorphic temperatures using the calibrations of Rahl et al. (2005) and Kouketsu et al. (2014). We used these calibrations since they are based on measurements likewise performed with a 532 nm laser. This is critically important since the D bands show an unusual and not yet fully understood dispersive behavior, whereby their frequency changes with the energy of the incident laser (e.g., Pimenta et al. 2007). We obtained geologically reasonable temperatures of 351 ± 30°C and 361 ± 30°C from the width of the D bands for carbon 1, and 333 ± 25°C from the intensity ratios (table 4.2), which agree well with the estimated upper temperature limit for metamorphism of the Hunsrück Slate (Wagner and Boyce 2006, and references therein). It is noteworthy that the FWHM of the D1 band of carbon 2 yields a very similar metamorphic temperature of 345 ± 30°C. On the other hand, the application of the calibration of Rahl et al. (2005) gives a geologically unrealistic temperature of 444 ± 25°C (table 4.2). This further demon-

Table 4.2.
Summary of Raman data for both identified carbon phases, crystallite size for carbon 1, and metamorphic temperatures estimated from two different Raman carbon geothermometers

	Raman shift (cm⁻¹)				FWHM (cm⁻¹)		Intensity ratios§		Size#	Temperatures (°C)		
Carb.	D3	G	D1	D2	D1	D2	R1	R2	(nm)	T^a	T^a	T^b
1	*Not obs*	1588.7(6)	1344.5(2)	1612.1(2)	54.2(2)	27.2(3)	2.20(1)	0.675(1)	8.2	361	351	333
2	1554.9(2)	1597.5(1)	1326.9(2)	*Not obs.*	62.0(1)	*Not obs.*	1.12(1)	0.518(1)	—	345	*443*	—

Notes: Numbers in brackets represent the 1-sigma standard error of the fitting procedure described in the text. Temperature value in italics is geologically implausible and can be discarded; — indicates not applicable.

§ Ratios are defined as R1 = I(D1) / I(G) and R2 = A(D1) / [A(D1) + A(D2) + A(G)], where I and A denote the intensity maximum (amplitude) respectively the integrated intensity (area).

Crystallite size estimated from a general equation from Pimenta et al. (2007).

[a] Temperatures are based on the full width at half maximum (FWHM) of D1 and D2 defect bands, respectively (Kouketsu et al. 2014). The calibration is valid for temperature between 150°C and 400°C and has an estimated error of ± 30°C.

[b] Temperature is based on bivariate polynomial calibration using both the R1 and R2 intensity ratio that is applicable from 100°C to 700°C with an estimated error of ± 25°C (Rahl et al. 2005).

strates that the Raman spectrum of this carbon phase is unusual compared to that of typical natural carbonaceous materials. In particular, it does not show a significant contribution of the G band, which is typically observed even in low grade carbonaceous matter (Beyssac et al. 2002; Rahl et al. 2005; Aoya et al. 2010; Kouketsu et al. 2014).

Fraction of Carbon 2 and DCP in the Bone Matrix

The fraction of amorphous carbon or any other phase in the analyzed volume can principally be taken from the CLS fit. Such values are accurate if (1) the Raman bands of the identified phases are not strongly polarized, that is, sensitive to crystal orientation effects, which is the case for most of the here identified phases, and (2) if the reference loadings for the CLS fit are directly extracted from the image/map from pure phase areas, which, however, was not possible for all phases. Since amorphous carbon and DCP phases inside the bone matrix are closely intergrown with each other, we could not find a location in the imaged area where pure Raman spectra of these phases were recorded. However, a computer-assisted search using the instrument software could identify spectra of DCP and carbon 1 with minimal contamination by bands from other phases in the excited and analyzed volume. These spectra were then used to generate synthetic reference spectra by fitting the individual bands with Gauss-Lorentz functions from which synthetic, "pure" DCP and carbon 1 spectra were generated (see also Hauke et al. 2019). We further corrected the spectra by the loss of intensity, which is indirectly caused by the fact that a part of the excited and analyzed volume is occupied by the "contamination" phase. This was done by adding the overall integrated intensity of all bands of the "contamination" phase to the overall integrated intensity of the synthetic spectrum (thereby neglecting any difference in the Raman scattering cross sections). With these reference loadings, we obtained an average amorphous fraction of 31 vol.% of DCP in the bone matrix from the measured distribution of the DCP fraction, with a standard deviation of ± 7 vol.%. Accordingly, the remaining 69 ± 7 vol.% of the matrix is occupied by amorphous carbon. In the future, this value will be compared with the organic fractions in other acanthodian bone samples in order to detect dependencies of the organic content on, for instance, bone histology, mineral paragenesis, and metamorphic grade, and, in turn, to better understand the fossilization process.

Estimation of the Crystallite Size of Carbon 1 from the R1 Intensity Ratio

Tuinstra and Koenig (1970) were the first to derive the in-plane crystallite size (L) of nanocrystalline graphite from the intensity ratio R1 as defined above. The authors found a good linear correlation between the inverse of the R1 ratio and the crystallite sizes. For many years since this pioneering work, this correlation has been used in numerous papers to estimate L in disordered carbon materials from Raman spectra, which often were measured with different laser energies. Due to the observed dispersive behavior of the D and G mode frequencies, however, this must have led to erroneous results. Only recently, Pimenta et al. (2007) experimentally derived a simple empirical relationship that allows the determination of L in nanographite systems for any excitation laser wavelength (λ_0) in the visible range:

$$L = (2.4 \cdot 10^{-10}) \cdot \lambda_0^4 \cdot R1^{-1} \tag{Eq. 4}$$

Using this expression and the measured R1 value for carbon 1, we obtain an average in-plane crystallite size in the order of 8 nm (table 4.2). Carbon 1 is thus identified as nanocrystalline graphitic carbon.

Discussion

The occurrence of DCP in the bone matrix of a ca. 405-million-year-old fossil fish spine is probably the most irritating finding of this study considering that the inorganic bone matrix of modern and fossil bone and teeth is usually composed of nanocrystalline, carbonated hydroxyapatite or fluora-patite (e.g., Thomas et al. 2011; Tütken and Vennemann 2011). However, several studies have reported some evidence of ACP in several mineralized tissues, such as in newly formed human bone (Eanes et al. 1965), newly formed fin bones of zebrafish (Mahamid et al. 2008), and in the radular teeth of chitons (Lowenstam and Weiner 1985). However, the occurrence of an ACP phase in such newly mineralized tissues of vertebrates is still debated (Grynpas et al. 2007). It would thus be even more surprising to detect a metastable, commonly highly reactive ACP phase in a ca. 405-million-year-old fossil.

Synthetic ACP is known to transform within hours or days into apatite in aqueous media at room temperature (e.g., Boskey and Posner 1974; Stammeier et al. 2018), although it can be stabilized to some extent by other cations (Boskey and Posner 1974; LeGeros et al. 2005). The survival of ACP therefore becomes even more puzzling considering that the vascular canals were found to be filled with quartz, ankerite, muscovite, nanocrystalline (ca. 8 nm) graphitic material (carbon 1), and a still unknown sulfide (or oxide) phase, which is unequivocal evidence that at some stage(s) in the history of the spine it was infiltrated by aqueous solutions along the vascular canals and cracks. Changes in the local physicochemical conditions, such as a change of the pH and redox conditions, must have triggered the precipitation of nanocrystalline graphite (carbon 1) and the other minerals inside the open space. Moreover, the hyperspectral Raman image shown in plate 4.2A reveals bone replacement domains that mainly consist of disordered graphite, well-ordered HAp, and in some areas quartz. The boarder of the replacement domains with the inner bone matrix is marked by relative sharp reaction fronts (white arrows in plate 4.2A) that progressed from the canals and cracks into the bone matrix. The Raman spectra across such a reaction front indicate gradual structural and/or chemical changes in the calcium phosphate away from the reaction front over length scales that by far exceed the spatial resolution (fig. 4.2). This indicates that the DCP phase partly reacted with the fluid ahead of the front and inside the bone matrix. The distribution of well-ordered HAp mainly in the outer and less in the inner parts of the vascular canals further suggests that most of the phosphate was possibly derived from partial dissolution of the DCP phase. The mineral paragenesis clearly points to a highly reducing fluid that was probably very Fe-rich, indicated by the Fe-rich ankerite (low #Mg value). If verified, the occurrence of troilite (stoichiometric FeS) and/or maghemite (Fe_2O_3) would constrain the oxygen and sulfur fugacity of the fluid.

Raman carbon geothermometry indicates that the fish bone has experienced temperatures of up to ca. 350°C. However, it should be noted that the laser energy may have caused a thermal broadening and frequency shifts (Tan et al. 1999) or structurally altered the carbon phases. We used a relatively high laser power of about 50 mW at the sample surface, which is significantly higher than the recommended upper power limit for carbon geothermometry of 3 mW (Beyssac et al. 2002). In addition, the preparation of the thin section by grinding and polishing may have changed the surface structure of the sample, which can also lead to erroneous temperatures (Byssac et al. 2002). On the other hand, the consistency of frequency and FWHM values throughout the mapped area does not indicate a heating or polishing effect, which would be expected to be different in different locations. In any case, we did not intend to determine the metamorphic temperature from carbon Raman spectra, but to get an initial overview of the mineral paragenesis, inorganic remnants, and texture of the fossil fish bone. For more reliable temperature estimates, measurements would have to be reproduced on unpolished surfaces with reduced laser power. Nevertheless, Raman carbon geothermometry may become an important tool in future fossilization studies because it provides a constraint on the temperature history of the fossil. Despite uncertainties in the absolute temperature estimate, relative differences in the temperature history of different fossil sites can be well investigated if the Raman spectra are carefully taken under identical analytical conditions.

Whatever the exact physicochemical conditions, the DCP phase represents the original mineralization of the acanthodian spine (at least in the spine section studied). Despite the observed spectral similarities with ADP, this apatite-like phosphate phase is more crystalline than amorphous and possibly characterized by a crystallite size in the lower nanometer range. Note that the DCP in our sample shows some Raman spectral differences when compared to synthetic ACP (fig. 4.3). Thermal broadening should be insignificant at 50mW surface power, considering laser heating tests on nanocrystalline HAp with crystallite sizes between 70 and 100 nm (Asjadi et al. 2019), but this may be different in even smaller crystallites. We finally note that the highly amorphous carbon phase (carbon 2), which was only detected within the bone matrix and thus most likely represents the remnant of the original extracellular organic material (e.g., collagen), did not mature to typical graphite-like carbon.

To solve these enigmas, future research will involve further Raman measurements with lower laser power to reexamine the carbon thermometry results and other micro- or nano-analytical techniques to study the chemical composition, structure, and nano-sized texture (intergrowth) of carbon and the nanocrystalline and disordered calcium phosphate phases, such as transmission electron microscopy, electron microprobe, nanoscale secondary ion mass spectrometry, and atomic-scale atom probe tomography.

Conclusions

In this chapter, we have introduced the reader to the main theoretical and practical principles of confocal Raman spectroscopy, as well as its potential paleontological applications, through a detailed analysis of a hyperspectral Raman data set from a ca. 405-million-year-old fin spine from an acanthodian fish from the Hunsrück Slate in Germany. This data set delivers a wealth of textural, mineralogical, and geological information, but at the same time raises a number of new questions, which will be addressed in future studies using micro- to nano-analytical techniques.

Raman spectroscopy has the potential to become a central analytical technique in vertebrate paleontology, invertebrate paleontology, and paleobotany. It is mostly nondestructive if the laser power is correctly chosen, can be applied to a variety of sample types and sample sizes (from micrometer to centimeter sizes), and can identify inorganic and organic substances. The various information obtained from hyperspectral Raman imaging/mapping may well be used to select suitable regions for paleoecological and paleobiochemical reconstructions using stable isotope tracers (e.g., Tütken and Vennemann 2011). Raman spectrometers are relatively easy to run and thus do not necessarily need an experienced operator. However, Raman spectra are not always easy to interpret and require a certain level of experience and theoretical knowledge. Furthermore, the exemplary study presented here highlights the need to apply Raman spectroscopy also in combination with other analytical methods and to carefully process and evaluate the data obtained.

ACKNOWLEDGMENTS

The authors wish to thank Jes Rust for providing and making us aware of this interesting sample, as well as Martin Sander and Carole Gee for their helpful comments on early drafts of the manuscripts. We appreciate the valuable suggestions and corrections made by an anonymous reviewer and thank the reviewer very much. This work was supported by the Deutsche Forschungsgemeinschaft (DFG, German Research Foundation), Project number 396710782 to P. Martin Sander for TM (University of Bonn). This is contribution number 18 of the DFG Research Unit 2685, "The Limits of the Fossil Record: Analytical and Experimental Approaches to Fossilization."

WORKS CITED

Aoya, M., Kouketsu, Y., Endo, S., Shimizu, H., Mizukami, T., Nakamura, D., and Wallis, S. 2010. Extending the applicability of the Raman carbonaceous—Material geothermometer using data from contact metamorphic rocks. *Journal of Metamorphic Geology*, 28: 895–914.

Asjadi, F., Geisler, T., Salahi, I., Euler, H., and Mobasherpour, I. 2019. Ti-substituted hydroxylapatite precipitated in the presence of titanium sulphate: A novel photocatalyst? *American Journal Chemistry and Applications*, 6: 1–10.

Atkinson, A., and Jain, S. 1999. Spatially resolved stress analysis using Raman spectroscopy. *Journal of Raman Spectroscopy*, 30: 885–891

Avril, C., Malavergne, V., Caracas, R., Zanda, B., Reynard, B., Charon, E., Bobocioiu, E., Brunet, F., Borensztain, S., Pont, S., Tarrida, M., and Guyot, F. 2013. Raman spectroscopic properties and Raman identification of CaS–MgS–MnS–FeS–Cr_2FeS_4 sulfides in meteorites and reduced sulfur-rich systems. *Meteoritics & Planetary Science*, 48: 1415–1426.

Barthel, H.J., Fougerouse, D., Geisler, T., and Rust, J. 2020. Fluoridation of a lizard bone embedded in Dominican amber suggests open-system behavior. *PLOS ONE*, 15(2): e0228843.

Berg, R.W., and Nørbygaard, T. 2006. Wavenumber calibration of CCD detector Raman spectrometers controlled by a sinus arm drive. *Applied Spectroscopy Reviews*, 41: 165–183.

Bersani, D., Andò, S., Vignola, P., Moltifiori, G., Marino, I.-G., Lottici, P.P., and Diella, V. 2009. Micro-Raman spectroscopy as a routine tool for garnet analysis. *Spectrochimica Acta Part A: Molecular and Biomolecular Spectroscopy*, 73: 484–491.

Beyssac, O., Goffe, B., Chopin, C., and Rouzaud, J.N. 2002. Raman spectra of carbonaceous material in metasediments: A new geothermometer. *Journal of Metamorphic Geology*, 20: 859–871.

Boskey, A.L., and Posner, A.S. 1974. Magnesium stabilization of amorphous calcium phosphate: A kinetic study. *Materials Research Bulletin*, 9: 907–916.

Bowie, B.T., Chase, D.B., Lewis, I.R., and Griffiths, P.R. 2006. Anomalies and artifacts in Raman spectroscopy. In: Chalmers, J.M., and Griffiths, P.R., eds., *Handbook of Vibrational Spectroscopy*. John Wiley & Sons, doi:10.1002/0470027320.s3103.

Böhme, N., Hauke, K., Neuroth, M., and Geisler, T. 2019. In situ Raman imaging of high-temperature solid-state reactions in the $CaSO_4$–SiO_2 system. *International Journal of Coal Science and Technology*, 6: 247–259.

Brasier, M.D., Green, O.R., Jephcoat, A.P., Kleppe, A.T., Van Kranendonk, M.J., Lindsay J.F., Steele, A., and Grassineau, N.V. 2002. Questioning the evidence for Earth's oldest fossils. *Nature*, 416: 76–81.

Breitenfeld, L.B., Dyar, M.D., Carey, C., Tague, T.J., Jr., Wang, P., Mullen, T., and Parente, M. 2018. Predicting olivine composition using Raman spectroscopy through band shift and multivariate analyses. *American Mineralogist*, 103: 1827–1836.

Brody, R.H., Edwards, H.G.M., and Pollard, A.M. 2001. A study of amber and copal samples using FT-Raman spectroscopy. *Spectrochimica Acta Part A: Molecular and Biomolecular Spectroscopy*, 57: 1325–1338.

Burke, E.A. 2001. Raman microspectrometry of fluid inclusions. *Lithos*, 55: 139–158.

Chou, I.-M., and Wang, A. 2017. Application of laser Raman micro-analyses to Earth and planetary materials. *Journal of Asian Earth Sciences*, 145: 309–333.

De Grauw C.J., de Bruijn J.D., Otto C., Greve, J. 1996. Investigation of bone and calcium phosphate coatings and crystallinity determination using Raman microspectroscopy. *Cells and Materials*, 6: 57–62.

Denison, R. 1979. Acanthodii. In: Schultze H.-P., ed. *Handbook of Paleoichthyology*, 5: 1–62.

Dieing, T., Hollricher, O., and Toporski, J. 2012. Confocal Raman Microscopy. *Springer Series in Optical Sciences*, 158: 1–292.

Drzewicz, P., Natkaniec-Nowak, L., and Czapla, D. 2016. Analytical approaches for studies of fossil resins. *TrAC Trends in Analytical Chemistry*, 85: 75–84.

Dubois, F., Mendibide, C., Pagnier, T., Perrard, F., and Duret, C. 2008. Raman mapping of corrosion products formed onto spring steels during salt spray experiments. A correlation between the scale composition and the corrosion resistance. *Corrosion Science*, 50: 3401–3409.

Eanes, E.D., Gillessen, I.H., and Posner, A.S. 1965. Intermediate states in the precipitation of hydroxyapatite. *Nature*, 208: 365–367.

Ehrlich, H., Rigby, J.K., Botting, J.P., Tsurkan, M.V., Werner, C., Schwille, P., Petrášek, Z., Pisera, A., Simon, P., Sivkov, V.N., Vyalikh, D.V., Molodtsov, S.L., Kurek, D., Kammer, M., Hunoldt, S., Born, R., Stawski, D., Steinhof, A., Bazhenov, V.V., and Geisler, T. 2013. Discovery of 505-million-year old chitin in the basal demosponge *Vauxia gracilenta*. *Scientific Reports*, 3: 3497.

Enami, M., Nishiyama, T., and Mouri, T. 2007. Laser Raman microspectrometry of metamorphic quartz: A simple method for comparison of metamorphic pressures. *American Mineralogist*, 92: 1303–1315.

Ensikat, H.J., Geisler, T., and Weigend, M. 2016. A first report of hydroxylated apatite as structural biomineral in Loasaceae—Plants' teeth against herbivores. *Scientific Reports*, 6: 26073.

Everall, N. 2004. Depth profiling with confocal Raman microscopy, part I. *Spectroscopy*, 19: 22–28.

Everall, N. 2010. Confocal Raman microscopy: Common errors and artefacts. *Analyst*, 135: 2512–2522.

Fateley, W.G., McDevitt, N.T., and Bentley, F.F. 1971. Infrared and Raman selection rules for lattice vibrations: The Correlation Method. *Applied Spectroscopy*, 25: 155–173.

Ferraro, J. 2002. *Introductory Raman Spectroscopy*, 2nd ed. Academic Press, San Diego.

France, C.A.M., Thomas, D.B., Doney, C.R., and Madden, O. 2014. FT-Raman spectroscopy as a method for screening collagen diagenesis in bone. *Journal of Archaeological Science*, 42: 346–355.

Fries, M., and Steele, A. 2012. Raman spectroscopy and confocal Raman imaging in mineralogy and petrography. In: Dieing, T., Hollricher, O., and Toporski, J., eds. *Confocal Raman Microscopy*. Springer Verlag, Springer Series in Optical Sciences, 158: 111–136.

Frost, R., Kloprogge, T., and Schmidt, J. 1999. Non-destructive identification of minerals by Raman microscopy. *The Internet Journal of Vibrational Spectroscopy*, 3: 40–44. www.irdg.org/ijvs.

Geisler, T. 2002. Isothermal annealing of partially metamict zircon: Evidence for a three-stage recovery process. *Physics and Chemistry of Minerals*, 29: 420–429.

Geisler, T., Burakov, B.E., Zirlin, V., Nikolaeva L., and Pöml, P. 2005. A Raman spectroscopic study of high-uranium zircon from the Chernobyl. *European Journal of Mineralogy*, 17: 883–894.

Geisler, T., Dohmen L., Lenting C., and Fritzsche, M.B. 2019. Real-time in situ observations of reaction and transport phenomena during silicate glass corrosion by fluid-cell Raman spectroscopy. *Nature Materials*, 18: 342–348.

Geisler, T., Perdikouri, C., Kasioptas, A., and Dietzel, M. 2012. Real-time monitoring of the overall exchange of oxygen isotopes between aqueous CO_3^{2-} and water by Raman spectroscopy. *Geochimica et Cosmochimica Acta*, 90: 1–11.

Geisler, T., Popa, K., Konings, R.J., and Popa, A.F. 2006. A Raman spectroscopic study of the phase transition of $BaZr(PO_4)_2$: Evidence for a trigonal structure of the high-temperature polymorph. *Journal of Solid State Chemistry*, 179: 1490–1496.

Geisler, T., Popa, K., and Konings R.J.M. 2016. Evidence for lattice strain and non-ideal behavior in the $(La_{1-x}Eu_x)PO_4$ solid solution from X-ray diffraction and vibrational spectroscopy. *Frontiers in Earth Science*, 4: 64.

Gillet, P. 1996. Raman spectroscopy at high pressure and high temperature. Phase transitions and thermodynamic properties of minerals. *Physics and Chemistry of Minerals*, 23: 263–275.

Gouadec, G., and Colomban, P. 2007. Raman Spectroscopy of nanomaterials: How spectra relate to disorder, particle size and mechanical properties. *Progress in Crystal Growth and Characterization of Materials*, 53: 1–56.

Grynpas, M.D., and Omelon, S. 2007. Transient precursor strategy or very small biological apatite crystals? *Bone*, 41: 162–164.

Guinet, Y., Sauvajol, J., and Muller, M. 1988. Raman studies of orientational disorder in crystals: The 1.fluoroadamantane plastic phase. *Molecular Physics*, 65: 723–738.

Hartl, C., Schmidt, A.R., Heinrichs, J., Seyfullah, L.J., Schäfer, N., Gröhn, C., Rikkinen, J., and Kaasalainen, U. 2015. Lichen preservation in amber: Morphology, ultrastructure, chemofossils, and taphonomic alteration. *Fossil Record*, 18: 127–135.

Hauke, K., Kehren, J., Böhme, N., Zimmer, S., and Geisler T. 2019. In situ hyperspectral Raman imaging: A new method to investigate sintering processes of ceramic materials at high temperature. *Applied Sciences*, 9: 1310.

He, J., and Bismayer, U. 2019. Polarized mapping Raman spectroscopy: Identification of particle orientation in biominerals. *Zeitschrift für Kristallographie–Crystalline Materials*, 234: 395–401.

Hernández, B., Pflüger, F., Kruglik, S.G., and Ghomi, M. 2013. Characteristic Raman lines of phenylalanine analyzed by a multiconformational approach. *Journal of Raman Spectroscopy*, 44: 827–833.

Herzberg G. 1945. *Molecular Spectra and Molecular Structure. II, Infrared and Raman Spectra of Polyatomic Molecules*. D. Van Nostrand Company (Reprinted edition 1991, Krieger Publishing, Malabar, Florida).

Hutsebaut, D., Vandenabeele, P., and Moens, L. 2005. Evaluation of an accurate calibration and spectral standardization procedure for Raman spectroscopy. *Analyst*, 130: 1204–1214.

Izraeli, E.S., Harris, J.W., and Navon, O. 1999. Raman barometry of diamond formation. *Earth and Planetary Science Letters*, 173: 351–360.

Jerve, A., Bremer, O., Sanchez, S., and Ahlberg, P.E. 2017. Morphology and histology of acanthodian fin spines from the late Silurian Ramsåsa E locality, Skåne, Sweden. *Palaeontologia Electronica*, 20.3.56A: 1–19.

Juang, C.B., Finzi, L., and Bustamante C.J. 1988. Design and application of a computer-controlled confocal scanning differential polarization microscope. *Review of Scientific Instruments*, 59: 2399–2408.

Karim, R.M., Kwon, O.-H., Park, C., and Lee, K. 2019. A study of colormaps in network visualization. *Applied Sciences*, 9: 4228.

Kazanci, M., Roschger, P., Paschalis, E.P., Klaushofer, K., and Fratzl, P. 2007. Bone osteonal tissues by Raman spectral mapping: Orientation–composition. *Journal of Structural Biology*, 156: 489–496.

King, H., and Geisler, T. 2018. Tracing mineral reactions using confocal Raman spectroscopy. *Minerals*, 8: 158.

Kiseleva, D., Shilovskyl, O., Shagalov, E., Ryanskaya, A., Chervyakovskaya, M., Pankrushina, E., and Cherednichenko, N. 2019. Composition and structural features of two Permian parareptile (Deltavjatia vjatkensis, Kotelnich Site, Russia) bone fragments and their alteration during fossilization. *Palaeogeography, Palaeoclimatology, Palaeoecology*, 526: 28–42.

Kitajima, M. 1997. Defects in crystals studied by Raman scattering. *Critical Reviews in Solid State and Material Sciences*, 22: 275–349.

Knight, D.S., and White, W.B. 1989. Characterization of diamond films by Raman spectroscopy. *Journal of Materials Research*, 4: 385–393.

Kouketsu, Y., Mizukami, T., Mori, H., Endo, S., Aoya, M., Hara, H., Nakamura, D., and Wallis, S. 2014. A new approach to develop the Raman carbonaceous material geothermometer for low-grade metamorphism using peak width. *Island Arc*, 23: 33–50.

Kovesi, P. 2015. Good colour maps: How to design them. *arXiv:1509.03700v1 [cs. GR]*.

Kuczumow, A., Nowak, J., Kuzioła, R., and Jarzębski, M. 2019. Analysis of the composition and minerals diagrams determination of petrified wood. *Microchemical Journal*, 148: 120–129.

Lafuente, B., Downs, R.T., Yang, H., and Stone, N. 2015. The power of databases: The RRUFF project, pp. 1–30. In: Armbruster, T., and Danisi, R.M., eds. *Highlights in Mineralogical Crystallography*, W. De Gruyter, Berlin.

LeGeros, R.Z., Mijares, D., Park, J., Chang, X.-F, Khairoun, I., Kijkowska, R., Dias, R., and LeGeros, J.P. 2005. Amorphous calcium phosphates (ACP): Formation and stability. *Key Engineering Materials*, 284–286: 7–10.

Leissner, L., Schlüter, J., Horn, I., and Mihailova, B. 2015. Exploring the potential of Raman spectroscopy for crystallochemical analyses of complex hydrous silicates: I. Amphiboles. *American Mineralogist*, 100: 2682–2694.

Leung, Y., Anton Walters, M., Blumenthal, N.C., Ricci, J.L., and Spivak, J.M. 1995. Determination of the mineral phases and structure of the bone-implant interface using Raman spectroscopy. *Journal of Biomedical Materials Research*, 29: 591–594.

Lohumi, S., Kim, M.S., Qin, J., Cho, B.-K. 2017. Raman imaging from microscopy to macroscopy: Quality and safety control of biological materials. *TRAC—Trends in Analytical Chemistry*, 93: 183–198.

Lönartz, M.I., Dohmen, L., Lenting, C., Trautmann, C., Lang, M., and Geisler, T. 2019. The effect of heavy ion irradiation on the forward dissolution rate of borosilicate glasses studied in situ and real time by fluid-cell Raman spectroscopy. *Materials*, 12: 1480.

Lowenstam, H., and Weiner, S. 1985. Transformation of amorphous calcium phosphate to crystalline dahllite in the radular teeth of chitons. *Science*, 227: 51–53.

Mahamid, J., Sharir, A., Addadi, L., and Weiner S. 2008. Amorphous calcium phosphate is a major component of the forming fin bones of zebrafish: Indications for an amorphous precursor phase. *Proceedings of the National Academy of Sciences*, 105: 12748–12753.

Marshall, A.O., and Marshall, C.P. 2015. Vibrational spectroscopy of fossils. *Palaeontology*, 58: 201–211.

Marshall, C.P., Edwards, H.G., and Jehlicka, J. 2010. Understanding the application of Raman spectroscopy to the detection of traces of life. *Astrobiology*, 10: 229–243.

Meier, R.J. 2005. On art and science in curve-fitting vibrational spectra. *Vibrational Spectroscopy*, 2: 266–269.

Menneken, M., Geisler, T., Nemchin, A.A., Whitehouse, M.J., Wilde, S.A., Gasharova, B., and Pidgeon R.T. 2017. CO_2 fluid inclusions in Jack Hills zircons. *Contribution to Mineralogy and Petrology*, 172: 66.

Nasdala, L., Brenker, F.E., Glinnemann, J., Hofmeister, W., Gasparik, T., Harris, J.W., Stachel, T., and Reese I. 2003. Spectroscopic 2D-tomography: Residual pressure and strain around mineral inclusions in diamonds. *European Journal of Mineralogy*, 15: 931–935.

Nasdala, L., Smith, D.C., Kaindl, R., and Ziemann, M.A. 2004. Raman spectroscopy: Analytical perspectives in mineralogical research. *Spectroscopic Methods in Mineralogy*, 6: 281–343.

Nikiel, L., and Jagodzinski, P.W. 1993. Raman spectroscopic characterization of graphites: A re-evaluation of spectra/structure correlation. *Carbon*, 31: 1313–1317.

Ou-Yang, H., Paschalis, E., Boskey, A., and Mendelsohn, R. 2000. Two-dimensional vibrational correlation spectroscopy of in vitro hydroxyapatite maturation. *Biopolymers*, 57: 129–139.

Pelletier, M. 2003. Quantitative analysis using Raman spectrometry. *Applied Spectroscopy*, 57: 20A–42A.

Pimenta, M., Dresselhaus, G., Dresselhaus, M.S., Cancado, L., Jorio, A., and Saito, R. 2007. Studying disorder in graphite-based systems by Raman spectroscopy. *Physical Chemistry Chemical Physics*, 9: 1276–1290.

Rahl, J.M., Anderson, K.M., Brandon, M.T., and Fassoulas C. 2005. Raman spectroscopic carbonaceous material thermometry of low-grade metamorphic rocks: Calibration and application to tectonic exhumation in Crete, Greece. *Earth and Planetary Science Letters*, 240: 339–354.

Rividi, N., van Zuilen, M., Philippot, P., Menez, B., Godard, G., and Poidatz, E. 2010. Calibration of carbonate composition using micro-Raman analysis: Application to planetary surface exploration. *Astrobiology*, 10: 293–309.

Rousseau, D.L., Bauman, R.P., and Porto, S. 1981. Normal mode determination in crystals. *Journal of Raman Spectroscopy*, 10: 253–290.

Salje, E. 1992. Hard mode spectroscopy: Experimental studies of structural phase transitions. *Phase Transitions*, 37: 83–110.

Salje, E., Schmidt, C., and Bismayer, U. 1993. Structural phase transition in titanite, $CaTiSiO_5$: A Raman spectroscopic study. *Physics and Chemistry of Minerals*, 19: 502–506.

Saloman, E.B. 2006. Wavelengths, energy level classifications, and energy levels for the spectrum of neutral mercury. *Journal of Physical and Chemical Reference Data*, 35: 1519–1548.

Saltonstall, C.B., Beechem, T.E., Amatya, J., Floro, J., Norris, P.M., and Hopkins, P.E. 2019. Uncertainty in linewidth quantification of overlapping Raman bands. *Review of Scientific Instruments*, 90: 013111.

Schlücker, S. 2012. Confocal Raman microscopy. In: Dieing, T., Hollricher, O., and Toporski, J., eds. *Microscopy and Microanalysis*, 18: 1494–1494.

Schopf, J.W., Kudryavtsev, A.B., Agresti, D.G., Wdowiak, T.J., and Czaja, A.D. 2002. Laser-Raman imagery of Earth's earliest fossils. *Nature*, 416: 73–76.

Schrof, S., Varga, P., Galvis, L., Raum, K., and Masic, A. 2014. 3D Raman mapping of the collagen fibril orientation in human osteonal lamellae. *Journal of Structural Biology*, 187: 266–275.

Sidorov, N., Yanichev, A., Palatnikov, M., and Gabain, A. 2014. Effects of the ordering of structural units of the cationic sublattice of LiNbO 3: Zn crystals and their manifestation in Raman spectra. *Optics and Spectroscopy*, 116: 281–290.

Skála, R., Císařová, I., and Drábek, M. 2006. Inversion twinning in troilite. *American Mineralogist*, 91: 917–921.

Smith, D.C. 2005. The RAMANITA© method for non-destructive and in situ semi-quantitative chemical analysis of mineral solid-solutions by multidimensional calibration of Raman wavenumber shifts. *Spectrochimica Acta Part A: Molecular and Biomolecular Spectroscopy*, 61: 2299–2314.

Socrates, G. 2004. *Infrared and Raman Characteristic Group Frequencies: Tables and Charts*, 3rd ed. John Wiley & Sons, Chichester, England.

Stammeier, J.A., Purgstaller, B., Hippler, D., Mavromatis, V., and Dietzel, M. 2018. In-situ Raman spectroscopy of amorphous calcium phosphate to crystalline hydroxyapatite transformation. *MethodsX*, 5: 1241–1250.

Stange, K., Lenting, C., and Geisler, T. 2018. Insights into the evolution of carbonate-bearing kaolin during sintering revealed by in situ hyperspectral Raman imaging. *Journal of the American Ceramic Society*, 101: 897–910.

Tamor, M.A., and Vassell, W. 1994. Raman "fingerprinting" of amorphous carbon films. *Journal of Applied Physics*, 76: 3823–3830.

Tan, P.H., Deng, Y.M., Zhao, Q., and Cheng, W.C. 1999. The intrinsic temperature effect of the Raman spectra of graphite. *Applied Physics Letters*, 74: 1818–1820.

Tanabe, K., and Hiraishi, J. 1980. Correction of finite slit width effects on Raman line widths. *Spectrochimica Acta Part A: Molecular Spectroscopy*, 36: 341–344.

Termine, J., and Lundy, D. 1974. Vibrational spectra of some phosphate salts amorphous to X-ray diffraction. *Calcified Tissue Research*, 15: 55–70.

Thomas, D.B., McGoverin, C.M., Fordyce, R.E., Frew, R.D., and Gordon, K.C. 2011. Raman spectroscopy of fossil bioapatite—A proxy for diagenetic alteration of the oxygen isotope composition. *Palaeogeography, Palaeoclimatology, Palaeoecology*, 310: 62–70.

Thomas, D.B., McGraw, K.J., Butler, M.W., Carrano, M.T., Madden, O., and James, H.F. 2014. Ancient origins and multiple appearances of carotenoid-pigmented feathers in birds. *Proceedings of the Royal Society B*, 281: 20140806.

Tuinstra, F., and Koenig, J.L. 1970. Raman spectrum of graphite. *The Journal of Chemical Physics*, 53: 1126–1130.

Tütken, T., and Vennemann, T.W. 2011. Fossil bones and teeth: Preservation or alteration of biogenic compositions? *Palaeogeography, Palaeoclimatology, Palaeoecology*, 310: 1–8.

Vandenabeele, P., Edwards, H., and Jehlička, J. 2014. The role of mobile instrumentation in novel applications of Raman spectroscopy: Archaeometry, geosciences, and forensics. *Chemical Society Reviews*, 43: 2628–2649.

Wagner, T., and Boyce, A.J. 2006. Pyrite metamorphism in the Devonian Hunsrück Slate of Germany: Insights from laser microprobe sulfur isotope analysis and thermodynamic modeling. *American Journal of Science*, 306: 525–552.

Wojdyr, M. 2010. Fityk: A general-purpose peak fitting program. *Journal of Applied Crystallography*, 43: 1126–1128.

Zhang, M. 2017. Raman study of the crystalline-to-amorphous state in alpha-decay-damaged materials, pp. 103–122. In: Maaz, K., ed. *Raman Spectroscopy and Applications.* InTech, Rijeka, Croatia.

Zhang, M., Salje, E.K.H., Farnan, I., Graeme-Barber, A., Danield, P., Ewing, R.C., Clark, A.M., and Leroux, H. 2000. Metamictization of zircon: Raman spectroscopic study. *Journal of Physics: Condensed Matter*, 12: 1915–1925.

CHAPTER **5**

From Ultrastructure to Biomolecular Composition
Taphonomic Patterns of Tissue Preservation in Arthropod Inclusions in Amber

H. JONAS BARTHEL, VICTORIA E. MCCOY, AND JES RUST

ABSTRACT | Amber and copal represent unique and important sources of fossil arthropods. Their extraordinary preservation of morphological characters and internal tissues is well known and unrivaled by any other type of *Konservat-Lagerstätte*. Thus, amber is regarded as an ideal source for ancient biomolecules such as lipids, proteins, and DNA. Here we review the state of knowledge of amber and copal preservation, placing particular emphasis on evidence for a dichotomous pathway of soft tissue preservation in amber. Hypotheses to explain the outstanding preservation of inclusions focus on the resin chemistry and specifically volatile compounds that perfuse the tissue. These volatile compounds have effects that range from the active inhibition of decay enzymes and microorganisms to the inert dehydration caused by the reactions of terpenes and sugars with water in the tissues. However, chemical analysis of inclusions from various deposits indicates that the process of preserving arthropods in amber does not always follow the same exact taphonomic pathway. Furthermore, the quality of the amber fossils varies morphologically, from hollow inclusions to three-dimensional bodies of fossil arthropods exquisitely preserved with all their internal organs. Very early after entrapment, fossilization follows one of two pathways: either decay is severely inhibited, resulting in the preservation of soft tissues complete with ultrastructure, or decay proceeds as usual and the inclusion loses all soft tissues. If the inclusion follows the first, decay-inhibited, pathway, there are then further slow changes in tissue chemistry and biomolecule preservation during diagenesis. In this review, we also summarize previous studies that document internal tissue preservation in amber inclusions and discuss several factors that are likely the major drivers of this phenomenon, such as resin chemistry and microbial gut composition of the entrapped arthropods. It is clear that more extensive taphonomic experimentation is necessary to understand the role of these major drivers in the fine ultrastructural preservation of fossil arthropods in amber. |

Introduction

Resin is a viscous plant exudate produced by trees primarily as a defense against herbivores and pathogens (Hinejima et al. 1992; Langenheim 1994; Lange 2015; Pichersky and Raguso 2018; McCoy et al., chap. 8). Fossil and ancient forms of resin (amber and copal, respectively) are well known and important archives of terrestrial arthropods, plants, and fungi from the Mesozoic to the present day. With their characteristic three-dimensional inclusions—some even fossilized during an ephemeral moment of behavior, such as a parasitic worm emerging from its host (Poinar and Buckley 2006)—and the preservation of minute, even subcellular, detail, the amber fossil record is a powerful tool for answering phylogenetic and paleoecological questions (Labandeira 2014). The uniquely detailed preservation of labile soft-bodied organisms, tissues, and tissue ultrastructure is so exceptional as to suggest that fossil organisms in amber or copal may even preserve ancient biomolecules such as DNA and proteins (an idea most famously suggested in the 1993 blockbuster movie *Jurassic Park*).

Most studies on amber focus on arthropods, which are the dominant group of organisms found as inclusions. The prevalence of arthropod inclusions is due to several factors, such as their ecology and small size, which make them more likely to be entrapped than other taxa like vertebrates (Martínez-Delclòs et al. 2004). However, there is also variation in entrapment frequencies within the arthropod clade that are based on further differences in ecology and size (Martínez-Delclòs et al. 2004; Solórzano Kraemer et al. 2015, 2018). Consequently, if fossilization can be considered a sieve that only allows certain organisms to pass from an ecosystem into the fossil record, amber is a sieve with a particularly fine mesh. The amber fossil record therefore represents only a very small, incomplete slice of a paleocommunity (Solórzano Kraemer et al. 2015, 2018). Fossil organisms in amber and copal have been extensively studied and reviewed in the past (Larsson 1978; Poinar 1992; Solórzano Kraemer 2007; Vávra 2009; Penney 2010; Soriano et al. 2010; Labandeira 2014; McCoy et al. 2018a; Seyfullah et al. 2018a, b), but studies emphasizing taphonomic aspects are relatively sparse compared to studies on taxonomy and ecology. Nevertheless, as with all exceptional fossils, the utility of inclusions in amber in any research context (taxonomic, ecological, etc.) depends on the quality of the preservation.

Chemical and microscopic investigations suggest that the quality of preservation in amber is not uniform and depends on several factors, such as the chemistry of the resin, the grade of dehydration of the inclusion, or the soil composition (Kohring 1998; Stankiewicz et al. 1998; Martínez-Delclòs et al. 2004; Koller et al. 2005; McCoy et al. 2018a, b). Many questions arise from this observation: How are the fossils preserved? What is the degree of alteration? Do original biomolecules get preserved? Are some structures (or tissues) generally "better" preserved than others? Are there differences in preservation between amber deposits of the same or different ages, or based on geographical origin and resin-producing tree origin? Understanding the process of fossilization in resin would allow us to answer these questions and therefore target the most suitable deposits for specific research questions, which may require the preservation of specific tissues or biomolecules.

Here we summarize the current state of research on the taphonomy of inclusions in amber and combine these data into a unified picture of taphonomic patterns in amber. While we focus primarily upon investigations of arthropods because these are the most widely studied taxa, we also occasionally include information on other taxa to strengthen our conclusions and clarify the patterns that we identify. Further, we present a hypothesis for a general dichotomous pathway of fossilization in amber that results in either the presence or absence of soft tissue preservation, and we focus particularly upon several factors that we think are likely to most strongly influence the quality of preservation when soft tissues remain. Finally, we suggest the most productive future pathways for research into amber taphonomy.

Tissue Ultrastructure

Ultrastructural preservation is absent in fossils from most amber deposits (Martínez-Delclòs et al. 2004; Soriano et al. 2010; Labandeira 2014; McCoy et al. 2018a). Recent large-scale investigations of amber taphonomy have revealed that the majority of amber deposits have no fossils, and the majority of fossiliferous amber sites are characterized by "empty" inclusions, that is, inclusions that are empty molds with just external tissue preservation (Martínez-Delclòs et al. 2004; Soriano et al. 2010; Labandeira 2014; McCoy et al. 2018a). However, for inclusions in amber with preserved tissues, the ultrastructural details of those tissues can be used as a valuable tool for understanding the process of fossilization in amber.

The earliest pioneers in the study of tissue ultrastructure in amber fossils looked at arthropods—the most common type of inclusion—and used traditional study methods such as thin-sectioning, light microscopy, scanning electron microscopy (SEM), and transmission electron microscopy (TEM). These studies usually utilized only one method, and at most a few specimens, and were focused on whether tissue ultrastructure is preserved in fossil animals in amber. As described by Grimaldi et al. (1994), Kornilow-itsch was the first to recognize the importance of studying tissue ultrastruc-ture; in 1903, he thin-sectioned legs from both neuropteran and dipteran inclusions in Baltic amber and identified striated muscle tissue in both types of insects. However, his results were not widely accepted due to the poor quality of his images and the fact that the piece exhibited surface crazing. At the time, surface crazing was thought to be specific to copal alteration, but it has since been realized that it occurs in poorly looked-after ambers as well, such as historic Baltic amber collections (e.g. Kettunen et al. 2019). Nineteen years after Kornilowitsch's study, von Lengerken (1922) success-fully extracted the tracheal parts of beetles from Baltic amber, although he was less interested in ultrastructural preservation and more interested in the chemical preservation of chitin.

Andrée and Keilbach (1936) and Voigt (1937) presented new results on ultrastructural preservation in amber and showed that it is not a rare occur-rence in inclusions with preserved soft tissues. The first set of authors re-ported the preservation of tissue ultrastructure in parts of the tracheal sys-tem, musculature, and "soft tissue" (gut material and ovarioles) (Andrée and Keilbach 1936); they also indicated that they had evidence for chitin preservation. However, they do not clearly state their methods, neither for extracting the inclusions from the amber nor for analyzing the inclusions, although Voigt (1937) notes that Andrée and Keilbach achieved their re-sults by dissolving the amber. Voigt (1937) himself applied the "lacquer film method," which he developed as an improvement over previous iterations of this technique in order to prepare insects, spiders, plants, and fecal pel-lets in numerous pieces of Baltic amber. Using light microscopy, Voigt iden-tified muscle tissue, trachea, and gland epithelia from the insects, and noted that while even badly preserved samples yielded ultrastructural details, small insects generally seemed to be better preserved than large ones.

After these two studies, there was a long hiatus of 39 years before the advent of newer technology, specifically SEM, facilitated the investigation of tissue ultrastructure in fossils. Mierzejewski (1976a, b) was one of the first to apply this new technology to amber, focusing on arthropod and wood inclusions in Baltic amber. He accessed the tissues by grinding and polishing the outer covering of amber until only a thin layer remained over the inclusion, which he then removed manually with a thin needle. With this technique, Mierzejewski documented the ultrastructural details of the eye of a dolichopodid fly, the elytra of a beetle, and the book lungs of a spi-der, as well as of the wood of *Pinus succinifera*. A few years later, Poinar and Hess (1982, 1985) and Poinar (1992) applied TEM to elucidate the ultrastructure of a single specimen of a fungus gnat. They were able to identify various cellular remains containing muscle fibers, nuclei, ribosomes, lipid droplets, endoplasmic reticula, and even mitochondria. Using the same method, Poinar (1992) also identified cellular-level structures in tissues of a braconid wasp in Cretaceous amber from Canada.

The era of ultrastructural studies that started in the 1970s is rounded out by a few analyses of mineralized arthropod inclusions in amber. Mineralized tissues are rare in amber but not unknown (Labandeira 2014); it is more common for minerals to replicate fine-scale tissue ultrastructure outside of amber (e.g., Martill 1990; Gee and Liesegang, chap. 6). Baroni-Urbani and Graeser (1987) used SEM to describe a pyritized cast of an ant in Baltic amber; the pyrite replicated the surface microsculptural features of the integument. A few years later, Schlüter (1989) also employed SEM to investigate the wing venation of three termites from the Cenomanian amber of France, whereby one sample was covered with marcasite and pyrite. He also showed images of two other mineralized specimens—an insect head that may have belonged to an ant, although definite identification was not possible, and the elytra of a beetle—both of which preserved surface microsculpture.

These studies unequivocally demonstrate the preservation of exceptional microscopic ultrastructural details in fossils from various amber sites, including Eocene Baltic amber, Cretaceous French amber, and Cretaceous Canadian amber. More importantly, all authors emphasized that the preserved ultrastructures were identical to those in extant groups, even though the documentation of extant samples was not included in these publications.

The next major advancement in the study of tissue ultrastructure in amber involved the first direct comparison of fossil and extant material by Henwood (1992a), who removed inclusions from amber by cutting or fracturing the specimen through the insect body and analyzing the inclusions with TEM or SEM. Henwood (1992a) also carried out decomposition experiments under varying conditions to develop an understanding of different decay sequences in the flight muscles of living blowflies. She then compared the end results with preserved muscle tissues from similar brachyceran flies in Dominican amber. Although muscle fibers and the finest detail of mitochondria (their internal membranes or cristae) were still preserved in the fossil fly in amber, they were not identical to those in freshly dead flies. However, they were comparable to these structures after some experimental decay in extant flies.

Specifically, Henwood noted that the experimental pathways involving dehydration resulted in the best-preserved tissues but also led to the shrinkage of muscle fibers and mitochondria. The fossil fly embedded in Dominican amber also exhibited these features, which may have been due to the dehydrating effects of resin. Henwood expanded her studies by using SEM to describe the internal tissue preservation of essentially all organ systems in a beetle in Dominican amber (Henwood 1992b). As would be expected, the ultrastructure of all tissues was well preserved. Thus, these results somewhat contradict the previous conclusions: although ultrastructure is recognizable and well preserved, it may not be exactly identical to the ultrastructure observed in the tissues of a living organism.

Shortly after Henwood's experiments, the most comprehensive study of internal tissue preservation in amber was published by Grimaldi et al. (1994). Grimaldi et al. addressed the question of consistency and patterns in soft tissue preservation in amber based on a large number of organisms embedded in two types of amber. Altogether, the sample set included 16 specimens of arthropods from Baltic and Dominican amber that varied in several aspects, such as size and degree of cuticle sclerotization (bees, beetles, termites, gnats, flies). Furthermore, the sample set also included one leaflet and seven anthers of the leguminous tree *Hymenaea* from Dominican amber. The inclusions were manually extracted by fracturing the amber: a groove was first circumscribed around the midsagittal line of the specimens, which in the end formed a circle less than 1 mm from the inclusion. With the help of a sharp blade, the internal edge of the groove was scored until slight leverage caused the piece to split, exposing the inclusions for SEM and TEM analysis. Soft tissue preservation was more common in Dominican than in Baltic amber; all the Dominican amber specimens had internal soft tissues, whereas half of the Baltic amber samples were just empty molds. However, in both ambers, when soft tissues were preserved, air sacs, tracheae, parts of the gut, remains of the brain, and muscle fibers were found in their original position and insertions.

Since then, a few other studies have demonstrated similarly exceptional preservation of tissue ultrastructure in a variety of inclusions from many sites. Kohring (1998) documented the preservation of various arthropod structures, such as alae, muscle tissue with myofibrils, cornea cells, dorsal vessels, and trachea, in inclusions in Baltic, Bitterfeld, and Dominican ambers. Using a method similar to that of Mierzejewski, Kohring exposed the inclusion by putting mechanical pressure on the amber piece, relying on the fact that inclusions represent discontinuities that favor the splitting of the amber along the main axis of the embedded organisms. Just over a decade later, Kowalewska and Szwedo (2009) combined SEM with energy-dispersive X-ray spectroscopy (EDS) analysis and examined the surface of a leafhopper from Baltic amber, revealing the presence of marcasite. In that same year, Tanaka et al. (2009) published SEM and TEM images of dolichopodid fly eyes. They examined two specimens trapped together in the same piece of Baltic amber and report a completely preserved visual system in both animals. A radial section through ommatidia revealed the presence of a basement membrane, rhabdomeres, trachea, and pigment cells (primary and secondary).

These studies reveal an interesting pattern of tissue preservation. First, they unambiguously demonstrate the exceptional preservation of tissue ultrastructure in fossil inclusions in amber. Second, they indicate a dichotomy in the preservation of tissue ultrastructure: either no soft tissue is preserved or tissue morphology is exquisitely preserved, even including subcellular structures (although they may be slightly altered in morphology due to shrinkage). Third, this dichotomy is apparent even within the same amber site; for example, Baltic amber inclusions are an even mix of empty molds and tissues preserved with ultrastructural details (Grimaldi et al. 1994). This suggests that there is a dichotomous pathway to tissue preservation: depending on fossilization variables (specific variables discussed in more detail below), there is either a rapid preservation or a rapid decay of soft tissues.

Biomolecular Composition

Fossil arthropods in amber have an exquisite life-like preservation, as if they were frozen in time, unchanged since the day they were first trapped in resin. As discussed above, the life-like nature of tissue preservation can even extend to microscopic ultrastructural details, which immediately prompts follow-up questions: Does the preservational fidelity of amber-embedded tissues extend beyond the ultrastructural scale into the molecular level? In other words, do amber-entombed tissues retain remnants of the original chemical composition?

The question of the chemical preservation of tissues was raised very early on in the study of amber. Alexander Tornquist, professor and head of the amber collection of the Albertus University Königsberg in the early 20th century, states in his monograph on the geology of East Prussia (Tornquist 1910) that no original material is preserved in amber (as mentioned earlier in this chapter, Kornilowitsch's success in extracting muscle tissue from Baltic amber was doubted by the majority of the scientific community at this time). In that same year, Klebs tested Tornquist's hypothesis by examining bone in Baltic amber to determine if original tissues in bone were still preserved (Klebs 1910). This study was carried out on the only lizard in amber known at that time (the "Königsberg specimen"), which had to be sacrificed for the cause; the mechanical opening of the amber to expose the inclusions resulted in the partial destruction of this unique specimen. Without the benefit of modern technology, his conclusions were limited. Klebs reported "coaly remains," but he did not check for muscle preservation; the rest of the inclusion was simply a hollow void.

Initial studies on the preservation of chitin in amber were carried out by von Lengerken (1913, 1922). In his first attempt, von Lengerken examined pieces that had been stored in alcohol for years. All of these inclusions showed no evidence of chitin preservation and were either hollow inside or contained coaly remains that could not be assigned to any specific organ (von Lengerken 1913). In contrast to this, an inclusion that was stored under dry conditions did yield chitin and had preserved partial remains of abdominal segments, including a ventral porus (von Lengerken 1913). It was reported that the chitin was not strongly altered in this specimen but had in some places transformed into brown coal. Later, von Lengerken successfully analyzed chitin of beetles from Baltic amber, which he extracted manually through needle preparation (von Lengerken 1922); in these specimens, the structure of the chitin could be observed under a microscope. In addition to the chitin, von Lengerken also extracted tracheal remains.

The geochemical basis of tissue fixation and preservation in inclusions in amber is not fully understood. It has been suggested that biomolecules in the fossil tissues might have undergone extensive polymerization, similar to nonenzymatic Maillard reactions (Bada et al. 1999; Briggs 1999), which would lead to chemical and mechanical fixation. This hypothesis was supported by the detection of volatile compounds released from ancient plants found in Egyptian graves, which are characteristic of Maillard reaction products (Evershed et al. 1997). High molecular weight Maillard polymers have been dissolved in in vitro experiments with the chemical phenacyl thiazolium bromide (PTB), which breaks specific crosslinks and allows for the release of the trapped proteins and DNA (Vasan et al. 1996). In accordance with this reaction, incubation with PTB released polymerase chain reaction-amplifiable DNA from insoluble polymers in ancient feces that were about 10,000 years old (Poinar et al. 1998). These results have led to the hypothesis that the carbohydrates, peptides, or short chains of DNA or RNA in the tissue of amber fossils may also be trapped in a three-dimensional insoluble network of geopolymers formed by Maillard reactions and therefore cannot be detected as native biological molecules (Bada et al. 1999). This hypothesis is supported by the observation that the color of amber-embedded tissues seen in thin section consists of various shades of brown, which is reminiscent of melanoidins, the insoluble products of the Maillard reaction (browning reaction).

Modern investigations of tissue composition of fossil inclusions in amber started with techniques such as pyrolysis–gas chromatography–mass spectrometry (Py-GC-MS), Raman spectroscopy (see Geisler and Menneken, chap. 4, for a more in-depth discussion of Raman spectroscopy), and histochemical staining to determine the major organic composition of fossils in amber and specifically to look for remnants of original tissue moieties. Stankiewicz et al. (1998) used Py-GC-MS to investigate cuticle preservation in arthropods (bees and beetles) and leaves from Dominican amber, as well as subfossil and ancient resin from Kenya. The younger (2–20 ka) samples from Kenya retained some of their original composition, including chitin (but not protein) in the arthropod cuticle and lignocellulose in the plants. The inclusions in the much older Dominican amber (ca.15–20 Ma; Iturralde-Vinent and MacPhee 1996) had no recognizable remnants of original chemistry. Edwards et al. (2007) used Raman spectroscopy to investigate insects in Mexican and Baltic ambers. They found evidence of severely degraded proteins in the insect in Mexican amber (22 Ma; de Lourdes Serrano-Sánchez et al. 2015) and no evidence of original biomolecules in the insect in Baltic amber (45 Ma). However, they suggest their Baltic amber specimen was hollow and that a specimen with preserved internal tissue would have yielded different results. These two amber studies of arthropod tissue chemistry give variable and contradictory results because they are based on multiple amber sites and methods, which makes it challenging to draw any conclusions about the patterns of chemical preservation of tissues in amber.

We will now briefly mention the results from studies on other tissues to help resolve these issues and clarify the patterns. Hartl et al. (2015) used Raman spectroscopy to analyze the tissue chemistry of lichens in amber and determined that the cortex and medulla can be distinguished on the basis of Raman spectra. However, these differences in the spectra could not be interpreted to determine if they retain some remnants of original chemistry, or if diagenesis and alteration had not yet completely obscured any original differences. Similarly, Koller et al. (2005) used histochemical staining on a cypress twig in Baltic amber to demonstrate the presence of some unidentified tissue biomolecules. They suggested that these represent original tissue moieties, but it is not clear to what extent the stains would also react to diagenetically altered biomolecules. In general, the results of Koller et al. (2005) suggest significant alteration or loss of original tissue chemistry during fossilization in amber, but also offer some hope that the inclusions may retain some trace remnants of original biomolecules.

Based on the likelihood of finding biochemical remnants, more recent studies have focused on targeted techniques to isolate and characterize specific molecules of interest. This was first attempted for DNA preservation in Dominican amber, using what was at the time the most up-to-date techniques for DNA purification, amplification, and sequencing (Cano et al. 1992; DeSalle et al. 1992; Poinar et al. 1996). Although the results initially looked exciting, later advances in our understanding of contamination and appropriate methods for ancient DNA analysis showed that the nucleotides recovered were contaminants (Gutiérrez et al. 1998). In addition, all later studies that tried to recover DNA from amber under strict laboratory conditions were unsuccessful (Austin et al. 1997a, b; Smith and Austin 1997; Walden and Robertson 1997; Hebsgaard et al. 2005). These results were further supported by a more recent study that failed to extract DNA from fossil inclusions in much younger copal, suggesting that DNA preserves very poorly in amber (Penney et al. 2013).

Protein and amino acid extraction from fossils in amber has been more successful, most likely due to the higher preservation potential of proteins relative to DNA. Bada et al. (1994) found amino acids in arthropod inclusions in copal, as well as in Dominican, Baltic, and Lebanese ambers, but not in Mexican amber. The amino acids were localized in the inclusion rather than in the amber matrix, and their degree of racemization indicated that the amino acids came from intact protein fragments. Smejkal et al. (2011) obtained ancient amino acids and sequenced protein fragments from fossil fungal inclusions in amber, giving more support to the idea that these inclusions can contain intact protein fragments. More recently, McCoy et al. (2019) found similar results for fossil feathers in Burmese and Baltic ambers: ancient amino acids with low degrees of racemization and a composition similar to that expected for feathers degraded in resin. These results suggest that trace amounts of original biomolecules can be preserved in fossils in amber and can be identified using targeted, highly sensitive chemical techniques.

Taphonomic Pattern: The Dichotomous Pathway to Preservation

The primary goal in investigating the patterns of preservation of tissue ultrastructure and tissue chemistry is to understand the process of fossilization in amber. This process can be divided into three phases (Martínez-Delclòs et al. 2004; Labandeira 2014): entrapment, in which an insect or other organism is trapped in resin; early diagenesis, which occurs immediately after entrapment, when the resin is still rich in original volatiles that interact with the tissues; and late diagenesis, after the resin has lost many of its volatiles but during which chemical changes related to polymerization may still be occurring. The boundary between early diagenesis and late diagenesis, like the boundary between copal and amber, is not clearly defined (Labandeira 2014). Entrapment is the best studied of these three phases, having been thoroughly investigated and summarized in a few recent studies (Martínez-Delclòs et al. 2004; Solórzano Kraemer et al. 2015, 2018). Entrapment biases are unlikely to be a major factor influencing the patterns of tissue preservation in fossil inclusions in amber; if the amber has an inclusion, then by default an organism has been trapped.

The second phase, early diagenesis, is most strongly characterized by the interaction of resin volatiles with the tissues and microbiota of any entrapped organism. Through chemical analysis of inclusions in resin, various resin specialists have noted that the tissues are perfused with resin compounds (Grimaldi et al. 1994; Stankiewicz et al. 1998), indicating that resin chemicals can reach and interact with all organic tissues (Poinar and Hess 1985). Moreover, Henwood (1992a) demonstrated the importance of resin chemistry in preserving tissues by evaluating the decay of flies entrapped in a variety of substances. The flies in wax decayed more quickly and had relatively poorer tissue preservation than a control group left open in the air, and the slowest decay and best preservation were found in flies embedded in maple syrup. In other words, physical sealing of organic matter from oxygen alone is not a guarantee of inhibited decay and high-fidelity tissue preservation, rather, there must be some chemical interaction with the embedding medium (Henwood 1992a). The exact mechanism of decay inhibition is unclear and likely includes some combination of tissue dehydration and inhibition of microbial or enzymatic activity.

These analyses and experiments demonstrate the importance of early diagenesis in fossilization in amber, as well as the role of resin volatiles in exceptional tissue preservation. We further hypothesize that interactions between the tissue, bacteria, and volatiles (e.g. dehydration, enzyme inhibition, growth inhibition of bacteria) control the early dichotomous branching on the fossilization pathway. These interactions will determine if soft tissues (with their attendant ultrastructures) preserve over geological timescales or degrade. Previous authors have noted this dichotomous pattern of tissue loss or preservation in ambers of all ages (Grimaldi et al. 1994; Labandeira 2014), and we have observed it in young resin from Madagascar as well. Thus, this split in the taphonomic pathway must happen fairly early in diagenesis. More support comes from McCoy et al. (2018b), who observed this dichotomy after less than two years of experimental decay in resin: fruit flies encased in *Pinus* resin resulted in empty molds, whereas fruit flies in *Wollemia* resin retained all their tissues. The analytical method used by Mc-Coy et al. (2018b)—synchrotron tomography—precluded the assessment of tissue ultrastructural preservation, but given the preservation of tissue ultrastructure in, for example, 45-million-year-old Baltic amber (Grimaldi et al. 1994), it was likely preserved in the fruit flies in *Wollemia* resin as well.

Based on our hypothesis that chemical interactions with resin determine the loss or preservation of tissues, the most obvious influential factor would be resin chemistry, which is complex and can differ significantly between resin-producing taxa and even within an individual tree (Anderson and Winans 1991; Anderson et al.1992; Langenheim 2003; Lambert et al. 2008, 2012, 2015). This also proved to be the most important factor in the experiments by McCoy et al. (2018b) that resulted in the dichotomous pattern of tissue loss or preservation analogous to what is observed in the amber fossil record. This factor alone would be sufficient to explain the pattern if the mechanism of decay inhibition is a straightforward interaction between tissues and volatile resin compounds, such as dehydration. However, a more complex scenario involving the interaction of resin compounds with the microbiota in the inclusions suggests alternative factors, such as the composition of the gut microbiota, which can be very different between taxa or even within a single individual. McCoy et al. (2018b) noted differences in the extent of decay based on the formation of decay gas bubbles between fruit flies with no gut biota and fruit flies with a gut biota embedded in *Wollemia* resin; however, all still retained their internal soft tissues.

There is extensive literature on the interaction of various resins and resin constituents with various bacteria (e.g., Lang and Buchbauer 2012; Shuaib et al. 2013), but there is no clear pattern: any particular resin may inhibit the activity of some bacteria and promote the growth of other bacteria, and these bacterial assemblages vary among different resins. Experiments involving resins that are known precursors to amber along with bacteria common in arthropod guts may help untangle this complicated set of interactions. Even less well-understood is the role that resin volatiles may play in inhibiting autolytic decay; there is some evidence that chemical compounds in plant tissues, including those such as terpenoids and terpenes, which are also the primary components of resin, can inhibit enzymatic activity (e.g., Miyazawa and Yamafuji 2005; Chang et al. 2006; Bustanji et al. 2011; Adamczyk et al. 2015), suggesting this interaction deserves more study. Other less well-supported factors have been suggested as well, for example, dehydration prior to entrapment (Coty et al. 2014), the degree of sclerotization of the cuticle (Poinar and Hess 1985; Grimaldi et al. 1994), and the physical prevention of external bacteria from interacting with the inclusions, which is related to resin permeability (Smith 1880). We expect factors arising from the depositional environment, such as sediment interactions, heat, and pressure, to play a more important role in late diagenesis, but of course they may also have an impact on early diagenesis.

Resin polymerization during late diagenesis is a prerequisite for producing a fossil inclusion in amber, but we expect that there are still changes in the preservation of inclusions during this stage of fossilization in amber. Specifically, although the preservation of tissues and their ultrastructures seems to be determined during early diagenesis, we hypothesize that this phase, late diagenesis, controls the loss or preservation of original tissue chemistry over long time periods. In other words, if an inclusion continues down the tissue preservation branch of the dichotomous pathway of fossil preservation, it may experience further biochemical changes. These do not represent further dichotomous branches, but rather continuous changes.

Original tissue chemistry is more commonly preserved in copal than in amber, but remnants may remain even in Cretaceous amber (e.g., Bada et al. 1994; McCoy et al. 2019). Factors influencing the degradation of tissue chemistry in amber have been less well-studied than variables influencing the preservation of tissues in amber; this field is still in the early stages, and most studies are focused simply on investigating the extent of original tissue chemistry in fossils in amber (Bada et al. 1994; Stankiewicz et al. 1998; Koller et al. 2005; Edwards et al. 2007; Penney et al. 2013; Thomas et al. 2014; Hartl et al. 2015; McCoy et al. 2019). However, the degradation of biomolecules, such as proteins, is strongly influenced by environmental factors, such as temperature and water, which correspond to variables such as burial depth, diagenetic history, sediment composition, and amber permeability (Martínez-Delclòs et al. 2004; McCoy et al. 2018b).

Amber permeability has been briefly investigated within the context of determining whether air and water inclusions in amber are original, resulting in the conclusion that amber is a somewhat open, permeable system (Beck 1988; Hopfenberg et al. 1988). Further investigations to understand variations in permeability throughout the process of polymerization and between different resins would be a valuable addition to the field. Similarly, there have been a few studies subjecting resin to different temperatures, pressures, and sediments to understand how these variables influence polymerization and amberization (Gold et al. 1999; Poinar and Mastalerz 2000;

Winkler et al. 2001; Scalarone et al. 2003a, b; Bisulca et al. 2012; Saitta et al. 2019); applying these methods to resin with inclusions would help understand the role of these particular factors in the fossilization of inclusions.

Future Perspectives

We must emphasize that all previous studies of tissue ultrastructure obtained access to the tissues through the sawing, cutting, or cracking of the amber—methods that are inherently destructive to the inclusion. It may be possible to extract complete, undamaged inclusions by dissolving the amber around them (fig. 5.1; plate 5.1), but this method is restricted to samples of a few

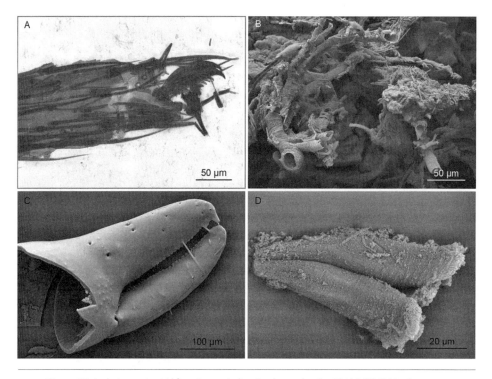

Figure 5.1. Inclusions extracted from Eocene Indian Cambay amber (ca. 53 Ma). (**A**) Light microscope image of an isolated insect leg. (**B**) SEM image showing parts of the tracheal system of an extracted beetle. (**C**) SEM image of a pedipalp of a pseudoscorpion. (**D**) SEM image showing the preservation of the muscular tissue that was removed from the pedipalp shown in **C**.

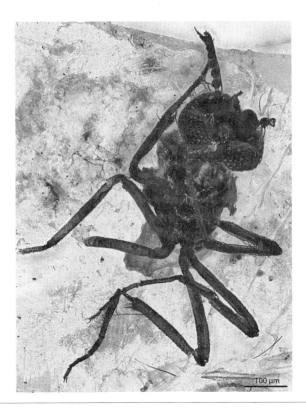

Plate 5.1. Nematoceran fly extracted from Indian Cambay amber (ca. 53 Ma). Except for the anten-nae, the anterior-most part of the body is nicely preserved and projects above the amber matrix. The abdomen is still enclosed by the resinous residue of the original amber.

localities and depends on the composition of the resin. Viehmeyer (1913), in his study of Sulawesi copal; Azar (1997), in his study of Lower Cretaceous Lebanese amber; Andrée and Keilbach (1936), in their study of Eocene Baltic amber; and Mazur et al. (2014), in their study of Eocene Indian amber, devel-oped protocols for dissolving their respective type of amber and were thereby able to recover more or less complete arthropod remains. Viehmeyer (1913), for example, exhumed complete, three-dimensional specimens and even large parts of insects by completely dissolving the amber matrix with differ-ent organic solvents; although the plant and insect remains were fragile, they could be mounted on microscope slides.

To date, all of these studies have only developed dissolution methods and observed the preservation of soft tissues; however, such extracted arthropod remains are an obvious target for future tissue ultrastructure studies. Younger ambers and copals, or just "ancient resins," are less polymerized than fossil resins such as amber and can be more easily dissolved, although some highly viscous nondissolvable fraction may remain in the vial (Viehmeyer 1913; Penney et al. 2013; Mazur et al. 2014; Barthel, own data). Over time, the resins start to polymerize and become amber, making it difficult or impossible to dissolve most ambers (Mazur et al. 2014). Dutta et al. (2010) and Dutta and Mallick (2017) have theorized that, due to their unique chemical composition, dammar resins undergo only minimal polymerization during fossilization. Therefore, these and similar class II resins originating from angiosperms (van Aarssen et al. 1990) are the most promising target for dissolution experiments.

Another approach is to apply newer methods such as high-resolution X-ray microcomputed tomography (micro-CT or CT scanning) and synchrotron X-ray tomographic microscopy (SRXTM, or synchrotron tomography) to amber inclusions (Dierick et al. 2007; Soriano et al. 2010; Kehlmaier et al. 2014; Sherratt et al. 2015; Pepinelli and Currie 2017). These two imaging techniques are nondestructive methods that can be used to develop three-dimensional models of the internal and external features of the inclusion. At the moment, these three-dimensional reconstructions are a powerful tool for taxonomy. It should be pointed out, however, that while synchrotron tomography can clearly show images of tissue ultrastructure (Zehbe et al. 2010), this method has not yet been applied to specimens in amber. Synchrotron tomography at these very detailed scales may prove to be an imaging tool as equally powerful as CT scanning for studying ultrastructure.

Further research is needed to understand the patterns of chemical preservation in fossils in amber. In particular, there are still several unresolved issues. Can remnants of original biomolecules preserve even when soft tissues have decayed away? Are there trace remnants of original chemical composition that characterize all arthropod fossils in amber with tissues preserved to the ultrastructure level? Or, are there one or more additional taphonomic filters that control the biomolecule preservation?

Fossilization of soft tissues in resin is most likely a complex process involving a number of factors. Further taphonomic experimentation is needed to understand the effects of these factors in isolation and in combination with one another. Previous experiments have been sufficient to confirm that some factors—specifically chemistry of the embedding resin and the composition of the gut microbiota of the entrapped arthropods—are likely to play a major role in tissue preservation and the early diagenetic phases of fossilization. These factors should be investigated in greater detail in a series of taphonomic experiments to develop a more complete understanding of their influence on decay and their interactions with one another.

As of yet, there have not been any experiments on the late diagenesis phase of fossilization in amber, and any attempts in this direction would be a major step forward. There are a number of experiments on the diagenesis of nonfossiliferous resin (Scalarone et al. 2003a, b; Pastorelli 2011; Beltran et al. 2017; Saitta et al. 2019) that could provide a good basis for experimentation on resin with inclusions.

Conclusions

Amber and its inclusions have been extensively studied for over a hundred years. Initially, most investigations were either chemical analyses of the amber or taxonomic and ecological investigations of the inclusions (ignoring the amber). In the past few decades, more emphasis has been placed on the unit of amber and its inclusions rather than focusing on one or the other in isolation, revealing a unique and interesting taphonomic pathway (Poinar and Hess 1985; Henwood 1992a; Grimaldi et al. 1994; Stankiewicz et al. 1998; McCoy et al. 2018a, b). Recent integrative investigations have revealed a robust taphonomic pattern in the fossilization of entrapped organisms: an early dichotomy between the loss or preservation of tissues and tissue ultrastructure, which corresponds to early diagenetic interactions with resin volatiles, and a later, more complex pattern of loss or preservation of original tissue chemistry, which corresponds to a variety of later diagenetic interactions.

While this sounds straightforward, the taphonomic pattern in amber fossilization may be influenced by a wide array of complex factors, such as resin chemistry, composition of the microbiota in the inclusions, and parameters of the depositional environment. A deeper analysis of localities with amber fossils in order to understand these taphonomic patterns would be useful and extremely interesting. However, concurrent taphonomic experiments in the laboratory to better understand and quantify the effect of the resin chemistry and microbiota would help with understanding the taphonomic processes of fossilization in amber even more. Finally, we encourage amber researchers to consider both the amber and the inclusions together, rather than viewing them in isolation.

ACKNOWLEDGMENTS

The authors thank Kerstin Koch and Axel Huth (Rhine-Waal University of Applied Sciences) for access to the Centre for Microscopy and supporting the senior author by taking images of the extracted inclusions. Furthermore, thank you to reviewers Leyla Seyfullah (University of Vienna) and an anonymous colleague for their helpful comments on the manuscript. This work was funded by the Deutsche Forschungs-gemeinschaft (DFG, German Research Foundation), Project number 396704064 to Jes Rust for HJB and Project number 396637283 to Jes Rust for VEM (both at the University of Bonn). This is contribution number 19 of DFG Research Unit FOR 2685, "The Limits of the Fossil Record: Analytical and Experimental Approaches to Fossilization."

WORKS CITED

Adamczyk, S., Adamczyk, B., Kitunen, V., and Smolander, A. 2015. Monoterpenes and higher terpenes may inhibit enzyme activities in boreal forest soil. *Soil Biology & Biochemistry*, 87: 59–66.

Anderson, K.B., and Winans, R.E. 1991. Nature and fate of natural resins in the geo-sphere. I. Evaluation of pyrolysis–gas chromatography mass spectrometry for the analysis of natural resins and resinites. *Analytical Chemistry*, 63: 2901–2908.

Anderson, K.B., Winans, R.E., and Botto, R.E. 1992. The nature and fate of natural res-ins in the geosphere—II. Identification, classification and nomenclature of resin-ites. *Organic Geochemistry*, 18: 829–841.

Andrée, K., and Keilbach, R. 1936. Neues über Bernsteineinschlüsse. *Schriften der Physikalisch-Ökonomische Gesellschaft zu Königsberg*, 69: 124–128.

Austin, J.J., Ross, A.J., Smith, A.B., Fortey, R.A., and Thomas, R.H. 1997a. Problems of reproducibility—Does geologically ancient DNA survive in amber-preserved insects? *Proceedings of the Royal Society B*, 264: 467–474.

Austin, J.J., Smith, A.B., and Thomas, R.H. 1997b. Palaeontology in a molecular world: The search for authentic ancient DNA. *Trends in Ecology & Evolution*, 12: 303–306.

Azar, D. 1997. A new method for extracting plant and insect fossils from Lebanese amber. *Palaeontology*, 40: 1027–1060.

Bada, J.L., Wang, X.S., and Hamilton, H. 1999. Preservation of key biomolecules in the fossil record: Current knowledge and future challenges. *Philosophical Transactions of the Royal Society of London B*, 354: 77–86.

Bada, J.L., Wang, X.S., Poinar, H.N., Paabo, S., and Poinar, G.O. 1994. Amino acid racemization in amber-entombed insects: Implications for DNA preservation. *Geochimica et Cosmochimica Acta*, 58: 3131–3135.

Baroni-Urbani, C., and Graeser, S. 1987. REM-Analysen an einer pyritisierten Ameise aus Baltischem Bernstein. *Stuttgarter Beiträge zur Naturkunde B*, 133: 1–16.

Beck, C.W. 1988. In reply: Is the air in amber ancient? *Science*, 241: 718–719.

Beltran, V., Salvadó, N., Butí, S., Cinque, G., and Pradell, T. 2017. Markers, reactions, and interactions during the aging of *Pinus* resin assessed by Raman spectroscopy. *Journal of Natural Products*, 80: 854–863.

Bisulca, C., Nascimbene, P.C., Elkin, L., and Grimaldi, D.A. 2012. Variation in the deterioration of fossil resins and implications for the conservation of fossils in amber. *American Museum Novitates*, 3734: 1–19.

Briggs, D.E.G. 1999. Molecular taphonomy of animal and plant cuticles: Selective preservation and diagenesis. *Philosophical Transactions of the Royal Society of London B*, 354: 7–17.

Bustanji, Y., Al-Masri, I.M., Mohammad, M., Hudaib, M., Tawaha, K., Tarazi, H., and Alkhatib, H.S. 2011. Pancreatic lipase inhibition activity of trilactone terpenes of *Ginkgo biloba*. *Journal of Enzyme Inhibition and Medicinal Chemistry*, 26: 453–459.

Cano, R.J., Poinar, H., and Poinar, G.O. 1992. Isolation and partial characterisation of DNA from the bee *Proplebeia dominicana* (Apidae: Hymenoptera) in 25–40 million year old amber. *Medical Science Research*, 20: 249–251.

Chang, T.K.H., Chen, J., and Yeung, E.Y.H. 2006. Effect of *Ginkgo biloba* extract on procarcinogen-bioactivating human CYP1 enzymes: Identification of isorhamnetin, kaempferol, and quercetin as potent inhibitors of CYP1B1. *Toxicology and Applied Pharmacology*, 213: 18–26.

Coty, D., Aria, C., Garrouste, R., Wils, P., Legendre, F., and Nel, A. 2014. The first antitermite syninclusion in amber with CT-scan analysis of taphonomy. *PLOS ONE*, 9: e104410.

de Lourdes Serrano-Sánchez, M., Hegna, T.A., Schaaf, P., Pérez, L., Centeno-García, E., and Vega, F.J. 2015. The aquatic and semiaquatic biota in Miocene amber from the Campo LA Granja mine (Chiapas, Mexico): Paleoenvironmental implications. *Journal of South American Earth Sciences*, 62: 243–256.

DeSalle, R., Gatesy, J., Wheeler, W., and Grimaldi, D. 1992. DNA sequences from a fossil termite in Oligo-Miocene amber and their phylogenetic implications. *Science*, 257: 1933–1936.

Dierick, M., Cnudde, V., Masschaele, B., Vlassenbroeck, J., Van Hoorebeke, L., and Jacobs, P. 2007. Micro-CT of fossils preserved in amber. *Nuclear Instruments & Methods in Physics Research. Section A, Accelerators, Spectrometers, Detectors and Associated Equipment*, 580: 641–643.

Dutta, S., and Mallick, M. 2017. Chemical evidence for dammarenediol, a bioactive angiosperm metabolite, from 54 Ma old fossil resins. *Review of Palaeobotany and Palynology*, 237: 96–99.

Dutta, S., Mallick, M., Mathews, R.P., Mann, U., Greenwood, P.F., and Saxena, R. 2010. Chemical composition and palaeobotanical origin of Miocene resins from Kerala-Konkan Coast, western India. *Journal of Earth System Science*, 119: 711–716.

Edwards, H.G.M., Farwell, D.W., and Villar, S.E.J. 2007. Raman microspectroscopic studies of amber resins with insect inclusions. *Spectrochimica Acta. Part A, Molecular and Biomolecular Spectroscopy*, 68: 1089–1095.

Evershed, R.P., Bland, H.A., van Bergen, P.F., Carter, J.F., Horton, M.C., and Rowley-Conwy, P.A. 1997. Volatile compounds in archaeological plant remains and the Maillard reaction during decay of organic matter. *Science*, 278: 432–433.

Gold, D., Hazen, B., and Miller, W.G. 1999. Colloidal and polymeric nature of fossil amber. *Organic Geochemistry*, 30: 971–983.

Grimaldi, D.A., Bonwich, E., Delannoy, M., and Doberstein, S. 1994. Electron microscopic studies of mummified tissues in amber fossils. *American Museum Novitates*, 3097: 1–31.

Gutiérrez, G., and Marín, A. 1998. The most ancient DNA recovered from an amber-preserved specimen may not be as ancient as it seems. *Molecular Biology and Evolution*, 15: 926–929.

Hartl, C., Schmidt, A.R., Heinrichs, J., Seyfullah, L.J., Schäfer, N., Gröhn, C., Rikkinen, J., and Kaasalainen, U. 2015. Lichen preservation in amber: Morphology, ultrastructure, chemofossils, and taphonomic alteration. *Fossil Record*, 2: 127–135.

Hebsgaard, M.B., Phillips, M.J., and Willerslev, E. 2005. Geologically ancient DNA: Fact or artefact? *Trends in Microbiology*, 13: 212–220.

Henwood, A. 1992a. Exceptional preservation of dipteran flight muscle and the taphonomy of insects in amber. *Palaios*, 7: 203–212.

Henwood, A. 1992b. Soft-part preservation of beetles in Tertiary amber from the Dominican Republic. *Palaeontology*, 35: 901–912.

Hinejima, M., Hobson, K.R., Otsuka, T., Wood, D.L., and Kubo, I. 1992. Antimicrobial terpenes from oleoresin of ponderosa pine tree *Pinus ponderosa*: A defense mechanism against microbial invasion. *Journal of Chemical Ecology*, 18 (10): 1809–1818.

Hopfenberg, H.B., Witchey, L.C., Poinar, G.O., Beck, C.W., Chave, K.E., Smith, S.V., Horibe, Y., and Craig, H. 1988. Is the air in amber ancient? *Science*, 241: 717–721.

Iturralde-Vinent, M.A., and MacPhee, R.D.E. 1996. Age and paleogeographic origin of Dominican amber. *Science*, 273: 1850–1852.

Kehlmaier, C., Dierick, M., and Skevington, J.H. 2014. Micro-CT studies of amber inclusions reveal internal genitalic features of big-headed flies, enabling a systematic placement of *Metanephrocerus* Aczél, 1948 (Insecta: Diptera: Pipunculidae). *Arthropod Systematics & Phylogeny*, 72: 23–36.

Kettunen, E., Sadowski, E.-M., Seyfullah, L.J., Dörfelt, H., Rikkinen, J., and Schmidt, A.R. 2019. Caspary's fungi from Baltic amber: historic specimens and new evidence. *Papers in Palaeontology*, 5: 365–389.

Klebs, R. 1910. Über Bernsteineinschlüsse im allgemeinen und die Coleopteren meiner Bernsteinsammlung. *Schriften der physikalisch-ökonomischen Gesellschaft zu Königsberg*, 51: 217–242.

Kohring, R. 1998. REM-Untersuchungen an harzkonservierten Arthropoden. *Entomologia Generalis*, 23: 95–106.

Koller, B., Schmitt, J.M., and Tischendorf, G. 2005. Cellular fine structures and histochemical reactions in the tissue of a cypress twig preserved in Baltic amber. *Proceedings of the Royal Society B*, 272: 121–126.

Kowalewska, M., and Szwedo, J. 2009. Examination of the Baltic amber inclusion surface using SEM techniques and X-ray microanalysis. *Palaeogeography, Palaeoclimatology, Palaeoecology*, 271: 287–291.

Labandeira, C.C. 2014. Amber. Reading and writing of the fossil record: Preservational pathways to exceptional fossilization. *Paleontological Society Papers*, 20: 163–216.

Lambert, J.B., Santiago-Blay, J.A., and Anderson, K.B. 2008. Chemical signatures of fossilized resins and recent plant exudates. *Angewandte Chemie*, 47: 9608–9616.

Lambert, J.B., Santiago-Blay, J.A., Wu, Y., and Levy, A.J. 2015. Examination of amber and related materials by NMR spectroscopy. *Magnetic Resonance in Chemistry*, 53: 2–8.

Lambert, J.B., Tsai, C.Y.-H., Shah, M.C., Hurtley, A.E., and Santiago-Blay, J.A. 2012. Distinguishing amber and copal classes by proton magnetic resonance spectroscopy. *Archaeometry*, 54: 332–348.

Lang, G., and Buchbauer, G. 2012. A review on recent research results (2008–2010) on essential oils as antimicrobials and antifungals. A review. *Flavour and Fragrance Journal*, 27: 13–39.

Lange, B.M. 2015. The evolution of plant secretory structures and emergence of terpenoid chemical diversity. *Annual Review of Plant Biology*, 66: 19.1–19.21.

Langenheim, J.H. 1994. Higher plant terpenoids: A phytocentric overview of their ecological roles. *Journal of Chemical Ecology*, 20 (6): 1–58.

Langenheim, J.H. 2003. *Plant Resins: Chemistry, Evolution, Ecology, and Ethnobotany*. Timber Press, Oregon, USA.

Larsson, S.G. 1978. *Baltic Amber: A Paleobiological Study*. Scandinavian Scientific Press, Klampenborg, Denmark.M;

Martill, D.M. 1990. Macromolecular resolution of fossilized muscle tissue from an elopomorph fish. *Nature*, 346: 171–172.

Martínez-Delclòs, X., Briggs, D.E.G., and Peñalver, E. 2004. Taphonomy of insects in carbonates and amber. *Palaeogeography, Palaeoclimatology, Palaeoecology*, 203: 19–64.

Mazur, N., Nagel, M., Leppin, U., Bierbaum, G., and Rust, J. 2014. The extraction of fossil arthropods from Lower Eocene Cambay amber. *Acta Palaeontologica Polonica*, 59: 455–460.

McCoy, V.E., Soriano, C., and Gabbott, S.E. 2018a. A review of preservational variation of fossil inclusions in amber of different chemical groups. *Earth and Environmental Science, Transactions of the Royal Society of Edinburgh*, 107 (2–3): 203–211.

McCoy, V.E., Soriano, C., Pegoraro, M., Luo, T., Boom, A., Foxman, B., and Gabbott, S.E. 2018b. Unlocking preservation bias in the amber insect fossil record through experimental decay. *PLOS ONE*, 13: e0195482.

Mierzejewski, P. 1976a. On application of scanning electron microscope to the study of organic inclusions from the Baltic amber. *Rocznik Polskiego Towarzystwa Geologicznego*, 46: 291–295.

Mierzejewski, P. 1976b. Scanning electron microscope studies on the fossilization of Baltic amber spiders (preliminary note). *Annals of the Medical Section of the Polish Academy of Sciences*, 21: 81–82.

Miyazawa, M., and Yamafuji, C. 2005. Inhibition of acetylcholinesterase activity by bicyclic monoterpenoids. *Journal of Agricultural and Food Chemistry*, 53: 1765–1768.

Pastorelli, G. 2011. A comparative study by infrared spectroscopy and optical oxygen sensing to identify and quantify oxidation of Baltic amber in different ageing conditions. *Journal of Cultural Heritage*, 12: 164–168.

Penney, D. 2010. *Biodiversity of Fossils in Amber from the Major World Deposits*. Siri Scientific Press, Manchester.

Penney, D., Wadsworth, C., Fox, G., Kennedy, S.L., Preziosi, R.F., and Brown, T.A. 2013. Absence of ancient DNA in sub-fossil insect inclusions preserved in "Anthropocene" Colombian copal. *PLOS ONE*, 8: e73150.

Pepinelli, M., and Currie, D.C. 2017. The identity of giant black flies (Diptera: Simuliidae) in Baltic amber: Insights from large-scale photomicroscopy, micro-CT scanning and geometric morphometrics. *Zoological Journal of the Linnean Society*, 181: 846–866.

Pichersky, E., and Raguso, R.A. 2018. Why do plants produce so many terpenoid compounds? *New Phytologist*, 220: 692–702.

Poinar, G.O. 1992. *Life in Amber*. Stanford University Press, Stanford, California.

Poinar, G.O., and Buckley, R. 2006. Nematode (Nematoda: Mermithidae) and hairworm (Nematomorpha: Chordodidae) parasites in Early Cretaceous amber. *Journal of Invertebrate Pathology*, 93: 36–41.

Poinar, G.O., and Hess, R. 1982. Ultrastructure of 40-million-year-old insect tissue. *Science*, 215: 1241–1242.

Poinar, G.O., and Hess, R. 1985. Preservative qualities of recent and fossil resins: Electron micrograph studies on tissue preserved in Baltic amber. *Journal of Baltic Studies*, 16: 222–230.

Poinar, G.O., and Mastalerz, M. 2000. Taphonomy of fossilized resins: Determining the biostratinomy of amber. *Acta Geológica Hispánica*, 35: 171–182.

Poinar, H.N., Hofreiter, M., Spaulding, W.G., Martin, P.S., Stankiewicz, B.A., Bland, H., Evershed, R.P., Possnert, G., and Pääbo, S. 1998. Molecular coproscopy: Dung and diet of the extinct ground sloth *Nothrotheriops shastensis*. *Science*, 281: 402–406.

Poinar, H.N., Höss, M., Bada, J.L., and Pääbo, S. 1996. Amino acid racemization and the preservation of ancient DNA. *Science*, 272: 864–866.

Saitta, E.T., Kaye, T.G., and Vinther, J. 2019. Sediment-encased maturation: A novel method for simulating diagenesis in organic fossil preservation. *Palaeontology*, 62: 135–150.

Scalarone, D., Lazzari, M., and Chiantore, O. 2003a. Ageing behaviour and analytical pyrolysis characterisation of diterpenic resins used as art materials: Manila Copal and Sandarac. *Journal of Analytical and Applied Pyrolysis*, 68–69: 115–136.

Scalarone, D., van der Horst, J., Boon, J.J., and Chiantore, O. 2003b. Direct-temperature mass spectrometric detection of volatile terpenoids and natural terpenoid polymers in fresh and artificially aged resins. *Journal of Mass Spectrometry*, 38: 607–617.

Schlüter, T. 1989. Neue Daten über harzkonservierte Arthropoden aus dem Cenomanium NW-Frankreichs. *Documenta Naturae*, 56: 59–70.

Seyfullah, L.J., Beimforde, C., Dal Corso, J., Perrichot, V., Rikkinen, J., and Schmidt, A.R. 2018a. Production and preservation of resins—Past and present. *Biological Reviews of the Cambridge Philosophical Society*, 93: 1684–1714.

Seyfullah, L.J., Roghi, G., Dal Corso, J., and Schmidt, A.R. 2018b. The Carnian pluvial episode and the first global appearance of amber. *Journal of the Geological Society*, 175: 1012–1018.

Sherratt, E., del Rosario Castañeda, M., Garwood, R.J., Mahler, D.L., Sanger, T.J., Herrel, A., de Queiroz, K., and Losos, J.B. 2015. Amber fossils demonstrate deep-time stability of Caribbean lizard communities. *Proceedings of the National Academy of Sciences of the United States of America*, 112: 9961–9966.

Shuaib, M., Ali, A., Ali, M., Panda, B.P., and Ahmad, M.I. 2013. Antibacterial activity of resin rich plant extracts. *Journal of Pharmacy & Bioallied Sciences*, 5: 265–269.

Smejkal, G.B., Poinar, G.O., Righetti, P.G., and Chu, F. 2011. Revisiting Jurassic Park: The isolation of proteins from amber encapsulated organisms millions of years old, pp. 925–938. In: Ivanov, A.R., and Lazarev, A.V., eds. *Sample Preparation in Biological Mass Spectrometry*. Springer, Dordrecht.

Smith, A.B., and Austin, J.J. 1997. Can ancient DNA be recovered from the fossil record? *GeoScientist*, 7: 58–61.

Smith, E.A. 1880. Concerning amber. *The American Naturalist*, 14: 179–190.

Solórzano Kraemer, M.M. 2007. Systematic, palaeoecology, and palaeobiogeography of the insect fauna from Mexican amber. *Palaeontographica Abteilung A*, 282: 1–133.

Solórzano Kraemer, M.M., Kraemer, A.S., Stebner, F., Bickel, D.J., and Rust, J. 2015. Entrapment bias of arthropods in Miocene amber revealed by trapping experiments in a tropical forest in Chiapas, Mexico. *PLOS ONE*, 10: e0118820.

Solórzano Kraemer, M.M., Martínez-Delclos, X., Clapham, M.E., Arillo, A., Peris, D., Jäger, P., Stebner, F., and Peñalver, E. 2018. Arthropods in modern resins reveal if amber accurately recorded forest arthropod communities. *Proceedings of the National Academy of Sciences*, 115: 6739–6744.

Soriano, C., Archer, M., Azar, D., Creaser, P., Delclòs, X., Godthelp, H., Hand, S., J Jo Jones, A., Nel, A., Néraudeau, D., Ortega-Blanco, J., Pérez-de la Fuente, R., Perrichot, V., Saupe, E., Solórzano Kraemer, M., and Tafforeau, P. 2010. Synchrotron X-ray imaging of inclusions in amber. *Comptes Rendus Palevol*, 9: 361–368.

Stankiewicz, B.A., Poinar, H.N., Briggs, D.E.G., Evershed, R.P., and Poinar, G.O. 1998. Chemical preservation of plants and insects in natural resins. *Proceedings of the Royal Society B*, 265: 641–647.

Tanaka, G., Parker, A.R., Siveter, D.J., Maeda, H., and Furutani, M. 2009. An exceptionally well-preserved Eocene dolichopodid fly eye: Function and evolutionary significance. *Proceedings of the Royal Society B*, 276: 1015–1019.

Thomas, D.B., Nascimbene, P.C., Dave, C.J., Grimaldi, D.A., and James, H.F. 2014. Seeking carotenoid pigments in amber-preserved fossil feathers. *Scientific Reports*, 4: 5226.

Tornquist, A.J.H. 1910. *Geologie von Ostpreussen*. Gebrüder Borntraeger, Berlin.

van Aarssen, B.G.K., Cox, H.C., Hoogendoorn, P., and de Leeuw, J.W. 1990. A cadinene biopolymer in fossil and extant dammar resins as a source for cadinanes and bicadinanes in crude oils from South East Asia. *Geochimica et Cosmochimica Acta*, 54: 3021–3031.

Vasan, S., Zhang, X., Zhang, X., Kapurniotu, A., Bernhagen, J., Teichberg, S., Basgen, J., Wagle, D., Shih, D., Terlecky, I., Bucala, R., Cerami, A., Egan, J., and Ulrich, P.WW 1996. An agent cleaving glucose-derived protein crosslinks *in vitro* and *in vivo*. *Nature*, 382: 275–278.

Vávra, N. 2009. The chemistry of amber—Facts, findings and opinions. *Annalen des Naturhistorischen Museums in Wien Serie A*, 111: 445–473.

Viehmeyer, H. 1913. Ameisen aus dem Kopal von Celebes. *Stettiner Entomologische Zeitung*, 74: 141–155.

Voigt, E. 1937. Paläohistologische Untersuchungen an Bernstein-Einschlüssen. *Palaeontologische Zeitschrift*, 19: 35–46.

von Lengerken, H. 1913. Etwas über den Erhaltungszustand von Insekten-Inklusen im Bernstein. *Zoologischer Anzeiger*, 41: 284–286.

von Lengerken, H. 1922. Über fossile Chitinstrukturen. *Verhandlungen der Deutschen Zoologischen Gesellschaft*, 27: 73.

Walden, K.O.O., and Robertson, H.M. 1997. Ancient DNA from amber fossil bees? *Molecular Biology and Evolution*, 14: 1075–1077.

Winkler, W., Kirchner, E.C., Asenbaum, A., and Musso, M. 2001. A Raman spectroscopic approach to the maturation process of fossil resins. *Journal of Raman Spectroscopy*, 32: 59–63.

Zehbe, R., Haibel, A., Riesemeier, H., Gross, U., Kirkpatrick, C.J., Schubert, H., and Brochhausen, C. 2010. Going beyond histology. Synchrotron micro-computed tomography as a methodology for biological tissue characterization: From tissue morphology to individual cells. *Journal of the Royal Society Interface*, 7: 49–59.

CHAPTER 6

Experimental Silicification of Wood in the Lab and Field

Pivotal Studies and Open Questions

CAROLE T. GEE AND MORITZ LIESEGANG

ABSTRACT | Silicification is one of the most important processes in the fossilization of plants. Plant cells in the primary plant body, as well as woody tissues in secondary growth, can be exquisitely preserved through this process. In particular, the silicification of wood can result in the precise three-dimensional preservation of the cells in the secondary xylem, down to the minute structures in the cell walls requisite for taxonomic identification. Understanding the process of silicification is thus central to understanding the formation of fossil wood. Here we review the history of experimental silicification in wood, in the laboratory and in the field. While the first trials in turning wood into stone took place five centuries ago, it has only been in the last five decades that attempts to mineralize wood have intensified. Synthetic agents such as tetraethyl orthosilicate were first applied in the laboratory, while silicification experiments were done in natural hot-spring waters. Recent laboratory experimentation has produced incipient silicification in wood using natural starting materials. These laboratory-controlled experiments show that silicification can be a rapid process. Moreover, modern imaging and analytical methods have allowed the documentation of the silicification process of wood tissues on the nanolevel. Questions regarding silicification in wood, such as the conditions under which it occurs in rooted trees in silica-rich substrates, the diagenetic pathway of silica and its interaction with secondary cell walls, or the permeability of bark to silicification, are promising avenues for future experimental research. |

Introduction

Silicification is the most important fossilization process resulting in the preservation of internal tissues in plants. While the permineralization of plant matter may also occur through calcium carbonate, calcium phosphate, manganese oxide, various iron and copper minerals, fluorite, barite, natrolite, and smectite clay (Mustoe 2018), the most common preserving agent is by far silica ($SiO_2 140 \cdot nH_2O$) (St. John 1927; Leo and Barghoorn 1976). Besides silicified wood, the best-known types of plant permineralization are calcified, pyritized, and phosphatized wood. However, it is the process of silicification that can preserve cells in exquisite detail, even down to the microstructure of the cell walls (Muir 1970; Buurman 1972; Jefferson 1987) and cell contents (Taylor et al. 2005). Without silicification, wood and other plant tissues would rapidly decompose in an oxygenated environment, and we would know very little about the anatomy, life history, and evolution of land plants. Indeed, a plenitude is known about one of the earliest terrestrial ecosystems, the 407-million-year-old plants and animals in the Rhynie Chert, Scotland, because of the rapid and detailed preservation through silica of the organisms in this conservation fossil lagerstätte (Selden and Nudds 2012 and references therein).

The silicification of woody plant tissue is a complex, multistage process. Plant cell walls have a natural affinity to silica, attracting dissolved silica $Si(OH)_4$ from aqueous solutions and precipitating it in cellular spaces as a noncrystalline solid (opal-A) (Channing and Edwards 2004; Hellawell et al. 2015). After the aqueous silica (silicic acid) enters the slightly acidic environment within a wood cell, if the pH is less than about 8, the silica will precipitate out of solution due to the pH change (Ballhaus et al. 2012). The silicic acid in the wood cell then bonds with the hydroxyl groups of lignin and holocellulose molecules of the cell wall (Leo and Barghoorn 1976; Jefferson 1987), represented here by the reaction in fig. 6.1.

The precipitate of monosilicic acid, $Si(OH)_4$, forms a water-permeable film on the cell wall that templates the inner surface of the wood cell (Leo and Barghoorn 1976; Jefferson 1987; Hellawell et al. 2015), thereby protecting it from the fungal decomposition that would otherwise befall all wood; bacteria play a lesser role in wood degradation and do not mediate wood

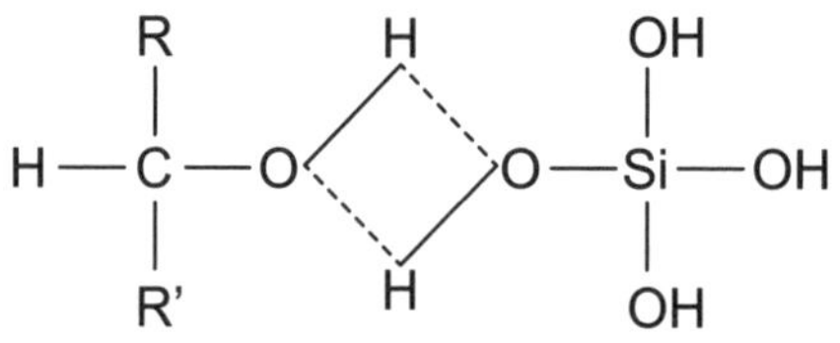

Figure 6.1. Bonding between silicic acid ($Si(OH)_4$ aq.) and the organic molecules (R, R') of the plant cell wall, resulting in a precipitate of monomeric silica on the inner cell walls of the wood cells. Redrawn from Jefferson (1987) by V.E. McCoy.

silicification (Zabel and Morrell 2012). The silica film in the wood cell preserves the minute structures in the cell wall, such as crossfield pits and circular bordered pits, which are so important for botanical identification of the wood. Continued intake of aqueous silica into the woody tissue may result in the infiltration and impregnation of the lignin and other organic molecules by silica in the cell walls, as well as in the infilling of the cell lumina. The initial silica phase is assumed to be noncrystalline opal-A (Channing and Edwards 2004; Ballhaus et al. 2012), which, with time and diagenesis, may transform into weakly crystalline opal (opal-CT), then into chalcedony or quartz (fig. 6.2A, B) (Mustoe 2016), so that the organic plant body is even-

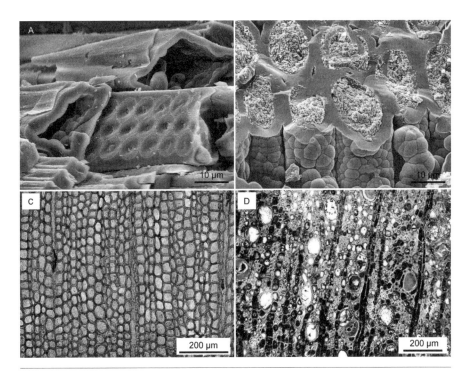

Figure 6.2. Silicified wood. (**A**) Scanning electron micrograph (SEM) of Miocene conifer wood from Churchill County, Nevada, USA, showing the radial tracheid walls with circular bordered pits and inner cell walls mineralized with opal-A. Photo courtesy of George Mustoe. (**B**) SEM of Miocene wood from Washoe County, Nevada, USA, in oblique transverse view, showing cell walls mineralized with botryoidal opal-A and lumen filled with granular opal-A. Photo courtesy of George Mustoe. (**C**) Transmitted light micrograph (LM) of conifer wood from Upper Jurassic from near Vernal, Utah, in transverse section, showing the preponderance of tracheids. Photo modified from Gee et al. (2019). (**D**) LM of angiosperm wood from middle Eocene from Laredo, Texas, in transverse section, showing the complex tissue of dicot wood; the largest cells are the vessel elements. Photo courtesy of Nicole Garten-Dölle.

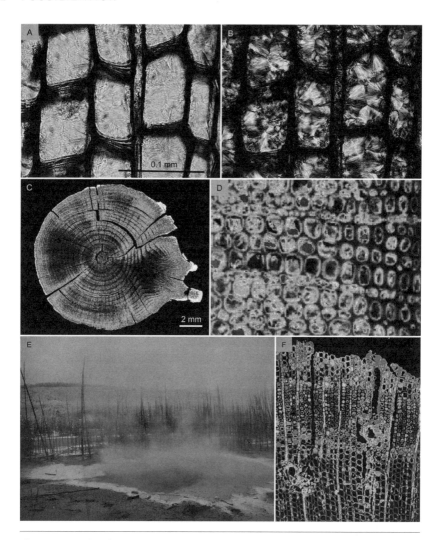

Plate 6.1. Examples of mineralized wood and a silica-rich hot-spring environment. (**A**) Early Permian lycopod wood of *Lepidodendron johnsonii* from Colorado, USA, in transverse section under transmitted plane-polarized light. (**B**) The same view as in **A**, but in cross-polarized light, showing gray-colored chalcedony infilling the cell lumina. **A** and **B** courtesy of George Mustoe. (**C**) EPMA map for silicon of a twig found lying atop a sinter apron of Cistern Spring, Yellowstone National Park, Wyoming. Relatively greater quantities of silicon are found along the periphery and in the desiccation cracks of the twig. (**D**) Close-up EPMA map for silicon of a single growth ring from **C**, showing differential amounts of silicon in the earlywood to the left than in the latewood to the right. Width of image = 0.23 mm. **C** and **D** adapted from Hellawell et al. (2015), reused with permission from Elsevier. (**E**) Cistern Spring, a silica-rich hot spring in Yellowstone National Park, Wyoming, USA. Photo: CTG. (**F**) EPMA map for silicon in wood from a sapling trunk standing upright on a sinter apron of Cistern Spring. Note the relative abundance of Si at the periphery of the trunk (warmer colors at top of image) and in the epithelial cells around the resin canals. Width of image = 650 μm. Image: ML.

tually rendered into a mineralized state (plate 6.1A, B). Most silicified wood consists of more than 90% SiO_2 by weight (Leo and Barghoorn 1976; Mustoe 2015, 2016, 2017). Any substrate that is able to bring elevated silica in solution (Ballhaus et al. 2012) is capable of silicifying plant material: a volcanic ash rich in glassy material (e.g., Läbe et al. 2012), a silica-rich hydrothermal solution from a hot spring (e.g., Hellawell et al. 2015, Liesegang and Gee 2020), or an immature sediment with abundant detrital feldspar.

Anatomical Structures and Water Conduction in Wood

Integral to understanding silicification in woody plants is the knowledge of anatomy and water conduction in the secondary tissues of a tree. Wood is a complex tissue, which means that it is composed of several cell types. In conifers, the axial system (cells running the length of tree trunk) is mostly made up of tracheids (fig. 6.2C), although scattered parenchyma cells and special structures such as resin canals may be present, too. In angiosperms, the axial system is composed of large-celled vessel elements, groups of parenchyma cells, and thick-walled fibers (fig. 6.2D). Water from the roots is primarily conducted upwards from the roots through the trunk via the tracheids in the conifers and via the vessel elements in the angiosperms. Tracheids and vessel elements, as well as the fibers, are stiffened by lignin by a secondary cell wall, making the plant cell rigid and stem woody. The vascular cells are also individually aligned so that water can flow continuously and mostly unimpeded from the roots to the leaves. When functional, tracheids and vessel elements are dead; thus, when these vascular cells are lined up with one another vertically, they form strong empty pipelines through which water is pulled up to the top of the tree.

In both conifers and angiosperms, water is also transported laterally along the radius of the tree trunk. This function is performed by the vascular rays, which are made up of ray cells. Unlike tracheids and vessel elements, ray cells are living, cytoplasm-filled cells when functional. Ray thickness and height are variable among various species, but conifers tend to have thin, short rays, while some angiosperms can have extremely thick and tall rays.

The wood cells in the axial system excel at conducting water upwards. Mature apple trees, for example, can consume more than 70 liters of water per day (Green et al. 2003). Even more impressive are the giants of the arborescent world, the coastal redwoods; a mere 45 m tall tree of *Sequoia sempervirens* uses even more water, needing up to 600 ± 145 liters a day (Dawson 1998). There are several forces at work necessary to pull water up the tree: root pressure, which draws in water into the roots from the soil;

capillary action, which results from the adhesion of water with the walls of the narrow, cylinder-shaped wood cells; cohesion, which occurs between water molecules; and transpiration, which releases water out of the leaves (Nobel 2009).

Pivotal Studies in Wood Silicification

Laboratory Wood Silicification Using Synthetic Starting Materials

Early experimental attempts at silicifying wood took place exactly five centuries ago. In 1520, alchemist Basilius Valentinus, a Benedictine monk in Erfurt, Germany, used powdered silica and *sal tartari* to try to turn wood into stone (Vail 1928; Zollfrank and Van Opdenbosch 2018). The *sal tartari*, presumably K_2CO_3, is thought to be the "subtartrate of potash" originating from grapes (Leo and Barghoorn 1976). Later in the 16th century, Kentmann (1518–1568) claimed to have made *lapides igne liquescentes* from alder wood by simmering the wood in a pot with brewing beer until the hops were done, then burying it in fresh sand or gravel in a cellar for three years. The source of the silica apparently came from beer mash itself, which is saturated with dissolved silica from the high amorphous silica content of grains (Leo and Barghoorn 1976). Afterwards, the wood was said to have made the best whetstone or flintstone (Meyer 1791; Leo and Barghoorn 1976).

In more recent times, others have used synthetic agents such as alkoxysilanes or sodium metasilicate to mineralize wood in the laboratory, often under vacuum (Drum 1968a, b; Buurman 1972; Leo and Barghoorn 1976; Saka and Ueno 1997; Shin et al. 2001; Persson et al. 2004; Götze et al. 2008). Internal cell casts, cell infilling, and cell wall penetration were preserved through these various laboratory silicification trials. In the case of Drum (1968a, b), siliceous casts were made from the inside of parenchyma cells by first soaking fresh birch twigs in sodium silicate solution for 12 to 24 hours, then wet-ashing the treated twigs with chromic acid for three days.

It was, however, the pivotal silicification experiments conducted by Leo and Barghoorn (1976) using tetraethyl orthosilicate (TEOS, formerly called tetraethoxysilane) that provided the basis for the now well-established concept of silica bonding to the reactive sites of the organic cell wall constituents during the early stage of wood silicification. To start, Leo and Barghoorn (1976) boiled a piece of wood in water until it was waterlogged and free of gas, then immersed it in alternating solutions of water and TEOS in sealed jars in an oven at 70°C for a few days to a month. Now and again, the jars were removed from the oven, uncapped, and placed under vacuum for about one half hour to aid diffusion of the liquid phase into the wood. Afterwards, removal of all organic material from the treated wood was achieved with oxidation using Schulze's solution, which is comprised of saturated aqueous potassium chlorate ($KClO_3$) and concentrated nitric acid (HNO_3) and is also applied to the maceration of leaf cuticle (e.g., Gee and McCoy, chap. 9). In the method developed by Leo and Barghoorn (1976), the silicification of wood occurred in solution of near neutral pH conditions and resulted in a greater degree of impregnation of the wood than with that of Drum (1968a, b).

According to Leo and Barghoorn (1976), their experimentally mineralized wood closely resembled naturally silicified wood from geologically young sediments ranging in age from the Miocene epoch to the Recent. Because the experimentally mineralized wood was too brittle for conventional petrographic thin-sectioning, the wood cells were teased apart and examined using scanning electron microscopy (SEM). In the scanning electron micrographs, delicate features in the wood, such as circular bordered pits in the conifer tracheids and scalariform perforation plates in the angiosperm vessel elements, could be observed.

From their work, Leo and Barghoorn (1976) concluded that silica is deposited in the early stages of mineralization on the cellular surfaces in the woody tissue, particularly on the inner cell wall and on the lining of the pit chambers (see also Ballhaus et al. 2012), then grows outwards, infilling void spaces within the cells. However, even in the earliest stages of silicification, some silica penetrates into the cell wall in close association with the organic matter. Hence, Leo and Barghoorn (1976) emphasized that the silicification of wood is basically a process of infiltration and impregnation, and not a molecule-by-molecule replacement of organic polymers in the cell walls by silica, as is commonly still believed today. They also posit that the surfaces within the wood tissue serve as an active template for silica deposition.

Following the methodology of Leo and Barghoorn (1976), Shin et al. (2001) also used TEOS as a silica source to replicate wood mineralization in the lab. However, instead of wood extraction, they used the surfactant cetyltrimethylammonium chloride. To prevent bulk precipitation, the rate of hydrolysis of the sol–gel process was kept under control by adjusting the amounts of solvent (ethanol) and acidity (hydrochloric acid) of the sol. These experiments produced detailed cellular wood structures of infilled cells and penetrated cell walls. Key in their experiments was the surfactant, which facilitated the formation of interconnected nanopores in the silica tracheid casts, because these interconnections became pathways for the volatile decomposition products of the organic matter during calcination.

A few years later, Persson et al. (2004a) modified the water-dispersed TEOS methods worked out by Leo and Barghoorn (1976) and Shin et al. (2001) by impregnating wood with TEOS dispersed in hexadecyltrimethylammonium chloride, ethanol, and hydrochloric acid. This was followed by calcination at 575°C, which produced brittle cell casts of silica. A second study by Persson et al. (2004b) was able to achieve completely infiltrated spaces between cellulose microfibrils in wood pulp. In 2008, Götze et al. explored the impregnation of wood with other sources of silica, such as sodium metasilicate, tetraethoxysilane, methyltriethoxysilane, colloidal silica suspension, and silica sols, to produce infilled tracheids.

A major goal of some of these early trials of wood mineralization in the lab was to produce biomimetic micro- and nanoscale ordered material that could be used as catalysts, ceramics, or water-repellent construction materials. For example, delignified wood could serve as a template to create an oxide ceramic with the natural porosity of wood. The use of wood as a template for the creation of bioinspired materials for engineering applications was recently reviewed by Dietrich et al. (2015) and Zollfrank and Van Opdenbosch (2018). It should be noted, however, that these trials in wood silicification do not truly reflect the process of mineralization in trees and wood in the natural world, for the synthetic compounds that were used as starting materials do not occur in nature.

Wood Silicification in the Lab and Field Using Natural Starting Materials

It was not until relatively recently that silica using natural starting materials was precipitated in plant tissue. In a long-term study, Akahane et al. (2004) immersed fresh wood pieces of alder wood for up to seven years in acidic hot-spring water at a temperature of ca. 50°C. Over this period of time, the silica concentration of the initially unsilicified pieces of wood increased to ca. 90% SiO_2 by weight. SEM and mapping of the silicon distribution with a SiKα X-ray scan showed that most of the silica occurs in the infilling of the cell lumina.

Channing and Edwards (2004) also took advantage of the naturally occurring, silica-rich conditions in a hot-spring environment to observe the early stages of silicification in plant tissue, although they used herbaceous plants with primary tissue and not woody plants with secondary tissue. As in the study by Akahane et al. (2004), Channing and Edwards (2004) placed the plant material in hot-spring water for a length of time, then documented and analyzed the resulting deposits with high-resolution instruments in the lab. They selected an herbaceous wetland plant, *Eleocharis rostellata*, as an analog for early land plants, and they immersed samples in two water-permeable, nylon mesh bags in the vent pool of Medusa Geyser in Yellowstone National Park, one for 30 and the other for 330 days. Later, in the lab, SEM and transmission electron microscopy (TEM) were used for the observation and documentation of the precipitates found in the plant tissue. Spot and area analysis with energy-dispersive X-ray spectrometry (EDS), as well as element mapping with EDS and wavelength-dispersive X-ray spectroscopy (WDS), were carried out to determine the source and extent of mineralization, while powder X-ray diffraction (XRD) determined mineral phases. Channing and Edwards (2004) found that in the plant tissue immersed for 30 days, silica deposits in the form of opal-A were found in the cell lumina, intercellular spaces, and within the cell walls. By 330 days, the deposition of opal-A had created a robust external and internal matrix that stabilized the soft plant tissue against collapse and had replicated cellular structure. Through their study, it was also shown that the soft primary tissue of an herbaceous plant, which would normally be susceptible to rapid decay or destruction in most terrestrial environments, had a preservation potential similar to plants with more robust secondary tissue, that is, with wood, in a hot-spring vent pool.

Nearly a decade later, a study carried out by Ballhaus et al. (2012) took a more comprehensive approach by determining the major driving forces that trigger silicification in wood by producing incipient silicification in wood in the laboratory using natural starting materials. To determine basic chemical parameters, dried conifer wood of Douglas fir (*Pseudotsuga menziesii*) was placed in tap water (pH 7.8); the pH dropped immediately to ca. 4 (fig. 6.3A), which reflects the slightly acidic microenvironment that develops in water-immersed wood during decomposition. Conversely, in another experiment, when powdered rhyolitic obsidian from Lipari Island, Italy, was added to tap water, the pH rose to ca. 9.5 within a minute (fig. 6.3B).

In short, when a basic aqueous solution of silica enters the acidic environment inside a wood cell, silica solubility decreases sharply (Fleming and Crerar 1982; Williams et al. 1985), and noncrystalline silica precipitates according to the reaction in equation 1. Thus, the most likely driving force behind the precipitation of the silica from solution entering a wood cell is a change in pH (Ballhaus et al. 2012), if the pH of the incoming fluid is larger than about 8. The inner cell wall of the wood cell serves as the precipitation surface for the silica.

$$H_3SiO_{4(aq)}{}^{1-} + H_3O^+ \text{ (wood)} + (n{-}3)H_2O \leftrightarrow SiO_2 \cdot nH_2O \text{ (opal)} \quad \text{(Eq. 1)}$$

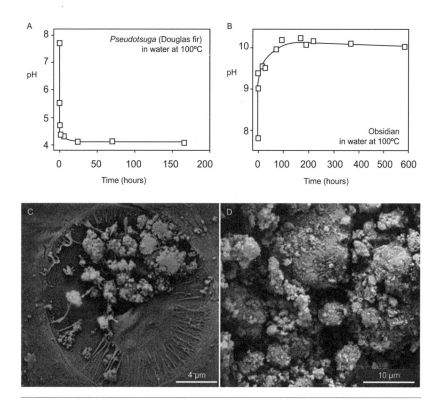

Figure 6.3. Exploratory experiments on wood silicification. **A, B**: Time-series experiments showing the evolution of the pH of tap water at an initial pH 7.8 after the addition of conifer wood or powdered obsidian (rhyolitic glass). (**A**) Tap water was reacted with *Pseudotsuga menziesii* wood at 100°C for ca. 160 hours. There was an immediate drop and leveling off of the pH to 4. (**B**) Tap water was reacted with powdered obsidian at 100°C for 584 hours. There was an immediate rise and subsequent equilibrium to a pH of ca. 10. **C, D**: Backscattered electron images of incipient silicification in recent *Pseudotsuga menziesii* wood after reaction with obsidian and water at 100°C for 10 days, illustrating the rapid deposition of silica precipitates (**C**) SEM of silica microspheres precipitated onto a circular bordered pit in the cell wall of treated wood. (**D**) SEM close-up of silica precipitates in the treated wood. Images adapted from Ballhaus et al. (2012), reused with permission from Elsevier.

With this chemical reaction in mind, Ballhaus et al. (2012) set out to produce incipient silicification in conifer wood in the lab at temperatures less than 100°C using natural starting materials found in natural settings. Cubes of *Pseudotsuga* wood were reacted with powdered obsidian and water, then heated to 100°C in a specially designed autoclave (see fig. 6.4B). Experiments were run for 112 days, although a white precipitate consisting of microspheres was soon observed in some tracheids (fig. 6.3C), especially on the circular bordered pits (fig. 6.3D), even after just a few days of treatment.

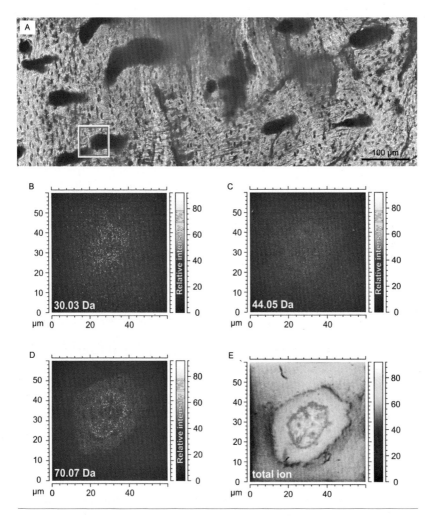

Plate 2.1. (A) Light microscope photo of bone with blood vessels of the archosauromorph reptile *Protanystropheus* sp. Rectangle marks area analyzed with ToF-SIMS. **B–E**: ToF-SIMS ion distribution maps (fast imaging mode) generated for selected masses and amino acids. (B) Glycine or proline. (C) Alanine. (D) Proline. (E) A blood vessel in cross section outlined by the total ion map based on amino acids. Modified from Surmik et al. (2016), Creative Commons Attribution License.

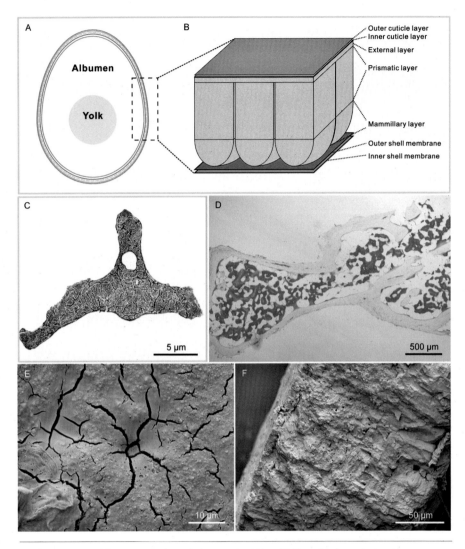

Plate 3.1. A, B: Schematic view through the egg of a chicken (*Gallus gallus domesticus*) and its egg-shell. Modified from Yang et al. (2018), Creative Commons Attribution License. (**A**) The anatomy of an egg. (**B**) The eggshell is composed of two crystalline layers, the prismatic layer and mammillary layer. The cuticle layer overlying the calcareous eggshell is further divided into two layers, a proteinaceous outer layer, and a hydroxyapatite inner layer. The shell membrane (*membrana testacea*) is also characterized by two layers. **C, D:** Histological images of medullary bone in a black swan, *Cygnus atratus* (CM-S16508). Modified from Canoville et al. (2019), Creative Commons Attribution License 4.0. (**C**) Medullary bone is evident in the cavities of the ground section of a caudal vertebra. (**D**) An Alcian blue-stained paraffin section shows the medullary bone as a blue area. **E, F:** Scanning electron microscope (SEM) images of a chicken eggshell. From Yang et al. (2018), Creative Commons Attribution License. (**E**) The cracked pattern of the cuticle layer on the surface of the eggshell. (**F**) A radial section through the eggshell.

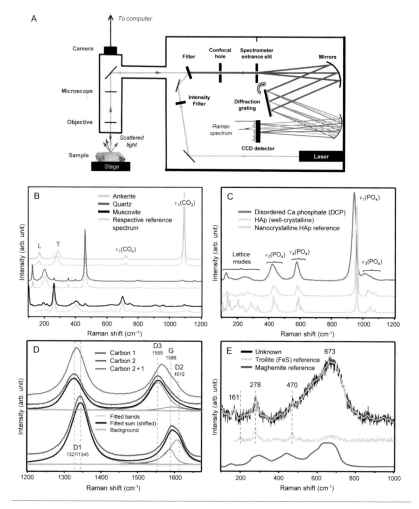

Plate 4.1. (**A**) Schematic sketch of a typical confocal Raman spectrometer setup (not to scale). See text for explanation. **B–E**: Representative Raman spectra of all phases detected within the mapped area of the acanthodian spine under study. (**B**) Ankerite, quartz, and muscovite, along with reference spectra from the RRUFF database (ankerite ID: R050181; quartz ID: R040031; muscovite ID: R040104). Ankerite bands are labeled according to the Herzberg notation and Rividi et al. (2010). (**C**) Calcium phosphate phases (DCP and HAp) along with a reference spectrum from nanocrystalline HAp (Asjadi et al. 2019). (**D**) Spectra of two different carbon phases (red and dark blue curves) and a mixed spectrum (light blue curve). The carbon spectra were fitted with three Gauss-Lorentz functions (gray curves) and a linear background (green line) for crystallite size and metamorphic temperature determination (see explanation in Chapter 4). The different Raman bands of the carbonaceous matter are labeled according to Kouketsu et al. (2014). (**E**) Raman spectrum of a yet unidentified phase that, however, shows some similarities with a RRUFF spectrum of troilite (RRUFF ID R070242) and maghemite (graphically extracted from Dubois et al. 2008). The yellow line represents a smoothed spectrum of the unknown phase using polynomial smoothing.

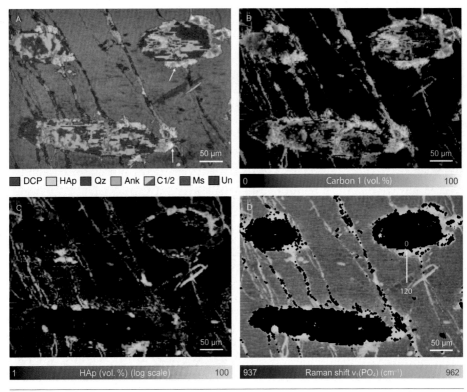

Plate 4.2. False-color hyperspectral Raman images of the fossil acanthodian spine. (**A**) Raman phase distribution image. In this image, the phase with the highest intensity contribution in a Raman spectrum is displayed in a color assigned to that particular phase. Note that this does not exclude that other phases are also present in a certain area (cf. **B** and **C**). Gray (**B**) and color scale (**C**) Raman images of the volume fractions of carbon 1 and HAp within the mapped area. Note that in **B**, the color-coding is in the log scale. (**D**) False-color Raman image of the $v_1(PO_4)$ frequency of calcium phosphate with location of the transect shown in fig. 4.2. Color-coding is based on the viridis colormap that is perceptually uniform consistently across the entire range of values. Abbreviations: DCP = disordered calcium phosphate; HAp = hydroxyapatite; Qz = quartz; Ank = ankerite; C1/2 = carbon 1 and 2 (cf. plate 4.1D); Ms = muscovite; Un = unidentified phase(s). White arrows in **A** point to replacement fronts that spread out from the vascular canals into the bone matrix.

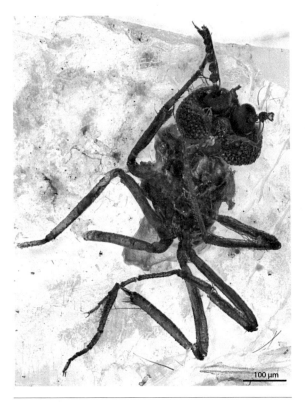

Plate 5.1. Nematoceran fly extracted from Indian Cambay amber (ca. 53 Ma). Except for the antennae, the anterior-most part of the body is nicely preserved and projects above the amber matrix. The abdomen is still enclosed by the resinous residue of the original amber.

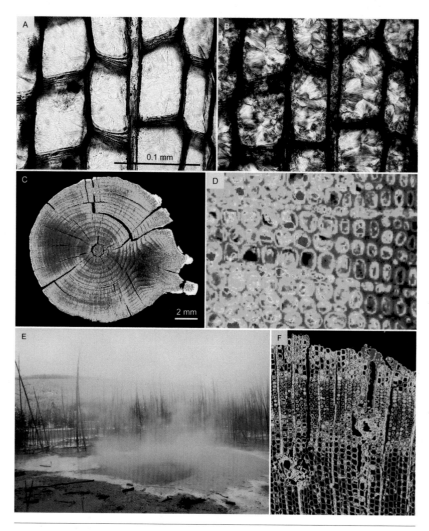

Plate 6.1. Examples of mineralized wood and a silica-rich hot-spring environment. (**A**) Early Permian lycopod wood of *Lepidodendron johnsonii* from Colorado, USA, in transverse section under transmitted plane-polarized light. (**B**) The same view as in **A**, but in cross-polarized light, showing gray-colored chalcedony infilling the cell lumina. **A** and **B** courtesy of George Mustoe. (**C**) EPMA map for silicon of a twig found lying atop a sinter apron of Cistern Spring, Yellowstone National Park, Wyoming. Relatively greater quantities of silicon are found along the periphery and in the desiccation cracks of the twig. (**D**) Close-up EPMA map for silicon of a single growth ring from **C**, showing differential amounts of silicon in the earlywood to the left than in the latewood to the right. Width of image = 0.23 mm. C and D adapted from Hellawell et al. (2015), reused with permission from Elsevier. (**E**) Cistern Spring, a silica-rich hot spring in Yellowstone National Park, Wyoming, USA. Photo: CTG. (**F**) EPMA map for silicon in wood from a sapling trunk standing upright on a sinter apron of Cistern Spring. Note the relative abundance of Si at the periphery of the trunk (warmer colors at top of image) and in the epithelial cells around the resin canals. Width of image = 650 μm. Image: ML.

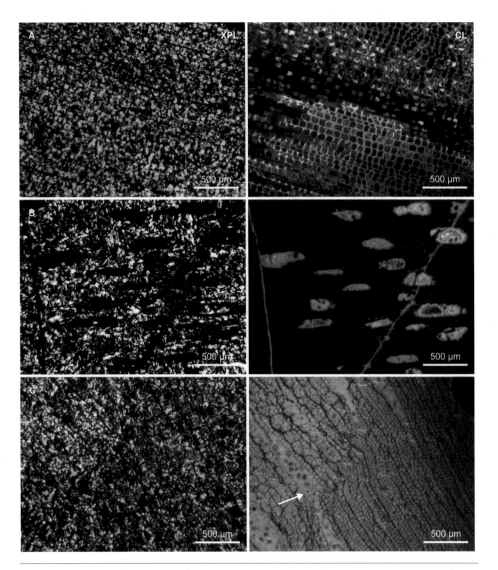

Plate 7.1. Transmitted light micrograph pairs of thin sections of silicified wood from various localities. Abbreviations: XPL = cross-polarized light; CL = cathodoluminescence. (**A**) Cordaite trunk (*Dadoxylon* sp.) from Chemnitz, Germany, showing well-preserved cells from primary silicification (yellow CL) and secondary hydrothermal overprint visible by transient blue CL. Cell walls have a different color than the lumina. (**B**) Miocene silicified wood from Limnos Island, Greece, showing weakly luminescent chalcedony and small voids filled with opal (bright blue CL). (**C**) Violet luminescent silicified wood thin section from the Thuringian Forest, Germany, with preserved cell structure, slight deformation features, and radiation halos (see arrow) around radioactive inclusions. Photos: JG.

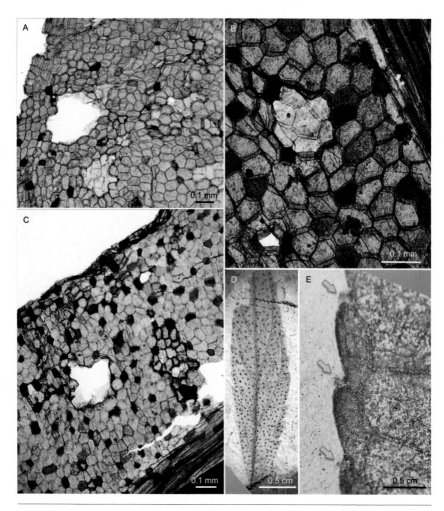

Plate 8.1. Chemical defenses of land plants in the fossil record. A–C: Thalli of *Metzgeriothallus sharonae* (Metzgeriales) showing the distribution of oil body cells as possible protection against microherbivores, as indicated by the herbivore damage avoiding the oil body cells, from the Middle Devonian (Givetian Stage) of the Cairo Quarry, Catskill Delta, Green County, New York, USA. Photos modified from Labandeira et al. (2014). (**A**) Thallus showing a sparse distribution of oil body cells and DT1 hole feeding; UCMP-20150.21. (**B**) Thallus exhibiting surface feeding (DT29) and herbivore avoidance of oil body cells; UCMP-250155.07. (**C**) Thallus showing a denser distribution of oil body cells and DT1 hole feeding; UCMP-250150.23. D, E: Cretaceous dicot leaves with trichomes or glands. Photos: VEM. (**D**) A leaf of morphotype HC81 (Urticaceae or Cannabaceae) showing capitate trichomes, each represented by an amber dot, from DMNS locality 2203, Williston Basin in Slope County, North Dakota, USA; latest Cretaceous (Maastrichtian Stage, ca. 66.5 Ma); specimen DMNS-19512 (see also Johnson 2002; Labandeira et al. 2002). (**E**) A leaf of an unidentified morphotype, most likely in the family Celastraceae, showing marginal tooth glands represented by a pink dot; from DMNS locality 3725, Kaiparowits Formation in southern Utah; Late Cretaceous, 76.6–74.5 Ma; specimen DMNS-43725 (see also Miller et al. 2013; Roberts et al. 2013). Glands in pink at the intersection of marginal veins and margin sinus are indicated by blue arrows. Repository abbreviations: DMNS, Denver Museum of Nature and Science, Colorado, USA; UCMP, University of California Museum of Paleontology, Berkeley, California, USA.

Plate 8.2. Physical defenses of land plants in the fossil record. **A–C**: Leaves of *Annularia carinata* (Equisetales: Calamitaceae), displaying linear to slightly curved foliar trichomes originating from the midveins of leaves; from the Midland Basin, NMNH locality 40013; Jack County, Texas, USA; latest Carboniferous (latest Pennsylvanian, Gzhelian Stage); specimen USNM-696247a. **(A)** Stem at a node showing the radiate arrangement of leaves. **(B)** Four leaves showing long trichomes originating from midvein region. **(C)** Enlargement of area outlined in **B**, showing insertion, disposition, and course of trichomes, four of which are indicated by blue arrows. **D, E**: Leaves of *Wielandiella villosa* (Bennettitales: Williamsoniaceae); from the Yanliao Biota, Daohugou 1 locality, Ningcheng County, Inner Mongolia, China; latest Middle Jurassic (Callovian Stage), showing spine-like trichomes along the midrib. **(D)** Upwardly oriented, spine-like trichomes originating along the midrib; CNU-PLA-NN-2006540P. **(E)** Midrib area, showing dark hued trichomes; CNU-PLA-NN-2006902. Repository abbreviations: CNU, Capital Normal University paleontological collections, Beijing, China; USNM, National Museum of Natural History, Washington, DC, USA. Photos modified from Xu et al. (2018).

Plate 8.3. More chemical defenses of land plants in the fossil record. **A–C**: Leaves of *Macroneuropteris scheuchzeri* (Medullosales: Neurodontopteridaceae), displaying resin tubules embedded in foliage that are part of the plant's resin duct system, from the Midland Basin, NMNH locality 40013 in Jack County, Texas, USA; latest Carboniferous (latest Pennsylvanian, Gzhelian Stage). **(A)** Leaf with margin at right showing the distribution of tissue embedded tubules oriented upper left to lower right; NMNH-606154a. **(B)** Enlargement of foliage showing tubules, some of which are indicated by blue arrows; note the different trajectories of tubules and leaf venation; NMNH-606298a. **(C)** Same specimen as in **B**, with some tubules indicated by blue arrows and piercing-and-sucking mark in center. Repository abbreviations: NMNH, National Museum of Natural History, Washington, DC, USA. Photos modified from Xu et al. (2018).

Plate 9.1. Structural color, bioluminescence, and the pigmentary colors of chlorophylls, anthocyanins, and carotenoids in plants. Cluster of dark blue berries (**A**) and a close-up (**B**) of *Elaeocarpus angustifolius* (commonly known as the blue quandong), showing its shiny indigo surface produced exclusively by structural color, at the Brisbane Botanic Gardens Mt. Coot-tha, Australia. Photos: CTG. (**C**) The sparkly blue "mirror" of the mirror orchid, *Orchis speculum* subspecies *speculum*, which results from an anthocyanin coupled with structural color. Photo: Esculapio, Creative Commons Attribution–ShareAlike 2.5 Generic license. (**D**) Blue-colored light from bioluminescent dinoflagellates during a red tide. Photo: Catalano82, Creative Commons Attribution 2.0 Generic License. (**E**) Marine stromatolite at Shark Bay, Australia, containing chlorophylls *a* and *f*, as well as the bluish pigment phycocyanobilin, releasing here numerous oxygen bubbles as a byproduct of photosynthesis. Photo courtesy of Louise Woo. (**F**) The spores of *Equisetum* are unusual in containing chlorophyll. They are typically 50 μm in diameter. Photo: CTG. (**G**) Anthocyanin (reds) and carotenoid (yellows) pigments in a grape leaf, which show up in the fall after the breakdown of chlorophyll (greens), in the Ahr Valley near Bonn, Germany, in early November. Photo: CTG.

Plate 9.2. Colors in leaves. (**A**) Molecular structure of chlorophyll *a*. Drawing: VEM. (**B**) Variegated fern frond of the Cretan brake, *Pteris cretica*, with pale yellow pinnules rimmed with green, against a background of kelly green-colored *Selaginella*, at the Phipps Conservatory in Pittsburgh, Pennsylvania, USA. Photo: VEM. (**C**) The fresh, light to medium green colors of various ferns, again at the Phipps Conservatory. Photo: VEM. (**D**) Blue-grayish green fronds of the cycad *Encephalartos arenarius* at the Huntington Gardens, California, USA. Photo: CTG. (**E**) Dark forest green conifers on Latemar mountain in the Dolomites, South Tyrol, Italy. Photo: CTG. (**F**) *Plectranthus* (*Solenostemon*) *scutellarioides* "Wizard Coral Sunrise," commonly known as coleus, is prized for its deep red and pink colors, which are produced by anthocyanins, and the striking color patterns in its leaves. Photo: Aleksandrs Balodis, Creative Commons Attribution–ShareAlike 4.0 International license.

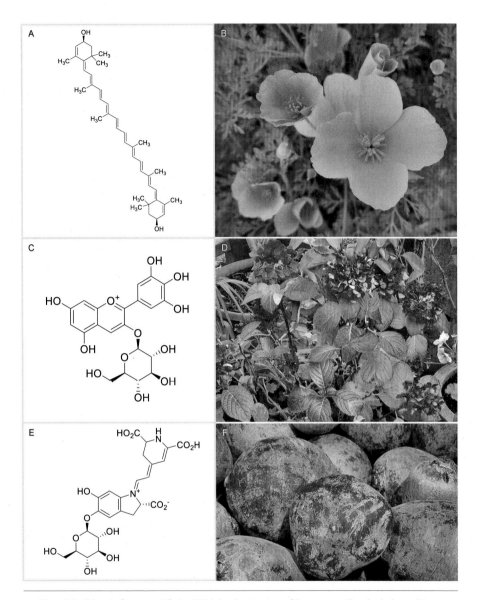

Plate 9.3. Colors in flowers and fruits. (**A**) Molecular structure of the carotenoid eschscholtzxanthin. (**B**) The bright orange color in the California poppy, *Eschscholzia californica*, is produced by carotenoids (including abundant eschscholtzxanthin) and a ridged ultrastructural surface. Photo: CTG. (**C**) Molecular structure of the anthocyanin delphinidin-3-glucoside. (**D**) *Hydrangea* is unusual because the red-violet color of its flower heads and autumnal crimson of its leaves come from a sole pigment, delphinidin-3-glucoside. Photo: CTG. (**E**) Molecular structure of the betalain pigment called betanin, which is the primary pigment in beetroots. (**F**) Even when raw, unwashed, and unpeeled, the beet-red color of *Beta vulgaris* is unmistakable. Photo: CTG.

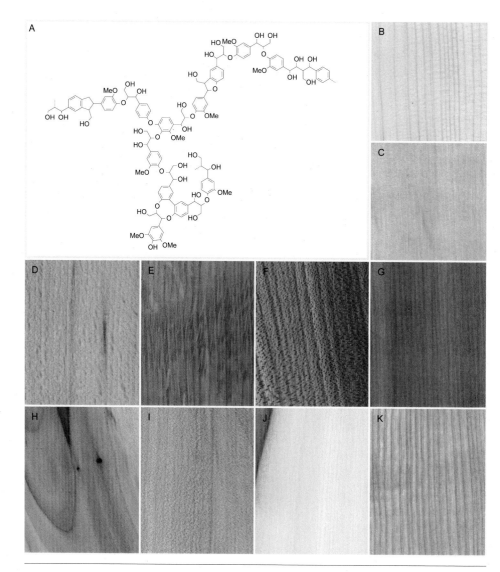

Plate 9.4. Colors in wood. (**A**) One possible molecular structure of lignin. **B–K**: Woods from Scandinavia. Photos: Anonimski, Creative Commons CC0 1.0 Universal Public Domain Dedication. (**B**) *Picea abies* (Norway spruce). (**C**) *Juniperus communis* (common juniper). (**D**) *Fagus sylvatica* (beech). (**E**) *Quercus robur* (pedunculate oak). (**F**) *Ulmus glabra* (wych elm). (**G**) *Prunus avium* (wild cherry). (**H**) *Pyrus communis* (pear). (**I**) *Acer platanoides* (Norway maple). (**J**) *Tilia cordata* (small-leaved lime). (**K**) *Fraxinus excelsior* (ash).

Plate 9.5. Brown colors in fossil plants attributable to carbonization of organic matter. **A–D**: 23-million-year-old fossil plants from the late Oligocene of Rott near Bonn, Germany. Photos: Georg Oleschinski. (**A**) Leaves of *Tremophyllum tenerrimum*, a member of the Ulmaceae. (**B**) An unknown fossil flower in which the intensity of brown colors differs between the delicate petals and the sturdier sepals and stalk. (**C**) Another unidentified flower in which the brown colors are darker and more uniform; nevertheless, darker stripes can be observed on the petals. (**D**) Fossil palm frond. **E, F**: 150-million-year-old wood from the Morrison Formation in northeast Utah, USA. Photos: CTG. (**E**) Fossil log in the field. The light beige, orange, and brown colors were likely produced by traces of iron precipitates. (**F**) Thin section in radial section through the fossil wood. The dark brown horizontal bars are resin plugs that have likely also undergone carbonization since the Late Jurassic.

Plate 9.6. Bright colors in fossil plants. (**A**) Fossil dicot leaf that is still green, presumably due to chlorophyll derivatives present even after 45 million years, collected from the Eocene lignites in Geiseltal but now stored in the collections of the Museum für Naturkunde Berlin. Photo: VEM. (**B**) Precious opal-replaced fossil wood from the middle Miocene Virgin Valley Formation, Nevada, USA. Photo: James St. John, Creative Commons Attribution 2.0 Generic license. (**C**) Tumbled bits of silicified wood from the Ginkgo Petrified Forest State Park, showing diagenetic colors ranging from red to pink, orange to light and dark brown, and even black and white. Photo: Georg Oleschinki. (**D**) The center of a 230-million-year-old tree trunk with a rainbow of diagenetic colors from the Petrified Forest National Park, Arizona, USA. Photo courtesy of Sidney R. Ash.

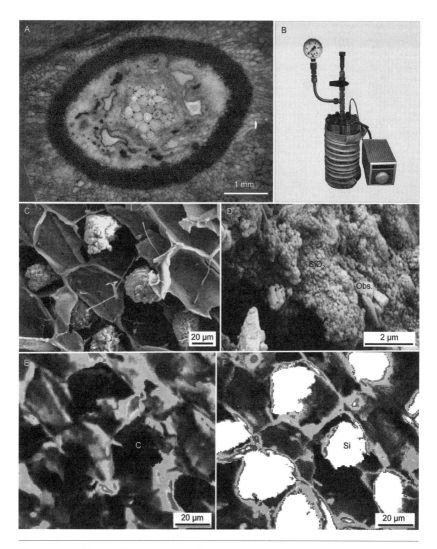

Figure 6.4. Silicification of primary tissues in plants. **(A)** The exquisitely preserved center of an adventitious root of *Psaronius* from the Permian of Chemnitz, Germany, as an example of silicified soft plant tissue. Specimen K4610 from the Naturkundemuseum Chemnitz, Germany. Photo: Georg Oleschinski. **B–F:** Experimental silicification of primary plant tissue in the lab. Original images courtesy of Sashima Läbe. **(B)** The autoclave specially designed by Chris Ballhaus at the University of Bonn for silicifying plant tissue; the sample of plant tissue is placed in the copper tube above the black-colored ball valve. **(C)** SEM image of white clumps of silica in the steam-treated soft tissue pith of the tree fern *Dicksonia antarctica*. **(D)** Detail of white clumps, showing the difference between the spheroids of precipitated silica (SiO_2) and the angular pieces of contaminant obsidian (Obs.). **(E)** Electron probe microanalyzer (EPMA) map for carbon (C) in the same area of tissue as in **C**, showing that the precipitated clumps in the parenchyma pith cells are not composed of carbon. **(F)** EPMA map for silicon (Si) in the same area of tissue as in **C**, showing that the precipitated clumps in the parenchyma pith cells do indeed contain silicon.

Electron probe microanalysis (EPMA) was carried out that identified the microspheres as precipitated silica.

The same set of questions regarding silicification in secondary tissues in wood can also be asked about the primary, nonwoody tissues in stems. The primary tissue in the stems and adventitious roots of *Psaronius* tree ferns from the early Permian of Chemnitz, Germany, for example, can be exquisitely preserved (fig. 6.4A). In this case, silicification in the Permian plants is thought to have occurred as a result of hydrothermal activity with silica-rich steam at temperatures over 150°C (Läbe et al. 2012).

To replicate the silicification conditions at Chemnitz some 290 million years ago, laboratory experiments were conducted in which the parenchyma-rich pith of a tree fern, *Dicksonia antarctica*, was treated with powdered obsidian and water (Läbe et al. 2012). In this case, the experiment was run at 150°C, with water in the vapor phase, to simulate the hydrothermal conditions during the volcanic eruption in the Chemnitz forest during the Permian (e.g., Rößler et al. 2012). Using the same autoclave construction (fig. 6.4B) as in the laboratory experiments by Ballhaus et al. (2012), Läbe et al. (2012) observed white clumps made up of microspheres in the parenchyma cells (fig. 6.4D). Element distribution maps with EPMA showed that these spheres do not contain carbon (fig. 6.4E) but rather silicon (fig. 6.4F). EPMA spot analysis of the material precipitated in the *Dicksonia* pith confirmed that it is possible to precipitate opal (pure hydrated SiO_2) out of silica-rich steam into parenchyma cells. Hence, the studies by Ballhaus et al. (2012) and Läbe et al. (2012) demonstrate that the experimental silicification of plants using natural starting materials in the lab is not only successful in secondary tissue (wood) but can be produced in primary stem tissue as well.

Open Questions

Tissue Pathway of Aqueous Silica in Autochthonous Trees

While there are many case studies on wood fossilization, nearly all of them concentrate on the geochemical and time sequences involved in the mineralization of wood after aqueous silica has started to enter the wood cell and lay down a film in the cell lumen. What is poorly understood is the tissue and cellular pathway of aqueous silica into the tree before silica precipitation has begun on the wood cell walls. The open question here is whether a living tree rooted in a silica-rich substrate, such as found in a hydrothermal setting, can take up aqueous silica from the substrate and transport it through its wood to the leaves, thereby depositing silica on the inside of its conducting cells. In vivo mineralization is not as implausible as it sounds, for trees have no means of excluding ions and complexes that are dissolved in the groundwater from their vascular system (Ma 2005). Many tropical trees are known to take up considerable quantities of aqueous silica species and deposit them in the form of opal in their tracheids and rays (Scurfield et al. 1974; Santana et al. 2013).

If it is possible for the wood cells of an in situ tree to be mineralized in vivo—that is, while the tree is still alive—then this has implications for the fossilization of ancient forests with rooted trees. Specifically, this suggests that standing fossil trees, for example, those growing in hydrothermal settings, may have been silicified in situ and in vivo. In essence, the trees would have unwittingly contributed to their own mineralization and eventual demise by the continued uptake of silica-laden ground water.

To this end, a decorticated (barkless) twig of *Pinus contorta* (lodgepole pine) that grew at Yellowstone National Park was collected from the sinter apron of a hot spring and analyzed using EPMA (plate 6.1C, D) to determine the tissue pathways of aqueous silica into the wood (Hellawell et al. 2015). It was shown that silicon occurs in differing amounts in the earlywood and latewood of conifer wood growth rings; in this case, the earlywood lumina has significantly more silicon than the latewood lumina (plate 6.1D). This unequal concentration of silicon, or in this case, precipitated silica, was attributed in this study to the botanical phenomenon that 90% of the water within a single growth ring is transported in the earlywood (Domec and Gartner 2002). It is thus implied that a differential concentration of silica in the different parts of a growth ring is characteristic of living, autochthonous trees—that is, in situ rooted trunks in growth position.

While the results of the Hellawell et al. (2015) study offer tantalizing evidence of in vivo silicification, the EPMA results showed that there was more than one pathway of aqueous silica into the twig from the sinter apron. The greatest concentration of silicon was just under the external surface of the decorticated twig and in the desiccation cracks extending into the wood from the periphery of the twig. The entry of aqueous silica through the twig's periphery most likely occurred while the twig lay atop the silica-rich water on the sinter apron of the hot spring. Thus, to separate out the different pathways of aqueous silica and to focus on the issue of self-silicification in standing trees, further analytical work should be carried out specifically on in situ trees that are rooted in original growth position on silica-rich substrates, such as on the sinter apron of Cistern Spring (plate 6.1E). The major question is then if in situ trees can take up and accumulate silica, and if the silica does penetrate and replace the organic compounds in the cell wall, if the cell lumina will become infilled with silica, or both. Preliminary studies show that uptake through the roots is possible but negligible in regard to preservation by silica (plate 6.1F) (Liesegang and Gee 2020). To what extent this pathway from the roots through the axial vascular cells of the tree is responsible for mineralizing cells is unclear. Other pathways, such as through the barkless outer surface of the trunk during flooding, may prove to be a more effective way to mineralize a standing tree (Liesegang and Gee 2020).

Silica Diagenesis

It is general consensus in plant silicification studies that, during the early stage of the silicification process, significant amounts of noncrystalline silica form, which impregnate and infill organic material (Scurfield and Segnit 1984; Mustoe 2008; Dietrich et al. 2013). During diagenesis, this early precipitate is commonly thought to transform into more crystalline phases, from opal-CT to opal-C, then to quartz, with each phase represented by different crystal sizes and growth morphologies (Liesegang et al., chap. 7). In contrast, the impact of silica diagenesis on the integrity of plant material has received little attention so far.

The volume loss and micromorphological transformation of silica phases from opal-A to quartz may severely modify the stability of cellular detail and thus dictate its preservation potential. To maintain an intact cellular structure, this dissolution–reprecipitation process requires a constant influx of silica-rich fluid to balance the inherent volume loss.

Further experimental work with different fluids at variable temperatures and pressures will be useful to understand the diagenetic pathway of silica and its interaction with secondary cell walls. The starting fluids could vary in their silicon concentration and amount of other relevant ions, such as sodium, calcium, or aluminum. Precise concentration measurements would allow us to determine exactly the fluid chemical composition before and after interaction with a wood substrate to determine what and how much is taken up into the wood. Silica phase solubility is significantly modified by differences in temperature, and experiments under controlled temperatures below 100°C, which are assumed to occur during wood silicification in natural settings, can be run in the laboratory as analogs to specific natural conditions. Such experimental work would not only shed light on the wood silicification process in general but would also provide important insights into the modification of organic matter over geological time.

The Role of Tree Bark in Silicification

In what is commonly called "petrified wood" in the rock record, the plant tissue under study is usually silicified secondary xylem, although recently primary plant tissues have received some attention as well (Channing and Edwards 2004, 2009; Läbe et al. 2012). However, as far as it is known, no experimental work has been done on the silicification of bark. Although there have been a number of reports on silicified or mineralized bark of conifers or other plants in the literature, these are mainly limited to anatomical descriptions (Scott 1902, 1924; Ramanujam and Stewart 1969; Taylor 1988; Galtier and Scott 1991; Meyer-Berthaud and Taylor 1991; Galtier et al. 1993; Artabe et al. 1999; Ash and Savidge 2004; Decombeix et al. 2010; Bomfleur et al. 2013; Decombeix 2013; Decombeix and Meyer-Berthaud 2013; Slater et al. 2015; Degani-Schmidt and Guerra-Sommer 2016; Yang et al. 2017).

In fact, the absence of bark on allochthonous woody axes—logs, branches, twigs—may prove to be the key to the preservation potential of anatomical structures in wood. *Bark* is a nontechnical, collective term for the peripheral tissues of a woody stem on the outer side of the vascular cambium. Starting at its interface with the vascular cambium, it consists of the phloem, cortex, phelloderm, cork cambium (phellogen), and cork (phellem).

Of particular interest to mineralization is the outermost tissue, the cork. The cell walls of cork contain suberin, a waxy substance that protects the stem against water loss, as well as against the invasion of insects into the stem and infection by bacterial and fungal spores. If the suberized cell walls of the cork prevent water from moving out of the wood into the environment, they similarly keep water from entering into the tree's core from the environment. Thus, water can only pass into the secondary xylem of a woody axis rapidly if it is decorticated (barkless).

Whether there is a difference in preservational quality in the wood of fossil conifer trees preserved with bark and those found without bark has not been studied. It is also unknown whether trunks with well-preserved bark and wood were initially silicified in their original growth position with intact root systems, or whether well-preserved bark and wood can also be found in allochthonous logs.

There are differences in the thickness of bark in different tree species. For example, total bark thickness in conifers tends to be thicker than in angiosperms. In a comparison of bark thickness in 11 tree species in two habitats, wetland bog and savannah, it was found that the *Pinus* species had thicker bark than the hardwood species in the same habitat (Schafer et al.

2015). There were also differences between bark thickness among the nine angiosperms. In this case, a thicker bark was thought to be a better adaptation for fire protection (Schafer et al. 2015). However, following this argument, thicker bark may also pose a greater impediment to silicification in allochthonous logs and branches than in decorticated woody limbs.

Thus, experimental work in the laboratory with highly concentrated aqueous silica flowing through woody axes with and without bark should be undertaken to decipher the role of cortication and decortication in wood silicification. The direction of flow through a woody axis, whether parallel or perpendicular to the axial cells, is also important. Since woods are so variable in the abundance, thickness, and arrangement of bark, a series of experimental taphonomic studies should be carried out on woody axes with different bark thicknesses and morphologies.

Conclusions

The experimental silicification of wood, whether it be in the field or in the laboratory, transcends a mere replication of the silicification process that occurs in nature. In being able to produce mineralization by silica in wood in the laboratory under specific and controlled conditions, such as under certain temperatures or with different combinations of starting materials, we can sort out experimentally the taphonomical processes that act on wood during the delicate balance between decay and preservation, as well as determine the optimal and suboptimal conditions for plant silicification. Because incipient silicification has been produced in both primary and secondary plant tissue in the lab in recent years, the thrust of experimentation should move on to other open questions, such as those focusing on the cellular and tissue pathways of aqueous silica in the plant body, the diagenesis of silica in cell walls, and the effect of bark on wood silicification. The ultimate goal of experimental silicification, whether in the lab or field, is to understand the sedimentological, mineralogical, chemical, and plant anatomical conditions influencing the varying degrees of cellular and subcellular preservation in wood in the fossil record, in allochthonous as well as autochthonous occurrences.

ACKNOWLEDGMENTS

The authors are grateful to reviewer George Mustoe (Western Washington University, USA) for helpful discussions, as well as for the use of several photos. This work was funded by the Deutsche Forschungsgemeinschaft (DFG, German Research

Foundation), Project number 39676817 to CTG and Chris Ballhaus (both at the University of Bonn). This is contribution number 20 of the DFG Research Unit 2685, "The Limits of the Fossil Record: Analytical and Experimental Approaches to Fossilization."

WORKS CITED

Akahane, H., Furuno, T., Miyajima, H., Yoshikawa, T., and Yamamoto, S. 2004. Rapid wood silicification in hot spring water: An explanation of silicification of wood during the Earth's history. *Sedimentary Geology*, 169: 219–228.

Artabe, A.E., Brea, M., and Zamuner, A.B. 1999. *Rhexoxylon brunoi* Artabe, Brea et Zamuner, sp. nov., a new Triassic corystosperm from the Paramillo de Uspallata, Mendoza, Argentina. *Review of Palaeobotany and Palynology*, 105: 63–74.

Ash, S.R., and Savidge, R.A. 2004. The bark of the Late Triassic *Araucarioxylon arizonicum* tree from Petrified Forest National Park, Arizona. *International Association of Wood Anatomists Journal*, 25: 349–368.

Ballhaus, C., Gee, C.T., Bockrath, C., Greef, K., Mansfeldt, T., and Rhede, D. 2012. The silicification of trees in volcanic ash: An experimental study. *Geochimica et Cosmochimica Acta*, 84: 62–74.

Bomfleur, B., Decombeix, A.-L., Escapa, I.H., Schwendemann, A.B., and Axsmith, B. 2013. Whole-plant concept and environment reconstruction of a *Telemachus* conifer (Voltziales) from the Triassic of Antarctica. *International Journal of Plant Sciences*, 174: 425–444.

Buurman, P. 1972. Mineralization of wood. *Scripta Geologica*, 12: 1–43.

Channing, A., and Edwards, D. 2004. Experimental taphonomy: Silicification of plants in Yellowstone hot-spring environments. *Transactions of the Royal Society of Edinburgh, Earth Sciences*, 94: 503–521.

Channing, A., and Edwards, D. 2009. Silicification of higher plants in geothermally influenced wetlands: Yellowstone as a Lower Devonian Rhynie analog. *Palaios*, 24: 505–521.

Dawson, T.E. 1998. Fog in the California redwood forest: Ecosystem inputs and use by plants. *Oecologia*, 117: 476–485.

Decombeix, A.-L. 2013. Bark anatomy of an Early Carboniferous tree from Australia. *International Association of Wood Anatomists Journal*, 34: 183–196.

Decombeix, A.-L., and Meyer-Berthaud, B. 2013. A *Callixylon* (Archaeopteridales, Progymnospermopsida) trunk with preserved secondary phloem from the Late Devonian of Morocco. *American Journal of Botany*, 100: 2219–2230.

Decombeix, A.-L., Taylor, E.L., and Taylor, T.N. 2010. Anatomy and affinities of permineralized gymnospermous trunks with preserved bark from the Middle Triassic of Antarctica. *Review of Palaeobotany and Palynology*, 163: 26–34.

Degani-Schmidt, I., and Guerra-Sommer, M. 2016. Charcoalified *Agathoxylon*-type wood with preserved secondary phloem from the lower Permian of the Brazilian Parana basin. *Review of Palaeobotany and Palynology*, 226: 20–29.

Dietrich, D., Lampke, T., and Rößler, R. 2013. A microstructure study on silicified wood from the Permian Petrified Forest of Chemnitz. *Paläontologische Zeitschrift*, 87: 397–407.

Dietrich, D., Viney, M., and Lampke, T. 2015. Petrifactions and wood-templated ceramics: Comparisons between natural and artificial silicification. *IAWA Journal*, 36: 17–185.

Domec, J.C., and Gartner, B.L. 2002. How do water transport and water storage differ in coniferous earlywood and latewood? *Journal of Experimental Botany*, 53: 2369–2379.

Drum, R.W. 1968a. Petrification of plant tissue in the laboratory. *Nature*, 218: 784–785.

Drum, R.W. 1968b. Silicification of *Betula* wood tissue in vitro. *Science*, 161: 175–176.

Fleming, B.A., and Crerar, D.A. 1982. Silicic acid ionization and calculation of silica solubility at elevated temperature and pH. *Geothermics*, 11: 15–29.

Galtier, J., Brown, R.E., Scott, A.C., Rex, G.M, and Rowe, N.P. 1993. A Late Dinantian flora from Weaklaw, East Lothian, Scotland. *Palaeontology Special Papers*, 49: 57–74.

Galtier, J., and Scott, A.C. 1991. *Stanwoodia*, a new genus of probable early gymnosperms from the Dinantian of East Kirkton, Scotland. *Transactions of the Royal Society of Edinburgh, Earth Sciences*, 82: 113–123.

Gee, C.T., Sprinkel, D.A., Bennis, M.B., and Gray, D.E. 2019. Silicified logs of *Agathoxylon hoodii* (Tidwell et Medlyn) comb. nov. from Rainbow Draw, near Dinosaur National Monument, Uintah County, Utah, USA, and their implications for araucariaceous conifer forests in the Upper Jurassic Morrison Formation. *Geology of the Intermountain West*, 6: 77–92.

Götze, J., Möckel, R., Langhof, N., Hengst, M., and Klinger, M. 2008. Silicification of wood in the laboratory. *Ceramics-Silikáty*, 52: 268–277.

Green, S., Vogeler, I., Clothier, B.E., Mills, T.M., and van den Dijssel, C. 2003. Modelling water uptake by a mature apple tree. *Australian Journal of Soil Research*, 41: 365–380.

Hellawell, J., Ballhaus, C., Gee, C.T., Mustoe, G.E., Nagel, T.J., Wirth, R., Rethemeyer, J., Tomaschek, F., Geisler, T., Greef, K., and Mansfeldt, T. 2015. Silicification of recent conifer wood at a Yellowstone hot spring. *Geochimica et Cosmochimica Acta*, 149: 79– 87.

Jefferson, T.H. 1987. The preservation of conifer wood: Examples from the Lower Cretaceous of Antarctica. *Palaeontology*, 30: 233–249.

Läbe, S., Gee, C.T., Ballhaus, C., and Nagel, T. 2012. Experimental silicification of the tree fern *Dicksonia antarctica* at high temperature with silica-enriched H_2O vapor. *Palaios*, 27: 835–841.

Leo, R.F., and Barghoorn, E.S. 1976. Silicification of wood. *Botanical Museum Leaflets, Harvard University*, 25: 1–33.

Liesegang, M., and Gee, C.T. 2020. Silica entry and accumulation in standing trees in a hot-spring environment: Cellular pathways, rapid pace and fossilization potential. *Palaeontology*, 63: 651–660.

Ma, J.F. 2005. Plant root responses to three abundant soil minerals: Silicon, aluminum and iron. *Critical Reviews in Plant Sciences*, 24: 267–281.

Meyer, F.A.A. 1791. Briefe über einige mineralogische Gegenstände an Herrn Peter Camper. Johann Christian Dieterich, Göttingen.

Meyer-Berthaud, B., and Taylor, T.N. 1991. A probable conifer with podocarpacean affinities from the Triassic of Antarctica. *Review of Palaeobotany and Palynology*, 67: 179–198.

Muir, M.D. 1970. A new approach to the study of fossil wood. *Proceedings of the Third Annual Electron Microscope Symposium*, Chicago, 129–136.

Mustoe, G.E. 2008. Mineralogy and geochemistry of late Eocene silicified wood from Florissant Fossil Beds National Monument, Colorado. In: Meyer, H.W., and Smith, D.M., eds. *Paleontology of the Upper Eocene Florissant Formation, Colorado. Geological Society of America Special Paper*, 435: 127–140.

Mustoe, G.E. 2015. Late Tertiary petrified wood from Nevada, USA: Evidence of multiple silicification pathways. *Geosciences*, 5: 286–309.

Mustoe, G.E. 2016. Density and loss on ignition as indicators of the fossilization of silicified wood. *International Association of Wood Anatomist Journal*, 37: 98–111.

Mustoe, G.E. 2017. Wood petrifaction: A new view of permineralization and replacement. *Geosciences*, 7: 1–17.

Mustoe, G.E. 2018. Mineralogy of non-silicified wood. *Geosciences*, 8: 11–32.

Nobel, P.S. 2009. *Physicochemical and Environmental Plant Physiology*, 4th ed. Academic Press, San Diego.

Persson, P.V., Fogden, A., Hafrén, J., Daniel, G., and Iversen, T. 2004a. Silica-cast replicas for morphology studies on spruce and birch xylem. *International Association of Wood Anatomists Journal*, 25: 155–164.

Persson, P.V., Hafrén, J., Fogden, A., Daniel, G., and Iversen, T. 2004b. Silica nanocasts of wood fibers: A study of cell-wall accessibility and structure. *Biomacromolecules*, 5: 1097–1101.

Ramanujam, C.G.K., and Stewart, W.N. 1969. Taxodiaceous bark from the Upper Cretaceous of Alberta. *American Journal of Botany*, 56: 101–107.

Rößler, R., Zierold, T., Feng, Z., Kretzschmar, R., Merbitz, M., Annacker, V., and Schneider, J.W. 2012. A snapshot of an early Permian ecosystem preserved by explosive volcanism: New results from the Petrified Forest of Chemnitz. *Palaios*, 27: 814–834.

Saka, S., and Ueno, T. 1997. Several SiO_2 wood-inorganic composites and their fire-resisting properties. *Wood Science and Technology*, 31: 457–466.

Santana, M.A.E., Rodrigues, L.C., Coradin, V.T.R., Okino, E.Y.A., and de Souza, M.R. 2013. Silica content of 36 Brazilian tropical wood species. *Holzforschung*, 67: 19–24.

Schafer, J.L., Breslow, B.P., Hohmann, M.G., and Hoffmann, W.A. 2015. Relative bark thickness is correlated with tree species distributions along a fire frequency gradient. *Fire Ecology*, 11: 74–87. doi:10.4996/fireecology.1101074.

Scott, D.H. 1902. On the primary structure of certain Palaeozoic stems with the *Dadoxylon* type of wood. *Transactions of the Royal Society of Edinburgh*, 40: 331–365.

Scott, D.H. 1924. Fossil plants of the *Calamopitys* type, from the Carboniferous rocks of Scotland. *Transactions of the Royal Society of Edinburgh*, 53: 569–596.

Scurfield, G., Anderson, C.A., and Segnit, E.R. 1974. Silica in woody stems. *Australian Journal of Botany*, 22: 211–229.

Scurfield, G., and Segnit, E.R. 1984. Petrifaction of wood by silica minerals. *Sedimentary Geology*, 39: 149–167.

Selden, P., and Nudds, J. 2012. *Evolution of Fossil Ecosystems*, 2nd ed. Manson Publishing Ltd., London.

Shin, Y., Liu, J., Chang, J.H., Nie, Z., and Exarhos, G.J. 2001. Hierarchically ordered ceramics through surfactant-templated sol-gel mineralization of biological cellular structures. *Advanced Materials*, 13: 728–732.

Slater, B.J., McLoughlin, S., and Hilton, J. 2015. A high-latitude Gondwanan lagerstätte:

The Permian permineralised peat biota of the Prince Charles Mountains, Antarctica. *Gondwana Research*, 27: 1446–1473.

St. John, R. 1927. Replacement vs. impregnation in petrified woods. *Journal of Geology*, 35: 729–739.

Taylor, E.L. 1988. Secondary phloem anatomy in cordaitean axes. *American Journal of Botany*, 75: 1655–1666.

Taylor, T.N., Kerp, H., and Hass, H. 2005. Life history biology of early land plants: Deciphering the gametophyte phase. *Proceedings of the National Academy of Science*, 102: 5892–5897.

Vail, J.G. 1928. *Soluble Silicates in Industry*. Chemical Catalog Co., New York, 443 pp.

Williams, L.A., Parks, G.A., and Crerar, D.A. 1985. Silica diagenesis, I. Solubility controls. *Journal of Sedimentary Petrology*, 55: 301–311.

Yang, J.Y., Shen, J.J., Xu, X., Chen, Y.X., Wei, H.B., Kerp, H., and Feng, Z. 2017. The bark anatomy of *Ningxiaites specialis* from the Permian of China. *Review of Palaeobotany and Palynology*, 240: 11–21.

Zabel, R.A., and Morrell, J.J. 2012. *Wood Microbiology: Decay and its Prevention*. Academic Press, San Diego.

Zollfrank, C., and Van Opdenbosch, D. 2018. Biotemplating principles, pp. 19–52. In: Yang, G., Xiao, L., and Lamboni, L., eds. *Bioinspired Materials Science and Engineering*. John Wiley & Sons, Inc., Hoboken, New Jersey, USA.

CHAPTER **7**

The Structure and Chemistry of Silica in Mineralized Wood

Techniques and Analysis

MORITZ LIESEGANG, FRANK TOMASCHEK, AND JENS GÖTZE

A B S T R A C T | Silicification of wood results in the precise preservation of its cellular details and provides an exceptional window into the anatomy, life history, and evolution of land plants across geological timescales. The wood silicification process can be accompanied by substantial changes in silica structure and chemistry that can record the environmental conditions during fossilization and diagenesis. Detailed knowledge about early fluid–wood interaction and all subsequent diagenetic steps is essential to understanding and reconstructing the complex mineralization process of wood by silica minerals. Here we review established and modern analytical methods that can be used to characterize silicified wood from bulk to atomic dimensions. Modern integrative approaches can potentially reveal insights into the silicification process that may proceed continuously from an amorphous precipitate to the crystalline end-product quartz, but can also include overprint episodes deviating from this classical diagenetic pathway. Using state-of-the-art instrumentation for silicified wood analysis is an effective means to reconstruct plant evolution in the geological past. |

Introduction

The silicification of plant material is the major preservation process for preserving delicate cellular detail in three dimensions. Despite its simple, idealized composition of $SiO_2 \cdot nH_2O$, silica has an extensive range of micromorphologies, structural characteristics, and chemical compositions (Iler 1979; Scurfield and Segnit 1984; Flörke et al. 1991; Graetsch 1994; Rodgers et al. 2004; Gaillou et al. 2008a, b; Mustoe 2008; Liesegang and Milke 2014; Liesegang et al. 2018). It is compositionally variable due to the amount of water bound to it and because of substitution of Si^{4+} by Al^{3+} or Fe^{3+}, which is compensated by the incorporation of mono- and divalent cations and OH groups (Flörke et al. 1982; Graetsch et al. 1985; Gaillou et al. 2008a; Liese-gang and Milke 2014; Chauviré et al. 2019). The high potential for the pres-ervation of plant structure and organic matter is largely due to the chemical and physical stability of the common crystalline end-product quartz of the silicification process in wood.

Silicification is a widespread process in surficial environments and, ac-cordingly, an in-depth understanding of the silicification of wood benefits from the extensive knowledge gained from decades of studies on natural and synthetic silica. The nature and terminology of the silica phases in silici-fied wood are, for example, identical to those formed in hot-spring and deep-sea deposits, and to silcrete and alteration products of volcanic rocks and deep-weathering environments (Jones et al. 1964; Jones and Segnit 1971; Flörke et al. 1976; Williams et al. 1985; Hesse 1989; Lynne et al. 2005; Thiry et al. 2006; Gaillou et al. 2008b; Liesegang and Milke 2014; Lowenstern et al. 2018). The description of silica minerals in mineralized wood is adopted from these previous studies without modification.

It has been pointed out that evidence for a silica mineralogical transfor-mation from opal-A to opal-CT to quartz during wood fossilization comes from siliceous sediments (e.g., Mizutani 1970, 1977; Kastner et al. 1977; Wil-liams et al. 1985) and hot-spring sinters (e.g., Rodgers et al. 2004 and refer-ences therein; Lynne et al. 2005), and is largely indirect (Mustoe 2008). Environmental conditions such as fluid chemistry, host-rock mineralogy, temperature, and pressure control may significantly alter this pathway (Mizu-tani 1970; Kastner et al. 1977; Isaacs 1982; Hinman 1998). The coexistence of different silica phases in silicified wood is most likely an expression of spatially and temporally variable concentrations of dissolved silica and the subsequent direct precipitation of silica minerals with different aqueous solubility (Scurfield and Segnit 1984; Mustoe 2008). A strict correlation be-tween age and silica maturity may not be universally valid, for example, in the light of the occurrence of opal-A silicified wood in Lower Cretaceous strata (Scurfield and Segnit 1984; Liesegang and Milke 2014). To date, no novel silica phase, growth morphology, or chemical composition unique to silicified wood has been identified.

Silicified wood is traditionally characterized and categorized by means of transmitted light microscopy. As pure silica phases are commonly trans-parent in thin section, plane-polarized light observation is of limited use for phase identification and deduction of the silicification process. However, observations between crossed polarizers reveal a multitude of characteristic birefringence phenomena and growth fabrics (Scurfield and Segnit 1984; Flörke et al. 1991; Graetsch 1994; Mustoe 2008; Nesse 2013; Liesegang and

Milke 2018). Other valuable analytical tools include observational methods such as cathodoluminescence and high-resolution scanning electron microscopy (SEM), as well as analytical methods such as X-ray diffraction, Raman spectroscopy, electron probe microanalysis, and laser ablation mass spectrometry. In other branches of the geosciences, these well-established analytical methods are regularly used to characterize the structural state and chemical properties of silica to reconstruct its formation pathway and the temporal and spatial changes of environmental conditions preserved in specific phases. Data from these analyses assist in reconstructing the processes, as well as the temporal and spatial changes in environmental conditions, that led to its formation. To date, most analytical methods in silicified wood research have been used to obtain only a qualitative fingerprint of the intricate silicification process and, in particular, the coupling between the wood organic substance and the silica precipitation and diagenesis. Incipient silicification can be studied with silicification experiments that result in the precipitation of noncrystalline silica (Leo and Barghoorn 1976; Götze et al. 2008; Gee and Liesegang, chap. 6). However, most of the silicified wood in the fossil record consists of α-quartz with variable structure and crystallite size, or poorly crystalline low-tridymite/cristobalite intergrowths (opal-CT) that may result from diagenetic silica maturation or a different initial precipitate (Scurfield and Segnit 1984; Mustoe 2008). Accordingly, any attempt to reconstruct the wood silicification process from these minerals requires the precise identification of their structural and chemical properties. The review presented here focuses on analytical techniques and the data they generate to characterize silicified wood and potentially reconstruct the paleoenvironmental conditions that formed these exceptional fossils.

Mineralogical Characterization

X-Ray Diffraction Analysis (XRD)

X-ray diffraction (XRD) analysis of silicified wood is the primary method for the bulk qualitative determination of silica phases, their structural state, and co-occurring minerals (Buurman 1972; Mitchell and Tufts 1973; Sigleo 1978; Scurfield 1979; Stein 1982; Scurfield and Segnit 1984; Jefferson 1987; Weibel 1996; Matysová et al. 2010; Hellawell et al. 2015; Mustoe 2015). X-rays are electromagnetic radiation with wavelengths of 0.5–2.5 Å. This wavelength is of the same order of magnitude as the interatomic distances in a crystal and, hence, the elastic scattering of X-rays can be used to study the atomic structure of a sample. In an ideal crystal, the distribution of atoms is perfectly periodic in three dimensions. The periodicity of this structure can be expressed as a lattice of geometric points, and planes in various geometric orientations can be inscribed into this three-dimensional arrangement. These planes are referred to as lattice planes, which are indexed using the Miller indices *hkl*. The distance between parallel equidistant planes is the *d*-value (d_{hkl}), which is usually given in angstroms (Å, 10^{-10} m).

A typical laboratory diffractometer used for XRD analysis is equipped with an X-ray tube containing a metal target (e.g., Cu, Mo, or W) emitting X-rays, a sample stage, and a detector. A variety of slits, filters, and monochromators are introduced into the incident and diffracted beam path to collimate, limit, and direct the beam, and to remove unwanted interfering components. When the incident beam hits the sample, the monochromatic X-rays of a known wavelength are scattered by the atoms in the crystal. These X-rays are scattered in all directions, but in some directions, constructive interference results in an increased intensity. The Bragg equation $n\lambda = 2d \cdot \sin(\theta)$ relates the X-ray wavelength λ of the incident beam, the interplanar distance *d* separating the lattice planes, the angle θ of incidence, and the positive integer value *n*, which is the order of reflection.

When the Bragg equation is satisfied, scattered X-ray beams are in phase and interfere constructively. At a constant wavelength of incident X-rays, a series of lattice planes produces diffraction peaks at a particular angle, θ. The resultant diffraction pattern (diffractogram) is unique for every substance and commonly serves as a fingerprint for mineral phases in silicified wood (fig. 7.1). The integrated mean size of the coherently scattering domains (crystallites) can be calculated using the Scherrer equation, which includes the diffraction angle, X-ray wavelength, peak width at half-maximum intensity, and a shape factor (Langford and Wilson 1978). Powdered samples are commonly prepared for the XRD analysis of silicified wood. Nondestructive XRD on polished surfaces is possible with focusing optics that allow the analysis of sample regions smaller than 100 μm (Liesegang et al. 2018).

The most frequently identified silica phase in silicified wood is trigonal α-quartz with a structure ranging from granular to fibrous (chalcedony). Micro- and macrocrystalline quartz can be readily distinguished from other low-temperature silica phases by their sharp and intense peaks at about 3.34, 4.26, and 1.82 Å. A quartz crystallinity index (CI) (Murata and Norman 1976) is occasionally calculated from the relative peak intensities of a multiplet located at ~1.37 Å in an attempt to identify the structural maturity of silicified wood (Stein 1982; Mustoe 2008). Diffractograms of chalce-

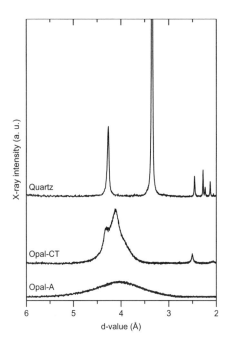

Figure 7.1. X-ray diffractograms of opal-A, opal-CT, and quartz—silica phases commonly observed in silicified wood. The diffractograms of opal phases display broad peaks, which result from peak overlap, structural disorder, and an overall small size of coherently scattering domains. Image: ML.

dony exhibit broader peaks at quartz positions and may include weak additional reflections of the silica phase moganite (Flörke et al. 1991; Graetsch et al. 1994). Due to the lower sensitivity of XRD analysis to short-range order, Raman spectroscopy is a more reliable technique to detect the amount and spatial distribution of the moganite component in chalcedony (Götze et al. 1998; Witke et al. 2004).

Compared with quartz, the diffractograms of opal phases show very broad peaks resulting from the overlap of multiple reflections, high structural defect density, and small size of coherently scattering domains. In quartz-dominated silicified wood, the high intensity quartz reflections and the XRD detection limit of ~1 vol.% may cover reflections of coexisting opal and prevent its identification. Jones and Segnit (1971) elaborated on the most commonly used mineralogical opal classification based on XRD analyses. This scheme distinguishes between three low-temperature typologies: amorphous opal (opal-A), disordered low-cristobalite/low-tridymite (opal-CT), and the more ordered low-cristobalite with minor low-tridymite stacking (opal-C).

The diffractogram of opal-A shows a single broad peak centered at approximately 4 Å, indicating a highly disordered structure. Occasionally, a broad peak of low intensity is located at ~2 Å. Systematic changes of the main peak position and asymmetry suggest nucleation and growth of low-tridymite-rich nanodomains in a noncrystalline matrix during early diagenesis (Liesegang et al. 2018). Opal-CT produces diffraction patterns that comprise broad overlapping reflections of low-tridymite at ~4.3 and 4.11 Å, low-cristobalite at 4.04 Å, and a disordered matrix (Jones and Segnit 1971; De Jong et al. 1987; Liesegang and Tomaschek 2020). Opal-CT is therefore commonly interpreted as representing a continuous series of disordered intergrowths between end-member low-cristobalite and low-tridymite stacking sequences. During the diagenetic transformation of opal-CT to opal-C, the main peak maximum shifts from a low-tridymite to a low-cristobalite position, reflecting the elimination of tridymite stacking in favor of cristobalite stacking. The XRD pattern of opal-C resembles low-cristobalite and shows more intense and narrow peaks than that of opal-CT. Opal-C is characterized by a major sharp peak between about 4.06 to 4.04 Å and by the presence of two other peaks at 2.85 and 3.14 Å.

Raman Spectroscopy

Raman spectroscopy is a versatile tool for the identification of mineral phases and their structural state or orientation at the microscale (Smith and Dent 2019; Geisler and Menneken, chap. 4). The technique is based on the Raman effect, that is, the inelastic scattering of light due to the interactions of light with matter. A typical instrument consists of a light source, a microscope with confocal optics, and a detector system. A spot on the sample is irradiated with a monochromatic laser beam, energy is transferred to elevate molecular vibrations in the material, and the inelastically scattered light is detected. The spot size is approximately 1 μm when a 100x objective is used. The recorded Raman spectra depict the shift in energy of the light, relative to that of the laser. Individual band positions are characteristic of the material; the shape and intensity of the bands depend on structural properties and crystal orientation. In Chapter 4 of this book, Geisler and Menneken provide a comprehensive review of the Raman techniques with a focus on investigations of fossilization processes. Raman spectroscopy is nondestructive and requires little sample preparation, however, polished slides are preferred.

Raman spectroscopy has been employed in the characterization of non- to microcrystalline silica (Kingma and Hemley 1994; Götze et al. 1998; Rodgers and Hampton 2003; Ilieva et al. 2007; Schmidt et al. 2013), as well as to the analysis of silicified wood (Dietrich et al. 2001; Witke et al. 2004; Saminpanya and Sutherland 2013; Viney et al. 2016).

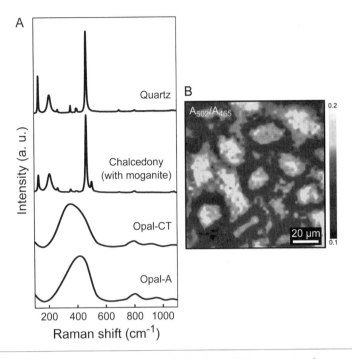

Figure 7.2. Raman single spectra and mapping of silica phases. (**A**) Raman spectra of common silica phases. Note that a moganite component in chalcedony leads to an additional band at 502 cm^{-1}, besides the main band of quartz at 465 cm^{-1}. (**B**) Raman map of silica phases in silicified wood from the Upper Jurassic Morrison Formation, Escalante Petrified Forest State Park in Utah, USA. The map displays the integral ratios of the main moganite versus quartz bands (A_{502}/A_{465}), tracing the cellular wood structure (dark cell walls). Images: FT.

The extensive range of silica phases and their structural state (Flörke et al. 1991; Graetsch 1994) can be distinguished based on their characteristic band positions, band width (high/low crystallinity), and band intensity ratios (moganite/quartz ratio). Representative spectra of silica phases commonly observed in silicified wood are illustrated in fig. 7.2A. Opal-A has a main broad band at ~400 cm^{-1}, which is located at ~330 cm^{-1} for opal-CT (Smallwood et al. 1997). The structural state of opal-CT, consisting of a disordered interstratification of low-cristobalite and tridymite layers, can be determined from fitting the bands of the respective end members (Ilieva et al. 2007). The most intense band of quartz is located at 465 cm^{-1}, while that of the SiO$_2$-polymorph moganite is located at 502 cm^{-1} (Götze et al. 1998). Moganite is commonly associated with microcrystalline silica varieties, and the moganite component in chalcedony can be quantified from intensity ratios of the main moganite versus quartz bands (Götze et al. 1998; Schmidt et al. 2013).

In addition to silica, inclusions or intergrown mineral phases, which provide clues about the diagenetic environment, can be readily identified from comparing the acquired spectra with reference data. The RRUFF Raman database (Lafuente et al. 2016) is a simple and easily accessible tool for mineral identification. Bands in the 1300–1600 cm^{-1} region identify carbonaceous material in silicified wood (Dietrich et al. 2001; Witke et al. 2004), which may be derived from degrading organic compounds present in recent plant material (Gierlinger et al. 2012). Raman spectroscopy is the only method of those summarized here that allows for a detailed characterization of the relict carbonaceous matter in fossil plant material. Infrared (IR) spectroscopy and pyrolysis–gas chromatography–mass spectrometry (Py-GC-MS) analyses on dissolved bulk material permit a more specific identification of remnant organic compounds in silicified wood (Sigleo 1978).

Raman spectral imaging is an extraordinarily powerful tool for generating detailed maps of the spatial distribution of mineral phases and their structural properties. Such a two-dimensional map is constructed from an extended array of single spectra that are acquired individually. Constructed maps can display the spatial arrangement of different phases, as well as quantitative structural properties, such as the relative proportion of moganite versus quartz in microcrystalline silica varieties. This information can be used to infer aspects of the fluid evolution, for example, the fluid chemistry controlling the degree of polymerization. In addition, the moganite content decreases with increasing structural ordering during progressive diagenesis (Heaney 1995). Accordingly, the moganite/quartz ratio potentially allows for distinguishing between different silica generations.

Fig. 7.2B shows a Raman map of silicified wood from the Upper Jurassic Morrison Formation (Escalante, Utah, USA). In this specimen, structurally distinct silica varieties outline the pattern of a relict wood structure. Notably, chalcedony in the cell walls is low in the moganite component compared to chalcedony in the lumina. Did the initial precipitation on an organic template control the silica structure, or have distinct silica structures formed in response to distinct fluid events? Such questions can be addressed if Raman spectral imaging is combined with other advanced analytical methods, such as cathodoluminescence-imaging and electron probe microanalysis element mapping, providing phase and chemical information with high spatial resolution.

Optical Microscopy

Cathodoluminescence Microscopy and Spectroscopy

Luminescence is a common light-emission phenomenon in solids that results from an emissive transition of electrons from an excited electronic state to a ground or other state with lesser energy (Marfunin 1979). Simply put, luminescence can be regarded as a process of converting various kinds of energy (e.g., ultraviolet, or UV, light = photoluminescence; electron beam = cathodoluminescence) into light. The luminescence process is the result of multiple interactions and can be described based on a scheme of the energy levels within a solid. In insulators and semiconductors, a band gap (forbidden zone) exists between the valence band and the conduction band. Additional energy levels in this forbidden zone can be created due to the existence of luminescence centers (defect centers = activators), such as impurity ions or lattice defects. During excitation with various kinds of energy, electrons can pass from the ground state to an excited state, which is attended by the appearance of an absorption band in the optical spectrum (excitation/absorption). When the electrons return from the excited state to the ground state or the activator level, they can emit the released energy in the form of light. The wavelength of the emitted light (photon energy) depends on the energy difference of the electron transition and is thus characteristic for a certain crystal structure and/or the type of the defect (Götze 2012).

Cathodoluminescence (CL), which is created by electron irradiation of a solid surface, can be observed on a wide variety of analytical instruments using electron beams (e.g., optical microscope, scanning electron microscope, electron microprobe; Götze and Kempe 2008, 2009). Based on different analytical arrangements, CL can be used both in a purely descriptive way to detect and distinguish different minerals or mineral generations by their variable CL colors or as an effective method for spatially resolved analysis of point defects in solids by spectral CL measurements. With this method, the bulk of emitted light is split into the certain wavelengths, and the maxima of emission intensities can be related to specific defects.

The use of special light microscopes equipped with a vacuum chamber and an electron gun (Neuser et al. 1995), as well as scanning and transmission electron microscopes, enables the detailed investigation of minerals and rocks at high magnification and reveals internal structures, different growth generations, or the distribution of trace elements at the microscale, which is not possible using conventional light microscopes (plate 7.1). The spatial resolution depends on the instrument used to excite the CL and is in the range of a few to tens of nanometers with SEM. Commonly, polished thin sections are used for CL microscopy, which allows for alternating observations and measurements in transmitted light or polarized light and CL. In addition, the combination with other advanced analytical methods (e.g., SEM, Raman spectroscopy, local chemical analysis, geochronology) provides useful information for the reconstruction of processes of mineral formation and alteration or the detection of microphases (Götze et al. 2013; Götze and Hanchar 2018).

Quartz and other SiO_2 phases, which are probably the most frequent mineral constituents in mineralized wood, are known to exhibit a multicolored luminescence with different shades of blue, violet, green, yellow, and red (plate 7.1). The type and frequency of luminescence emissions can vary between silica phases of different occurrences. This is due to the fact that the defects causing the different CL emissions often reflect the specific physicochemical conditions of formation. The visible CL colors are caused by varying intensities of several CL emission bands from the blue to the red spectral region and can be related to different structural defects (e.g., oxygen/silicon vacancies or broken bonds) and/or trace elements (e.g., Al, Li, Fe) in the lattice.

The most common emission bands at 450 and 650 nm occur in almost all silica minerals and can be related to lattice defects (e.g., Ramseyer et al. 1988; Götze et al. 2001; Stevens-Kalceff 2009). Another defect-related luminescence is yellow CL (emission band at 570 nm), which appears in macrocrystalline quartz and chalcedony formed under low-temperature hydrothermal conditions (Götze et al. 2015a). In contrast, certain trace elements can activate short-lived blue CL in hydrothermal quartz (Al^{3+}, emission band at 390 nm), green CL in different SiO_2 modifications (uranyl ion UO_2^{2+}; multiband emission around 500 nm), and the typical transient bottle-green CL of pegmatite quartz (Al^{3+}/Li^+; emission band at ca. 500 nm) (Ramseyer et al. 1988; Götze et al. 2001, 2005, 2015a, b; Stevens-Kalceff 2009).

The growing interest in luminescence studies of mineralized wood is based on the fact that information not available by other analytical methods can be obtained. Several studies have been performed in the last two decades emphasizing the advantages of CL in the investigation of silicified wood (e.g., Götze and Rößler 2000; Witke et al. 2004; Götze et al. 2008, 2015a; Matysová et al. 2008, 2010, 2016; Viney et al. 2016; Trümper et al. 2018). These investigations illustrate the utility of CL imaging for the visualization of primary cell structures and the identification of residual phases

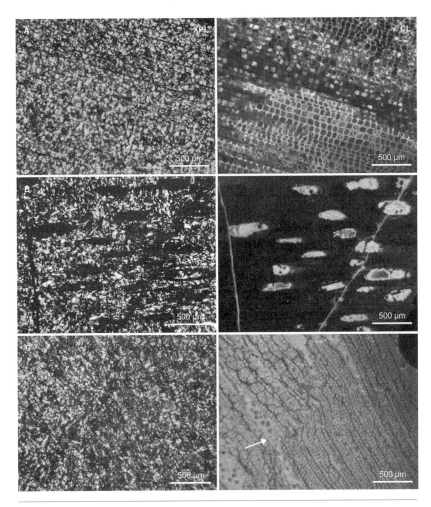

Plate 7.1. Transmitted light micrograph pairs of thin sections of silicified wood from various localities. Abbreviations: XPL = cross-polarized light; CL = cathodoluminescence. (**A**) Cordaite trunk (*Dadoxylon* sp.) from Chemnitz, Germany, showing well-preserved cells from primary silicification (yellow CL) and secondary hydrothermal overprint visible by transient blue CL. Cell walls have a different color than the lumina. (**B**) Miocene silicified wood from Limnos Island, Greece, showing weakly luminescent chalcedony and small voids filled with opal (bright blue CL). (**C**) Violet luminescent silicified wood thin section from the Thuringian Forest, Germany, with preserved cell structure, slight deformation features, and radiation halos (see arrow) around radioactive inclusions. Photos: JG.

and structures, as well as mineral microphases. Moreover, different silica phases could sometimes be identified, resulting in the recognition of multiple processes and episodes of silicification and replacement; one example was the coexistence of chalcedony and opal (plate 7.1B). All these data help to decipher the complex process of wood silicification and show the great potential of CL imaging and spectroscopy in future studies. Best results will be obtained by the combination of CL and other advanced analytical methods that have high sensitivity and local resolution.

Scanning Electron Microscopy (SEM)

SEM is a routinely used observational method to study the micromorphological relationships of silica phases in silicified wood and to determine the preservation of anatomical detail (Buurman 1972; Sigleo 1978; Scurfield 1979; Scurfield and Segnit 1984; Akahane et al. 2004; Dietrich et al. 2013; Hellawell et al. 2015; Mustoe 2017; Pe-Piper et al. 2019; Liesegang and Gee 2020). The spatial relationship of silica minerals, their accompanying phases, and the wood cellular structure can provide evidence of conditions during the silicification process (Matysová et al. 2010; Dietrich et al. 2013; Hellawell et al. 2015; Mustoe 2015; Liesegang and Gee 2020).

The basic components of a scanning electron microscope are an electron gun that generates an electron beam, a column with condenser lenses and scanning coils that produce a fine beam, a high vacuum sample chamber, and specialized detectors. The incident electrons interact with atoms in the sample at different depths, generating the two most-used signals for silicified wood analysis—secondary electrons (SE) from inelastic scattering and backscattered electrons (BSE) from elastic scattering. The interaction volume increases with the voltage and decreases with the atomic number Z of the studied material.

Low-energy SE are emitted from the sample surface and a few nanometers below. When the incident beam hits the sample, electrons penetrate the electron cloud around atoms of the sample. An energy transfer between electrons from the beam and electrons around atoms causes the ejection of an outer-shell electron that becomes a detectable secondary electron. The resultant SE signal yields information about the sample morphology and surface topography.

BSE result from elastic collisions of electrons with atoms and probe a larger sample volume than SE. The incident electrons pass close to atomic nuclei, change their trajectory, and some are scattered back to the sample surface. The number of the BSE reaching the detector correlates positively with the atomic number. Hence, mineral grains containing heavier atoms appear brighter in BSE images. This dependence allows differentiation between mineral phases in silicified wood, for example, impurity phases such as iron oxides or fluorite (Matysová et al. 2010; Dietrich et al. 2013), and visual identification of compositional heterogeneity within single grains.

For conventional imaging under vacuum, the sample surface must be electrically conductive to prevent build-up of electrostatic charge. This is commonly accomplished with a thin conductive coating of carbon, gold, or palladium, for example. Carbon is a suitable coating material for BSE imaging, whereas heavier elements can cover this sample signal. When a coating is not desired, samples can be analyzed with an environmental scanning electron microscope or a variable pressure scanning electron microscope. These SEMs permit usage of a gaseous environment at variable pressure to analyze uncoated, insulating, or wet samples (Goldstein et al. 2018).

For silica phase identification with secondary electron imaging, typical growth morphologies are commonly selected (fig. 7.3). Characteristic growth

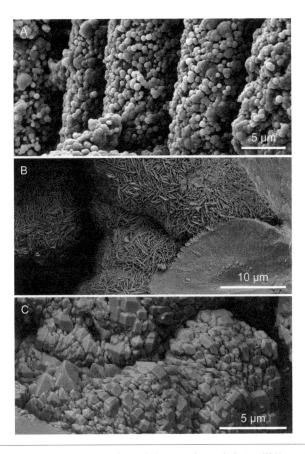

Figure 7.3. Secondary electron images of typical silica growth morphologies. (**A**) Non-uniform opal-A spheres covering and replacing the tracheids of coniferous wood from the Miocene Virgin Valley Formation in northern Nevada, USA. Photo courtesy of George Mustoe. (**B**) Opal-CT lepispheres grown on a partially dissolved quartz substrate (lower right) in chertified limestone from Andamooka, South Australia. Photo: ML. (**C**) Microcrystalline quartz in a cavity in the tightly cemented, microcline-rich Ordovician sandstone of the Mintabie beds, South Australia. Photo: ML.

fabrics can be identified on cell walls and other structures. In this microenvironment, quartz usually forms subhedral to euhedral prismatic crystals. Fine fibers form the bundles and radiating agate-like structures of chalcedony in silicified wood. Both chalcedony and granular quartz have been reported to traverse several cells (Scurfield and Segnit 1984). Inside cell walls, the fiber orientation of chalcedony may coincide with that of cellulose microfibrils (Scurfield and Segnit 1984).

Opaline silicas can also be characterized by their growth fabrics. Opal-A commonly occurs as aggregates of silica spheres with variable size distribution and degree of three-dimensional stacking. Coherently ordered sphere stacks, such as in precious opal (Jones et al. 1964; Liesegang et al. 2017), are rarely observed. Opal-CT is commonly deposited as mass aggregate or parallel bundles of fibers, whereas unobstructed growth can result in spherical aggregates (lepispheres) of discrete intergrown platelets (Flörke et al. 1976; Scurfield and Segnit 1984). Despite the often-distinct silica micromorphology that develops in silicified wood, definite identification requires a mineralogical characterization with analytical methods such as XRD or Raman spectroscopy.

Chemical Characterization

Electron Probe Microanalysis (EPMA) and Wavelength- and Energy-Dispersive X-ray Spectroscopy (WDS/EDS)

Electron probe microanalysis (EPMA) is used for the chemical characterization of inclusion-free sample areas with high spatial resolution (Goldstein et al. 2018). This analysis is usually combined with BSE imaging. Two analytical modes are commonly utilized for silicified wood analysis and spot analysis and, more rarely, element distribution mapping (Sigleo 1978; Boyce et al. 2001; Matysová et al. 2010, 2016; Hellawell et al. 2015; Pe-Piper et al. 2019; Liesegang and Gee 2020). With conventional EPMA, the chemical composition of silica and its associated minerals can be determined at a beam diameter as low as one micrometer. Structures or mineral grains too small to measure with a conventional setup can be analyzed with a field-emission gun (FEG) electron microprobe where the spatial resolution is superior by about a factor of five (Rinaldi and Llovet 2015).

Modern electron microprobes are commonly equipped with up to five spectrometers that simultaneously collect sample data. For wavelength-dispersive X-ray spectroscopy (WDS), a crystal with known d-spacing is positioned in each spectrometer. In the sample, the incident electron beam interacts with the atoms, resulting in the ejection of an inner shell electron. When the atoms relax to their ground state, they emit X-rays with characteristic wavelengths for each element. At an element-specific reflection angle of the analyzing crystals, the Bragg equation is satisfied and the X-rays are directed to and counted by the respective X-ray detectors. Comparison with well-characterized reference material of known composition allows the quantitative analysis of the sample with unknown chemical composition. Adequate measurements require flat, conductive sample surfaces that are usually obtained by polishing thin sections or round mounts. Nondestructive chemical analysis (without sawing or cracking) yielding similar results, which is particularly useful for holotype material, is possible with other methods, such as synchrotron rapid scanning X-ray fluorescence (SRS-XRF) (Edwards et al. 2014).

While WDS is the preferred method for quantitative trace and light element analysis, energy-dispersive X-ray spectroscopy (EDS) analysis is a very fast means to qualitatively and semiquantitatively determine the chemical composition of silicified wood at high spatial resolution (Jefferson 1987; Konhauser et al. 1992; Witke et al. 2004; Mustoe 2008; Dietrich et al. 2013). This method uses the ability of incident electrons to scatter the bound electrons from an inner electron shell of the atoms, thereby leaving a vacancy. This vacancy is filled when an electron drops down from an outer shell. The difference in binding energy between individual shells is characteristic for every element and might be emitted as X-rays with specific energy. This energy is detected by an energy-dispersive spectrometer and different elements in a bulk analysis are discriminated accordingly. The detection limit is commonly about 0.1 weight percent (1000 ppm) and, hence, of limited use for trace element analysis (Karowe and Jefferson 1987; Scott and Collinson 2003; Mustoe 2008).

Compared with EDS analysis, WDS provides a significantly higher energy/wavelength resolution and better peak-to-background ratio. The WDS detection limit of trace elements in quartz, such as Al and Ti, can be as low as about 0.001 weight percent (10 ppm), depending on the sample composition and analytical conditions (Donovan et al. 2011). The low detection limit and high spatial resolution of WDS analysis may outweigh the longer acquisition time required for spot analysis and element mapping, when compared to EDS analysis. The identification of the concentration and distribution of trace elements in silica at the micrometer scale provides deeper insight into the physical pathway and chemical characteristics of silica-precipitating fluids (Flörke et al. 1991; Gaillou et al. 2008b; Rusk et al. 2008; Liesegang and Milke 2014; Hellawell et al. 2015; Liesegang and Gee 2020). Element distribution mapping proves to be an especially suitable method to identify the two- and three-dimensional correlation of individual plant tissue and a silica-rich fluid (Boyce et al. 2001; Hellawell et al. 2015; Liesegang and Gee 2020). Element mapping shows that trace elements, such as sulfur or calcium, are commonly associated with organic material (Sigleo 1978; Boyce et al. 2001), while other elements, such as sodium, potassium, or aluminum, likely record the fluid environment.

Laser Ablation Inductively Coupled Plasma Mass Spectrometry (LA-ICP-MS)

Laser ablation inductively coupled plasma mass spectrometry (LA-ICP-MS) is widely used for the spatially resolved analysis of the elemental or isotopic composition of earth materials (Jenner and Arevalo 2016; Woodhead et al. 2016). Detailed accounts on the range of instruments, methodologies, and applications are given by Sylvester (2008). Commonly, a thin section or polished mount is prepared from the sample to provide a flat surface. A pulsed laser beam is focused on the region of interest to ablate a small volume of sample material. The produced aerosol is transported in a gas stream to the ICP-mass spectrometer. Here, at the ion source, inductively coupled argon plasma converts the aerosol particles to ions. The ions are extracted into the high vacuum of the mass spectrometer, separated according to their mass/charge ratio, and counted by a detector.

Element concentrations are quantified by reference to an external standard of known composition, in addition to a known abundance of one element as an internal reference (e.g., the concentration of Si, determined by EPMA analysis or estimated from stoichiometry). Accurate determination of isotopic ratios relies on a matrix-matched reference material, of similar composition and ablation properties, to correct for and minimize analytical bias. Data are usually acquired within the size of a laser spot (ca. 10 to 100 µm in diameter), can be collected along profiles by ablating a continuous line, or can even be compiled to elemental images (Ubide et al. 2015). Compared to EPMA, the spatial resolution of LA-ICP-MS is more limited; the much higher sensitivity (ca. 1 ppm level), however, allows any geological interpretations to be based on a much wider range of trace elements.

Trace element data have been collected in case studies on agate (Götze et al. 2009, 2016) and opal (Rondeau et al. 2012; Dutkiewicz et al. 2015; Chauviré et al. 2019) with the aim of characterizing and differentiating the properties and sources of silica-precipitating fluids. Of prime interest are trace elements (e.g., Sr, Ca, Ba) and rare earth element patterns, to test for potential source rocks of the fluid, and redox-sensitive elements (U, V, Ceanomaly), as a probe into the oxidation state. The first applications in silici-fied wood research include the study of coloring agents (Mustoe and Acosta 2016) and the use of trace element patterns to discriminate between an ini-tial silicification and a secondary overprint within an extended diagenetic evolution (Matysová et al. 2016; Jiang et al. 2018).

LA-ICP-MS has become an important tool for isotopic analyses (e.g., Woodhead et al. 2016), particularly for U-Pb geochronology (Schoene 2014). Since non- to microcrystalline silica can incorporate considerable amounts of uranium, causing green fluorescence in short-wave UV light (Amelin and Back 2006; Götze et al. 2015a), the U-Pb system of silica phases aids in constraining the timing of paleohydrologic events under near-surface dia-genetic conditions (Neymark 2014). Jiang et al. (2018) recently explored this system in a case study on mineralized wood from Arizona, USA. The ages obtained allow discrimination between an initial silicification and second-ary alteration events.

Case Study: Opal-A-Replaced Wood from the Australian Cretaceous

In order to reconstruct the silicification process in detail, it is of fundamen-tal importance to precisely identify the structural and chemical properties of silica and its accompanying phases in silicified wood in order to poten-tially characterize the silica-precipitating fluids and their evolution over timescales. The fluid properties and the temporal succession of diagenetic silica phases have a direct impact on the preservation potential of organic matter and need to be considered to establish a consistent framework for the multistage wood silicification process. An integrative analytical approach combines structural and chemical characterization to overcome the specific limitations of each analytical method.

In the following case study, we use SEM, XRD, Raman spectroscopy, and EPMA to identify and characterize noncrystalline silica (opal-A) and its ac-companying minerals in silicified coniferous wood from the Lower Creta-ceous Bulldog Shale in Andamooka, South Australia. The unsilicified host rock is an extensively bleached, iron-stained mudstone with a modal mineral content (in vol.%) of about 50% kaolinite, 35% quartz, 10% alunite, and minor amounts of illite, hematite, anatase, and barite (Liesegang and Milke 2014).

Secondary electron images obtained from freshly fractured material at an acceleration voltage of 5 kV and a beam current of 10 nA show that sil-ica, which infills and replaces plant material, consists of non-uniform spheres with a diameter from 130 to 400 nm (fig. 7.4A). The smooth surface of lumen infillings suggests the formation of a dense film on intact cell walls and circular bordered pit surfaces during the early stage of the silicification process. Cell lumina are occasionally filled with uniform ordered spheres similar to those observed in precious opal (Jones et al. 1964; Liesegang et al. 2017). The sphere micromorphology of opal-A clearly indicates that the pore space in the template material exceeded the sphere diameter during the replacement process, which is essential for spheres to form as free units. The formation of up to 1 μm large, randomly distributed pores in the opaline material during the replacement process highlights the importance of later silica-rich fluid ingress to compensate the volume loss and to preserve dimensional stability during crystallization of an opal-A precursor.

BSE images of polished thin sections show that minute mineral grains with a different grayscale are located in and between the cell walls (fig. 7.4B). These grains, which measure up to 150 nm in size, are a light brown color in transmitted light microscopy and have been identified as TiO_2 and kaolinite using Raman spectroscopy and electron microprobe spot analysis. The occurrence of TiO_2 can be interpreted as an indicator for temporary low pH conditions due to the higher fluid mobility of titanium at low pH (Summerfield 1983). This clearly shows that mineral impurities are not just an undesired byproduct of the analysis but provide deeper insight into the diagenetic environment of silicified wood. In fact, authigenic minerals can reflect fluid compositions, and host-rock alteration episodes may be discernible (Liesegang and Milke 2014; Matysová et al. 2016; Pe-Piper et al. 2019).

Bulk powder XRD shows that the wood sample consists of opal-A (fig. 7.4C). The main peak position at 4.04 Å and asymmetric shape indicate that tridymite-rich nanodomains formed in this micromorphologically intact, noncrystalline material (Liesegang et al. 2018). The XRD detection limit of ~1 vol.% prevents the identification of minor constituents that, accordingly, have been visualized and analyzed with BSE imaging, EPMA, and Raman spectroscopy.

Raman spectroscopy at a spot size of 1 μm and 532 nm excitation wavelength was used to locate and characterize remnant organic matter in the cell wall. The carbon distribution in silicified wood can also be mapped using EPMA (Boyce et al. 2001). However, opal-A is unstable under the electron beam owing to its high water content and disordered structure. Short counting times and low beam currents must be employed to preserve the structural integrity of the material over the duration of the analysis. Under these electron microprobe conditions, the number of counts obtained at a carbon Kα peak position may not exceed the background counts.

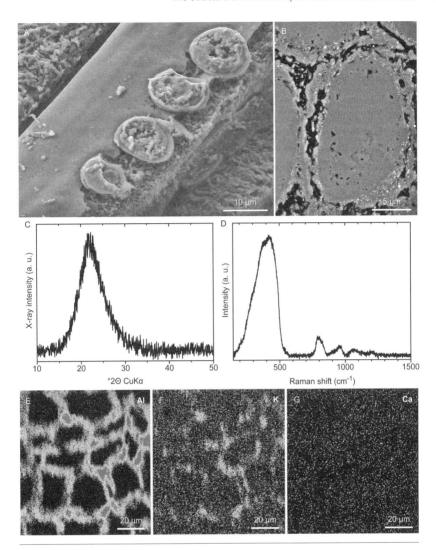

Figure 7.4. Silicified wood from the Lower Cretaceous Bulldog Shale in Andamooka, South Australia. (**A**) Secondary electron image of the freshly fractured sample with smooth lumen surface and silica-filled circular bordered pits. (**B**) BSE image of opal-A-replaced wood in cross section in a polished thin section. Lumina are filled with ordered and disordered spheres. Pores are abundant at the former cell wall position and at the interface of sphere stacks with different sphere size distribution. Bright spots are anatase (TiO$_2$). (**C**) X-ray diffractogram of bulk powdered wood, recorded using CuKα radiation. The broad asymmetric peak is centered at ~22°2θ (4.04 Å) and typical of opal-A. (**D**) Unpolarized Raman spectrum of an inclusion-free cell wall. The broad band at ca. 400 cm^{-1} is a multiplet composed of bands attributed to 4-, 5-, and 6-membered SiO$_4$ ring structures. The spectrum is consistent with opal-A. **E–G**: Elemental maps of inclusion-free earlywood in cross section. The maps of aluminum (**E**), potassium (**F**), and calcium (**G**) were recorded at a dwell time of 100 ms and pixel size of 0.5 μm. Black represents the lowest, and white the highest, abundance. Images: ML.

At a spot size of 1 μm, Raman spectroscopy allows more accurate carbon analysis than EPMA due to its lower destructive potential and high sensitivity for low concentrations of various organic matter. Analysis should generally be performed prior to any sample coating.

The Raman spectra obtained from inclusion-free cell wall domains show that the characteristic bands of carbon in the wavenumber region above 1300 cm^{-1} are absent, indicating that carbon was removed from the organic template during the silicification process (fig. 7.4D). This is consistent with results from EPMA element mapping of inclusion-free domains that show negligible concentrations of calcium and sulfur that could be associated with organic material (fig. 7.4E–G) (Sigleo 1978; Boyce et al. 2001). Conversely, the cell wall regions contain elevated amounts of aluminum and potassium— elements that are usually unrelated to the organic template and record the fluid environment. EPMA spot analysis shows that the sodium/potassium ratio of uniform spheres in lumina is lower than that of non-uniform spheres in cell walls. The low Na/K ratio indicates that the uniform spheres formed at low pH conditions (Liesegang and Milke 2014).

Future Research Directions

A detailed understanding of the wood silicification process opens a window into the past of individual plants, their respective ecosystems, and the fate of organic matter during diagenesis. The processes taking place at the cellular scale, interactions of silica-rich fluid with the organic substrate, and postdepositional silica diagenesis are key to the preservation of delicate subcellular features. While classical studies have focused on light microscopy and bulk analysis, modern analytical instrumentation can facilitate the study of the silica phases and reconstruction of the silicification process on the micro- to nanometer scale.

The physicochemical conditions of the fluid (e.g., fluid chemistry, temperature, pressure) dictate the silica structure and morphology. A detailed analysis of wood, silicified with quartz and chalcedony, can reconstruct the silicification process from the final, crystalline end of the diagenetic spectrum (e.g., Sigleo 1978; Matysová et al. 2010; Trümper et al. 2018), while research on noncrystalline silica precipitated during the incipient silicification of wood identifies the other, initial end (e.g., Hellawell et al. 2015; Liesegang and Gee 2020). The range between these two states may be characterized by large variations of fluid compositions, organic matter degradation, authigenic mineral formation, and general diagenetic overprint. Analysis of the chemical and structural properties of silica beyond a quantitative fingerprint will provide deeper insight into these factors. Further information can be obtained from texturally and chemically distinct domains that formed during multiple events (fig. 7.3). Combined CL analysis and EPMA mapping are a good starting point for tracing the physical pathway of silicifying fluid into plant material, as well as the intraplant microenvironments in which the silica phases precipitate. Assuming that the silica chemical composition reflects the initial fluid composition, host-rock alteration episodes may be discernible (Pe-Piper et al. 2019). Accordingly, it is essential to include detailed host-rock information in future studies.

Understanding of the silica diagenesis in the specific geological context is mandatory for identifying potential bias in the fossil wood record and will help to constrain the fossilization potential in different environments, which may, for example, change the long-standing view of the initially precipitated silica phase (i.e., noncrystalline versus crystalline). A coherent model should also address the sequence of silica-precipitation events and the compensation of volume loss during crystallization from a potentially noncrystalline precursor to quartz. Experimental diagenesis may contribute to filling this knowledge gap (Gee and Liesegang, chap. 6). Further, detailed knowledge about the fossilizing fluids may resolve the enigmatic topic as to why fossil bone and plants are so rarely preserved together.

Another notable gap in knowledge is the link between silica diagenesis and organic matter structure and preservation. Sigleo (1978) used IR spectroscopy and Py-GC-MS analyses on dissolved bulk material to identify remnant lignin compounds in silicified wood from the Triassic Chinle Formation in Arizona, USA. The compositional complexity of organic matter in silicified wood may be identified in the original context using modern in situ methods with high sensitivity and spatial resolution, such as matrix-assisted laser desorption/ionization–time-of-flight mass spectrometry (MALDI-ToF MS) or synchrotron-based X-ray absorption near-edge structure spectroscopy (STXM-based XANES) (Strullu-Derrien et al. 2019). Modern paleobotanical research may thus provide important information about the complex interplay between organic matter preservation and silica diagenesis.

Conclusions

The silicification of wood preserves delicate cellular detail in three dimensions. After initial precipitation, silica phases may undergo a diagenetic overprint that results in a multitude of distinct micromorphologies, as well as structural and chemical changes at the atomic level. These characteristics can be used to reconstruct the pathway of silica-rich fluid into and inside the wood substrate, the coupling between the intricate silicification process and the fate of wood organic substance, and the diagenetic history of wood transforming into a stable fossil. Silica phases evolve over geological time in response to the physicochemical properties of natural fluids and potentially in response to substantial changes in the organic matter composition of the plant material. Most of our understanding of the interaction of wood and silica and their coupled diagenetic fate has been based on qualitative data from light microscopy, electron microscopy, and bulk powder XRD. The modern instrumental tool kit currently available to earth scientists allows characterization of the silica structure and chemistry at the micro- to nanometer scale in an integrative manner.

ACKNOWLEDGMENTS

The authors thank George Mustoe for his helpful review and the use of his photograph, as well as an anonymous reviewer for detailed and constructive comments. Thanks should also go to book editors Carole T. Gee and Victoria E. McCoy, as well as to Mariah Howell, for helpful comments. This work was funded by the Deutsche Forschungsgemeinschaft (DFG, German Research Foundation), Project number 396706817 to Carole T. Gee for ML and Project number 396709532 to Thorsten Geisler for FT (both at the University of Bonn). This is contribution number 21 of the DFG Research Unit 2685, "The Limits of the Fossil Record: Analytical and Experimental Approaches to Fossilization."

WORKS CITED

Akahane, H., Furuno, T., Miyajima, H., Yoshikawa, T., and Yamamoto, S. 2004. Rapid wood silicification in hot spring water: An explanation of silicification of wood during the Earth's history. *Sedimentary Geology*, 169: 219–228.

Amelin, Y., and Back, M. 2006. Opal as a U-Pb geochronometer: Search for a standard. *Chemical Geology*, 232: 67–86.

Boyce, C.K., Hazen, R.M., and Knoll, A.H. 2001. Nondestructive, in situ, cellular-scale mapping of elemental abundances including organic carbon in permineralized fossils. *Proceedings of the National Academy of Sciences of the United States*, 98: 5970–5974.

Buurman, P. 1972. Mineralization of fossil wood. *Scripta Geologica*, 12: 1–43.

Chauviré, B., Rondeau, B., Alexandre, A., Chamard-Bois, S., La, C., and Mazzero, F. 2019. Pedogenic origin of precious opals from Wegel Tena (Ethiopia): Evidence from trace elements and oxygen isotopes. *Applied Geochemistry*, 101: 127–139.

De Jong, B.H.W.S., Van Hoek, J., Veeman, W.S., and Manson, D.V. 1987. X-ray diffraction and ^{29}Si magic-angle-spinning NMR of opals; incoherent long- and short-range order in opal-CT. *American Mineralogist*, 72: 1195–1203.

Dietrich, D., Lampke, T., and Rößler, R. 2013. A microstructure study on silicified wood from the Permian Petrified Forest of Chemnitz. *Paläontologische Zeitschrift*, 87: 397–407.

Dietrich, D., Witke, K., Rößler, R., and Marx, G. 2001. Raman spectroscopy on *Psaronius* sp.: A contribution to the understanding of the permineralization process. *Applied Surface Science*, 179: 230–233.

Donovan, J.J., Lowers, H.A., and Rusk, B.G. 2011. Improved electron probe microanalysis of trace elements in quartz. *American Mineralogist*, 96: 274–282.

Dutkiewicz, A., Landgrebe, T.C., and Rey, P.F. 2015. Origin of silica and fingerprinting of Australian sedimentary opals. *Gondwana Research*, 27: 786–795.

Edwards, N.P., Manning, P.L., Bergmann, U., Larson, P.L., van Dongen, B.E., Sellers, W.I., Webb, S.M., Sokaras, D., Alonso-Mori, R., Ignatyev, K., Barden, H.E., van Veelen, A., Anné, J., Egerton, V.M., and Wogelius, R.A. 2014. Leaf metallome preserved over 50 million years. *Metallomics*, 6: 774–782.

Flörke, O.W., Graetsch, H., Martin, B., Röller, K., and Wirth, R. 1991. Nomenclature of micro- and non-crystalline silica minerals, based on structure and microstructure. *Neues Jahrbuch für Mineralogie—Abhandlungen*, 163: 19–42.

Flörke, O.W., Hollmann, R., Von Rad, U., and Rösch, H. 1976. Intergrowth and twinning in opal-CT lepispheres. *Contributions to Mineralogy and Petrology*, 58: 235–242.

Flörke, O.W., Köhler-Herbertz, B., Langer, K., and Tönges, I. 1982. Water in microcrystalline quartz of volcanic origin: Agates. *Contributions to Mineralogy and Petrology*, 80: 324–333.

Gaillou, E., Delaunay, A., Rondeau, B., Bouhnik-le-Coz, M., Fritsch, E., Cornen, G., and Monnier, C. 2008a. The geochemistry of gem opals as evidence of their origin. *Ore Geology Reviews*, 34: 113–126.

Gaillou, E., Fritsch, E., Aguilar-Reyes, B., Rondeau, B., Post, J., Barreau, A., and Ostrooumov, M. 2008b. Common gem opal: An investigation of micro- to nanostructure. *American Mineralogist*, 93: 1865–1873.

Gierlinger, N., Keplinger, T., and Harrington, M. 2012. Imaging of plant cell walls by confocal Raman microscopy. *Nature Protocols*, 7: 1694–1708.

Goldstein, J.I., Newbury, D.E., Michael, J.R., Ritchie, N.W., Scott, J.H.J., and Joy, D.C. 2018. *Scanning Electron Microscopy and X-Ray Microanalysis*, 4th ed. Springer, New York.

Götze, J. 2012. Application of cathodoluminescence (CL) microscopy and spectroscopy in geosciences. *Microscopy and Microanalysis*, 18: 1270–1284.

Götze, J., Gaft, M., and Möckel, R. 2015a. Uranium and uranyl luminescence in agate/chalcedony. *Mineralogical Magazine*, 79: 983–993.

Götze, J., and Hanchar, J.M. 2018. Atlas of cathodoluminescence textures. *GAC Miscellaneous Publication*, 10. Geological Association of Canada.

Götze, J., Hanchar, J., Schertl., H.-P., Neuser, D.K., and Kempe, U. 2013. Optical microscope–cathodoluminescence (OM–CL) imaging as a powerful tool to reveal internal textures of minerals. *Mineralogy and Petrology*, 107: 373–392.

Götze, J., and Kempe, U. 2008. A comparison of optical microscope (OM) and scanning electron microscope (SEM) based cathodoluminescence (CL) imaging and spectroscopy applied to geosciences. *Mineralogical Magazine*, 72: 909–924.

Götze, J., and Kempe, U. 2009. Physical principles of cathodoluminescence and its applications to geosciences, pp. 1–22. In: Gucsik, A., ed., *Cathodoluminescence and Its Application in the Planetary Sciences*. Springer, Berlin.

Götze, J., Möckel, R., Kempe, U., Kapitonov, I., and Vennemann, T. 2009. Origin and characteristics of agates in sedimentary rocks from the Dryhead area, Montana, USA. *Mineralogical Magazine*, 73: 673–690.

Götze, J., Möckel, R., Langhof, N., Hengst, M., and Klinger, M. 2008. Silicification of wood in the laboratory. *Ceramics-Silikáty*, 52: 268–277.

Götze, J., Möckel, R. Vennemann, T., and Müller, A. 2016. Origin and geochemistry of agates in Permian volcanic rocks of the Sub-Erzgebirge basin, Saxony (Germany). *Chemical Geology*, 428: 77–91.

Götze, J., Nasdala, L., Kleeberg, R., and Wenzel, M. 1998. Occurrence and distribution of "moganite" in agate/chalcedony: A combined micro-Raman, Rietveld, and cathodoluminescence study. *Contributions to Mineralogy and Petrology*, 133: 96–105.

Götze, J., Pan Y., Stevens-Kalceff, M., Kempe, U., and Müller, A. 2015b. Origin and significance of the yellow cathodoluminescence (CL) of quartz. *American Mineralogist*, 100: 1469–1482.

Götze, J., Plötze, M., and Habermann, D. 2001. Cathodoluminescence (CL) of quartz: Origin, spectral characteristics and practical applications—A review. *Mineralogy and Petrology*, 71: 225–250.

Götze, J., Plötze, M., and Trautmann, T. 2005. Structure and luminescence characteristics of quartz from pegmatites. *American Mineralogist*, 90: 13–21.

Götze, J., and Rößler, R. 2000. Kathodolumineszenz-Untersuchungen an Kieselhölzern—I. Silifizierungen aus dem versteinerten Wald von Chemnitz (Perm, Deutschland). *Veröffentlichung Museum für Naturkunde Chemnitz*, 23: 35–50.

Graetsch, H. 1994. Structural characteristics of opaline and microcrystalline silica minerals. In: Heaney, P.J., Prewitt, C.T., and Gibbs, G.V., eds., *Silica. Reviews in Mineralogy*, 29: 209–232.

Graetsch, H., Flörke, O.W., and Miehe, G. 1985. The nature of water in chalcedony and opal-C from Brazilian agate geodes. *Physics and Chemistry of Minerals*, 12: 300–306.

Graetsch, H., Gies, H., and Topalović, I. 1994. NMR, XRD and IR study on microcrystalline opals. *Physics and Chemistry of Minerals*, 21: 166–175.

Heaney, P.H. 1995. Moganite as an indicator for vanished evaporites: A testament reborn? *Journal of Sedimentary Research*, 65: 633–638.

Hellawell, J., Ballhaus, C., Gee, C.T., Mustoe, G.E., Nagel, T.J., Wirth, R., Rethemeyer, J., Tomaschek, F., Geisler, T., Greef, K., and Mansfeldt, T. 2015. Incipient silicification of recent conifer wood at a Yellowstone hot spring. *Geochimica et Cosmochimica Acta*, 149: 79–87.

Hesse, R. 1989. Silica diagenesis: Origin of inorganic and replacement cherts. *Earth-Science Reviews*, 26: 253–284.

Hinman, N.W. 1998. Sequences of silica phase transitions: Effects of Na, Mg, K, Al, and Fe ions. *Marine Geology*, 147: 13–24.

Iler, R.K. 1979. *The Chemistry of Silica: Solubility, Polymerization, Colloid and Surface Properties, and Biochemistry*. Wiley, New York.

Ilieva, A., Mihailova, B., Tsintov, Z., and Petrov, O. 2007. Structural state of microcrystalline opals: A Raman spectroscopic study. *American Mineralogist*, 92: 1325–1333.

Isaacs, C.M. 1982. Influence of rock composition on kinetics of silica phase changes in the Monterey Formation, Santa Barbara area, California. *Geology*, 10: 304–308.

Jefferson, T.H. 1987. The preservation of conifer wood: Examples from the Lower Cretaceous of Antarctica. *Palaeontology*, 30: 233–249.

Jenner, F.E., and Arevalo, Jr., R.D. 2016. Major and trace element analysis of natural and experimental igneous systems using LA–ICP–MS. *Elements*, 12: 311–316.

Jiang, H., Lee, C.-T., and Parker, W.G. 2018. Trace elements and U-Pb ages in petrified wood as indicators of paleo-hydrologic events. *Chemical Geology*, 493: 266–280.

Jones, J.B., Sanders, J.V., and Segnit, E.R. 1964. Structure of opal. *Nature*, 204: 990–991.

Jones, J.B., and Segnit, E.R., 1971. The nature of opal I. Nomenclature and constituent phases. *Journal of the Geological Society of Australia*, 18: 37–41.

Karowe, A.L., and Jefferson, T.H. 1987. Burial of trees by eruptions of Mount St Helens, Washington: Implications for the interpretation of fossil forests. *Geological Magazine*, 124: 191–204.

Kastner, M., Keene, J.B., and Gieskes, J.M., 1977. Diagenesis of siliceous oozes—I. Chemical controls on the rate of opal-A to opal-CT transformation—An experimental study. *Geochimica et Cosmochimica Acta*, 41: 1041–1059.

Kingma, K.J., and Hemley, R.J. 1994. Raman spectroscopic study of microcrystalline silica. *American Mineralogist*, 79: 269–273.

Konhauser, K.O., Mann, H., and Fyfe, W.S. 1992. Prolific organic SiO_2 precipitation in a solute-deficient river: Rio Negro, Brazil. *Geology*, 20: 227–230.

Lafuente, B., Downs, R.T., Yang, H., and Stone, N. 2016. The power of databases: The RRUFF project, pp. 1–30. In: Armbruster, T., and Danisi, R.M., eds., *Highlights in Mineralogical Crystallography*. Walter De Gruyter, Berlin.

Langford, J.I., and Wilson, A.J.C. 1978. Scherrer after sixty years: Survey and some new results in the determination of crystallite size. *Journal of Applied Crystallography*, 11: 102–113.

Leo, R.F., and Barghoorn, E.S. 1976. Silicification of wood. *Botanical Museum Leaflets, Harvard University*, 25: 1–33.

Liesegang, M., and Gee, C.T. 2020. Silica entry and accumulation in standing trees in a hot-spring environment: Cellular pathway, rapid pace and fossilization potential. *Palaeontology*, 63: 651–660.

Liesegang, M., and Milke, R. 2014. Australian sedimentary opal-A and its associated minerals: Implications for natural silica sphere formation. *American Mineralogist*, 99: 1488–1499.

Liesegang, M., and Milke, R. 2018. Silica colloid ordering in a dynamic sedimentary environment. *Minerals*, 8: 12.

Liesegang, M., Milke, R., and Berthold, C. 2018. Amorphous silica maturation in chemically weathered clastic sediments. *Sedimentary Geology*, 365: 54–61.

Liesegang, M., Milke, R., Kranz, C., and Neusser, G. 2017. Silica nanoparticle aggregation in calcite replacement reactions. *Scientific Reports*, 7: 14550.

Liesegang, M., and Tomaschek, F. 2020. Tracing the continental diagenetic loop of the opal-A to opal-CT transformation with X-ray diffraction. *Sedimentary Geology*, 398: 105603.

Lowenstern, J.B., van Hinsberg, V., Berlo, K., Liesegang, M., Iacovino, K., Bindeman, I.N., and Wright, H.M. 2018. Opal-A in glassy pumice, acid alteration, and the 1817 phreatomagmatic eruption at Kawah Ijen (Java), Indonesia. *Frontiers in Earth Science*, 6: 11.

Lynne, B.Y., Campbell, K.A., Moore, J.N., and Browne, P.R.L. 2005. Diagenesis of 1900-year-old siliceous sinter (opal-A to quartz) at Opal Mound, Roosevelt Hot Springs, Utah, USA. *Sedimentary Geology*, 179: 249–278.

Marfunin, A.S. 1979. Spectroscopy, luminescence and radiation centres in minerals. Springer, Berlin.

Matysová, P., Götze, J., Leichmann, J., Škoda, R., Strnad, L., Drahota, P., and Grygar, T.M. 2016. Cathodoluminescence and LA-ICP-MS chemistry of silicified wood enclosing wakefieldite–REEs and V migration during complex diagenetic evolution. *European Journal of Mineralogy*, 28: 869–887.

Matysová, P., Leichmann, J., Grygar, T., and Rößler, R. 2008. Cathodoluminescence of silicified trunks from the Permo–Carboniferous basins in eastern Bohemia, Czech Republic. *European Journal of Mineralogy*, 20: 217231.

Matysová, P., Rößler, R., Götze, J., Leichmann, J., Forbes, G., Taylor, E.L., Sakala, J., and Grygar, T. 2010. Alluvial and volcanic pathways to silicified plant stems (Upper Carboniferous–Triassic) and their taphonomic and palaeoenvironmental meaning. *Palaeogeography, Palaeoclimatology, Palaeoecology*, 292: 127–143.

Mitchell, R.S., and Tufts, S. 1973. Wood opal—A tridymite-like mineral. *American Mineralogist*, 58: 717–720.

Mizutani, S. 1970. Silica minerals in the early stage of diagenesis. *Sedimentology*, 15: 419–436.

Mizutani, S. 1977. Progressive ordering of cristobalitic silica in the early stage of diagenesis. *Contributions to Mineralogy and Petrology*, 61: 129–140.

Murata, K.J., and Norman, M.B. 1976. An index of crystallinity for quartz. *American Journal of Science*, 276: 1120–1130.

Mustoe, G.E. 2008. Mineralogy and geochemistry of late Eocene silicified wood from Florissant Fossil Beds National Monument, Colorado. In: Meyer, H.W., and Smith, D.M., eds., Paleontology of the Upper Eocene Florissant Formation, Colorado. *Geological Society of America Special Paper*, 435: 127–140.

Mustoe, G.E. 2015. Late Tertiary petrified wood from Nevada, USA: Evidence of multiple silicification pathways. *Geosciences*, 5: 286–309.

Mustoe, G.E. 2017. Wood petrifaction: A new view of permineralization and replacement. *Geosciences*, 7: 119.

Mustoe, G.E., and Acosta, M. 2016. Origin of petrified wood color. *Geosciences*, 6: 25.

Nesse, W.D. 2013. *Introduction to Optical Mineralogy*. Oxford University Press.

Neuser, R.D., Bruhn, F., Götze, J., Habermann, D., and Richter, D.K. 1995. Kathodolumineszenz: Methodik und Anwendung. *Zentralblatt für Geologie und Paläontologie Teil I*, 1: 287–306.

Neymark, L. 2014. Uranium–lead dating, opal, pp. 858–863. In: Rink, W., and Thompson, J., eds., *Encyclopedia of Scientific Dating Methods*. Springer, Dordrecht.

Pe-Piper, G., Imperial, A., Piper, D.J., Zouros, N.C., and Anastasakis, G. 2019. Nature of the hydrothermal alteration of the Miocene Sigri Petrified Forest and host pyroclastic rocks, western Lesbos, Greece. *Journal of Volcanology and Geothermal Research*, 369: 172–187.

Ramseyer, K., Baumann J., Matter, A., and Mulis, J. 1988. Cathodoluminescence colours of alpha-quartz. *Mineralogical Magazine*, 52: 669–677.

Rinaldi, R., and Llovet, X. 2015. Electron probe microanalysis: A review of the past, present, and future. *Microscopy and Microanalysis*, 21: 1053–1069.

Rodgers, K.A., Browne, P.R.L., Buddle, T.F., Cook, K.L., Greatrex, R.A., Hampton, W.A., Herdianita, N.R., Holland, G.R., Lynne, B.Y., Martin, R., Newton, Z., Pastars, D., Sannazarro, K.L., and Teece, C.I.A., 2004. Silica phases in sinters and residues from geothermal fields of New Zealand. *Earth-Science Reviews*, 66: 1–61.

Rodgers, K.A., and Hampton, W.A. 2003. Laser Raman identification of silica phases comprising microtextural components of sinters. *Mineralogical Magazine*, 67: 1–13.

Rondeau, B., Cenki-Tok, B., Fritsch, E., Mazzero, F., Gauthier, J.-P., Bodeur, Y., Bekele, E., Gaillou, E., and Ayalew, D. 2012. Geochemical and petrological characterization of gem opals from Wegel Tena, Wollo, Ethiopia: Opal formation in an Oligocene soil. *Geochemistry: Exploration, Environment, Analysis*, 12: 93–104.

Rusk, B.G., Lowers, H.A., and Reed, M.H. 2008. Trace elements in hydrothermal quartz: Relationships to cathodoluminescent textures and insights into vein formation. *Geology*, 36: 547–550.

Saminpanya, S., and Sutherland, F.L. 2013. Silica phase-transformations during diagenesis within petrified woods found in fluvial deposits from Thailand–Myanmar. *Sedimentary Geology*, 290: 15–26.

Schmidt, P., Bellot-Gurlet, L., Leá, V., and Sciau, P. 2013. Moganite detection in silica rocks using Raman and infrared spectroscopy. *European Journal of Mineralogy*, 25: 797–805.

Schoene, B. 2014. U–Th–Pb geochronology. In: Holland, H.D., and Turekian, K.K., eds., *Treatise on Geochemistry*, 4: 341–378. Elsevier, Amsterdam.

Scott, A.C., and Collinson, M.E. 2003. Non-destructive multiple approaches to interpret the preservation of plant fossils: Implications for calcium-rich permineralizations. *Journal of the Geological Society*, 160: 857–862.

Scurfield, G. 1979. Wood petrifaction: An aspect of biomineralogy. *Australian Journal of Botany*, 27: 377–390.

Scurfield, G., and Segnit, E.R. 1984. Petrifaction of wood by silica minerals. *Sedimentary Geology*, 39: 149–167.

Sigleo, A.C. 1978. Organic geochemistry of silicified wood, Petrified Forest National Park, Arizona. *Geochimica et Cosmochimica Acta*, 42: 1397–1405.

Smallwood, A.G., Thomas, P.S., and Ray, A.S. 1997. Characterisation of sedimentary opals by Fourier transform Raman spectroscopy. *Spectrochimica Acta Part A: Molecular and Biomolecular Spectroscopy*, 53: 2341–2345.

Smith, E., and Dent, G. 2019. *Modern Raman Spectroscopy—A Practical Approach*. John Wiley & Sons, Manchester.

Stein, C.L. 1982. Silica recrystallization in petrified wood. *Journal of Sedimentary Research*, 52: 1277–1282.

Stevens-Kalceff, M.A. 2009. Cathodoluminescence microcharacterization of point defects in α-quartz. *Mineralogical Magazine*, 73: 585–606.

Strullu-Derrien, C., Bernard, S., Spencer, A.R.T., Remusat, L., Kenrick, P., and Derrien, D. 2019. On the structure and chemistry of fossils of the earliest woody plant. *Palaeontology*, 62: 1015–1026.

Summerfield, M.A. 1983. Geochemistry of weathering profile silcretes, southern Cape Province, South Africa. *Geological Society, London, Special Publications*, 11: 167–178.

Sylvester, P., ed. 2008. Laser ablation ICP–MS in the earth sciences: Current practices and outstanding issues. *Mineralogical Association Canada Short Course Series* 40: 1–356.

Thiry, M., Milnes, A.R., Rayot, V., and Simon-CoinÇon, R. 2006. Interpretation of palaeoweathering features and successive silicifications in the Tertiary regolith of inland Australia. *Journal of the Geological Society*, 163: 723–736.

Trümper, S., Rößler, R., and Götze, J. 2018. Deciphering silicification pathways of fossil forests: Case studies from the late Paleozoic of Central Europe. *Minerals*, 8: 432.

Ubide, T., McKenna, C.A., Chew, D.M., and Kamber, B.S. 2015. High-resolution LA-ICP-MS trace element mapping of igneous minerals: In search of magma histories. *Chemical Geology*, 409: 157–168.

Viney, M., Dietrich, D., Mustoe, G., Link, P., Lampke, T., Götze, J., and Rößler, R. 2016. An opalized tree from Gooding County, Idaho—Re-examination of an 1895 discovery. *Geosciences*, 6: 21.

Weibel, R. 1996. Petrified wood from an unconsolidated sediment, Voervadsbro, Denmark. *Sedimentary Geology*, 101: 31–41.

Williams, L.A., Parks, G.A., and Crerar, D.A. 1985. Silica diagenesis, I. Solubility controls. *Journal of Sedimentary Research*, 55: 301–311.

Witke, K., Götze, J., Rößler, R., Dietrich, D., and Marx, G. 2004. Raman and cathodoluminescence spectroscopic investigations on Permian fossil wood from Chemnitz—A contribution to the study of the permineralization process. *Spectrochimica Acta Part A*, 60: 2947–2956.

Woodhead, J.D., Horstwood, M.S.A., and Cottle, J.M. 2016. Advances in isotope ratio determination by LA-ICP-MS. *Elements*, 12: 317–322.

CHAPTER 8

Exceptional Fossilization of Ecological Interactions

Plant Defenses during the Four Major Expansions
of Arthropod Herbivory in the Fossil Record

VICTORIA E. MCCOY, TORSTEN WAPPLER,
AND CONRAD C. LABANDEIRA

A B S T R A C T | Plant–insect interactions play a major role in modern and fossil ecosystems. The fossil record of plant–insect interactions includes four major expansions of insect herbivory, which are apparent in the extensive and relatively well-known record of insect damage on fossil leaves. Here we examine the less well-studied, but equally important, fossil record of plant defenses against insect herbivory to determine how plants responded during each of these four expansions. We found that each expansion of insect herbivory is matched by an expansion of plant defenses. In the Early Devonian, both insect herbivory and plant defenses were sparse. During the mid-Carboniferous, plants developed diverse and abundant defensive hairs, called trichomes, and resin, the precursor to amber. In the Triassic, there was a greater diversity and abundance of trichomes and amber, as well as the first record of an accommodationist strategy of disposable, undefended leaves. The Cretaceous signaled the diversification of angiosperms, after which there was rapid development of the full catalog of defensive strategies, including a major expansion of phytoliths. Both insect herbivory and plant defenses follow a general pattern of increased abundance and diversity through time, and plant groups tend to develop defenses or other responses shortly after they become hosts for insect herbivores. Most of the information about fossil plant defenses comes from fossilized structural defenses or morphological evidence of chemical defenses. Future investigation of fossil chemical defenses using recently developed techniques will help us further understand the evolutionary history of plant–insect interactions. |

Introduction

Fossilization is typically considered exceptional when it encompasses an unusual density of organisms (concentration deposits) or unusually labile organisms (conservation deposits), therefore preserving significant paleobiological data (Seilacher 1970). However, other sources of exceptional paleobiological data are those fossils that preserve a record of behavior and ecological interactions (Boucot and Poinar 2010).

Plant–arthropod interactions are a product of hundreds of millions of years of antagonisms, commensalisms, and mutualisms. Such interactions include some of the most important and complex interorganismic relationships in modern ecosystems (Bryant et al. 1991; Labandeira et al. 1994; Labandeira 2007a; Howe and Jander 2008; War et al. 2012; Labandeira and Currano 2013), and plant–herbivore interactions in particular were a major component of ecosystems during the deep past (Labandeira and Currano 2013). Fossil evidence indicates that plants and insects were involved in major interactions soon after macroscopic ecosystems originated on land, and herbivory—the consumption of living plant tissues—represents an ancient feeding strategy that extends to the Early Devonian, about 410 million years ago (Labandeira et al. 2014; Misof et al. 2014; Condamine et al. 2016). Almost the entire fossil history of this interaction involved arthropod damage done to plants in a myriad of ways (Labandeira 2006a; Wappler et al. 2009, 2015). However, evidence of interactions in the opposite direction, specifically the prevention of plant hosts from herbivory by defense mechanisms, is sparse (Wilf et al. 2001; Wappler 2010; Xu et al. 2018). This evidence primarily includes morphological defenses such as trichomes, spines, thick cuticles, and plant architecture (Beck and Labandeira 1998; Labandeira 1998; Collinson and van Bergen 2004; Pott et al. 2012; Xu et al. 2018).

Nevertheless, modern studies of plant antiherbivore defenses indicate that plants have evolved two major strategies for contending with arthropod herbivores (Wilf et al. 2001; Labandeira 2007b; Schuman and Baldwin 2016). The accommodationist strategy allows photosynthetic tissue to be accessed by insect herbivores from the production of palatable, deciduous, thin-leaved foliage with short lifespans (Labandeira 2007b). The defensive strategy employs evergreen, sclerophyllous leaves with long lifespans that are highly defended by an eclectic spectrum of physical and chemical deterrents (Walling 2000; Labandeira 2007b; Wappler et al. 2012). Such physical defenses consist of attributes such as thick, structurally hardened leaves, abundant development of thick cuticles and epicuticular waxes, presence of spines or stinging trichomes, and massive structural tissues beyond what is architecturally sufficient to sustain leaf vigor (Hanley et al. 2007; Pott et al. 2012; Xu et al. 2018). Chemical defenses include compounds that can be deployed in a variety of plant tissues that can be directly toxic or unpalatable, interfere with reproduction or development, or attract predators, among other effects (Mithöfer and Boland 2012; War et al. 2012; Labandeira 2013).

To develop a more complete understanding of the patterns of plant–animal interactions through time, we summarize here the literature on basic patterns of plant defenses against arthropod herbivores and compare this to the major temporal patterns of evolution and diversification of arthropod feeding strategies. Specifically, we consider plant responses to the four main expansions of insect herbivory in the fossil record (Labandeira 2006a). For the purposes of this chapter, we assume the literature provides an accurate reflection of the fossil record. However, it is also possible that the patterns we report here are somewhat influenced by variations in the preservation of plant cuticle, including defensive structures, or variations in the degree to which certain time periods are studied. Specific examples of the fossil plant defenses and accommodations that we discuss in detail are listed in table 8.1 and put into a geological perspective in the timeline in figure 8.1.

Table 8.1.
Plant defensive structures and strategies developed during the four major expansions of arthropod herbivory and plant defenses. See fig. 8.1 for a geological perspective and timeline representation of these examples.

Key to fig. 8.1	Plant defenses	Lithostratigraphy, fossil flora	Location	Geological age	References
A	First major expansion	Various	Global	Early Devonian to mid-Carboniferous	Labandeira 2006a
1	Possible trichomes	Lower Old Red Sandstone	UK	419–410 Ma	Edwards 1986
2	Spines	Trout River Valley Formation	USA	407–393 Ma	Gensel et al. 1969; Andrews et al. 1977
3	Oil body cells	Catskill Delta	USA	388 Ma	Labandeira et al. 2014
4	Phytoliths	Aztec Siltstone	Antarctica	387–382 Ma	Carter 1999
5	Spines and trichomes	Wutubulake Formation	China	387–382 Ma [age disputed]	Cai and Chen 1996
6	Possible trichomes	Millboro Shale Formation	USA	387–372 Ma	Stein et al. 1983
7	Spines	Leigutai Member, Wutong Formation	China	372–369 Ma	D. Wang et al. 2014; D.-M. Wang et al. 2016; Liu et al. 2017
8	Possible trichomes	Xikuangshan Formation	China	372–358 Ma	Gerrienne et al. 2018
B	Second major expansion	Various	Global	Mid-Carboniferous to Middle Triassic	Labandeira 2006a
9	Resin rodlets	Tradewater Formation	USA	320 Ma	Bray and Anderson 2009
10	Hooked and glandular trichomes	Blanzy–Montceau Basin	France	304–299 Ma	Krings et al. 2002, 2003a
11	Resin rodlets	Williamson Drive	USA	303–298 Ma	Labandeira 2014a; Xu et al. 2018

(continued)

Table 8.1. (*continued*)

Key to fig. 8.1	Plant defenses	Lithostratigraphy, fossil flora	Location	Geological age	References
B	Second major expansion	Various	Global	Mid-Carboniferous to Middle Triassic	Labandeira 2006a
12	Phytoliths	Weller Coal Measures	Antarctica	298–272 Ma	Carter 1999
13	Trichomes, glands, glandular trichomes, spines	La Golondrina	Argentina	272–259 Ma	Cariglino 2018
14	Trichomes, glands, glandular trichomes, spines	Bainmedart Coal Measures	Antarctica	272–259 Ma	Pigg and Trivett 1994
C	Third major expansion	Various	Global	Middle Triassic to Present	Labandeira 2006a
15	Phytoliths	Lashly Formation	Antarctica	251–201 Ma	Carter 1999
16	Sclerophylly, trichome bases	Madygen Formation	Kyrgyzstan	242–227 Ma	Moisan et al. 2011
17	Accommodationist strategy	Yangcaogou Formation	China	205 Ma	Ding et al. 2015
18	Spines	Durikai fossil plant assemblage	Australia	182 Ma	McLoughlin et al. 2015
19	Trichomes	Daohugou Formation	China	168–163 Ma	Pott et al. 2012, 2015
20	Accommodationist strategy, fibrous bracts/scales, trichomes, sclerophylly, mimicry	Jiulongshan Formation	China	165 Ma	Y. Wang et al. 2012; Ding et al. 2015
21	Spines	Walloon Coal Measures	Australia	163–145	McLoughlin et al. 2015
22	Resin	Lebanese amber	Lebanon	135–125 Ma	Labandeira 2014a
23	Accommodationist strategy	Yixian Formation	China	125 Ma	Ding et al. 2015
24	Resin	Myanmar amber	Myanmar	99 Ma	Labandeira 2014a
25	Resin	Baltic amber	Europe, various countries	45 Ma	Labandeira 2014a
26	Defensive compounds	Clarkia flora	USA	16–15 Ma	Otto et al. 2003, 2005
D	Fourth major expansion	Various	Global	Middle part of Early Cretaceous to Present	Labandeira 2006a
27	Phytoliths	Zhonggou Formation	China	113–110 Ma	Wu et al. 2018
28	Trichomes	Puddledock flora	USA	113–100 Ma	Crane et al. 1994
29	Accommodationist strategy, sclerophylly	Fort Union Formation	USA	56 Ma	Wilf et al. 2001
30	Accommodationist strategy, sclerophylly	Wasatch Formation	USA	53 Ma	Wilf et al. 2001
31	Resin	Cambay amber	India	50 Ma	Labandeira 2014a
32	Latex	Geiseltal fossil lagerstätte	Germany	47.5–42.5 Ma	Mahlberg and Störr 1989
33	Extrafloral nectaries	Florissant Formation	USA	34 Ma	Pemberton 1992
34	Resin	Dominican amber	Dominican Republic	25 Ma	Labandeira 2014a

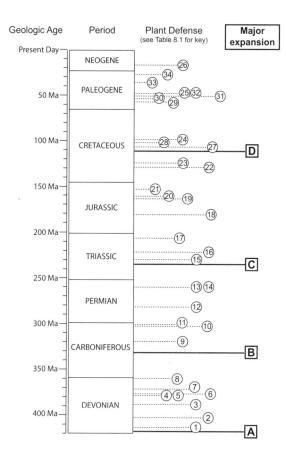

Figure 8.1. Timeline of examples of the fossil plant defenses mentioned in the text. See table 8.1 for the key to numbers and letters.

First Herbivore Expansion: Early Devonian

Plants are widely believed today to have colonized land sometime in the late Silurian, and arthropods, specifically hexapods, arachnids, and myriapods, were present by the Early Devonian (Ward et al. 2006). Shortly thereafter, arthropods began feeding upon plants, primarily upon their spores and stems (Labandeira 2006a, 2007a; Labandeira and Currano 2013). There is evidence of sporivory from various sites from the late Silurian to Early Devonian (417 to 407 Ma) (Edwards et al. 1995; Habgood et al. 2003), and piercing-and-sucking damage occurs on plants from the Early Devonian (407.5 Ma) (Mark et al. 2011; Krings and Harper 2019). These associations are best demonstrated in the Rhynie Chert flora (Kevan et al. 1975) and Gaspé flora (Trant and Gensel 1985; Banks and Colthart 1993) of the Early

Devonian (407–394 Ma) (McGregor 1973). More diverse assemblages of insect damage on a more diverse array of plant organs can be found by the Middle Devonian (388 Ma) Catskill Delta flora (Hernick et al. 2008), which includes piercing and sucking as well as galling and external foliage feeding on liverwort thalli (Labandeira et al. 2014). However, despite these examples of damaged plants, evidence for insect herbivory is rare, albeit also understudied, in the early Paleozoic (Labandeira 2006a; Labandeira and Currano 2013); many floras from this time are apparently devoid of insect damage (Shear and Kukalová-Peck 1990; Edwards and Selden 1992). Plant defenses are similarly sparse among early Paleozoic floras, attributable to a combination of rarity and lack of study.

Many of the damaged fossil plants mentioned above show one of their basic defense systems: wound reaction tissue (Kevan et al. 1975; Trant and Gensel 1985; Scott et al. 1992; Banks and Colthart 1993; Labandeira et al. 2014). Wound reaction tissue, such as callus, is produced in direct response to insect damage and often involves both physical and chemical defenses that include the local production of defensive proteins and other defensive compounds, as well as rapid cell growth, cell size enlargement, and thickening of cell walls to plug and seal the wounded area (Edwards and Wratten 1983; Baron and Zambryski 1995). This shows up in fossil plant matter as a dark rim around a damaged region and is commonly found in damaged fossil plants of all ages (Scott 1991, 1992). Other fossil evidence for possible chemical defenses include dark cells, which have been interpreted as oil body cells, in the Middle Devonian liverworts with diverse insect feeding traces (Labandeira et al. 2014) (plate 8.1A–C). In extant liverworts, oil body cells contain an array of defensive chemicals (up to 700 different compounds) to deter arthropod herbivores (He et al. 2013). Although the original chemistry of the dark cells in fossil liverworts has not been preserved, which means that they cannot directly be interpreted as chemical defenses, their morphology and distribution in the fossil thalli is consistent with extant liverwort oil body cells (Labandeira et al. 2014) (plate 8.1A–C). Other physical defenses include possible trichomes, spines, and phytoliths.

Definitive trichomes have not been identified from Devonian or Early Carboniferous plants, but some fossil plants from these time intervals exhibit cuticular outgrowths that resemble trichomes (Taylor et al. 2009). Trichomes are microscopic epidermal outgrowths (Levin 1973), commonly known as hairs, that can consist of one or more cells. They most likely evolved for a nondefensive purpose, such as preventing water loss, blocking ultraviolet (UV) radiation, or regulating plant temperature (Levin 1973; Gutschick 1999; Manetas 2003; Hanley et al. 2007). Trichomes also provide physical

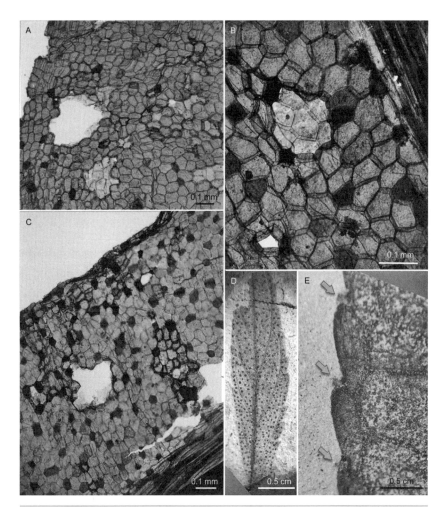

Plate 8.1. Chemical defenses of land plants in the fossil record. **A–C**: Thalli of *Metzgeriothallus sharonae* (Metzgeriales) showing the distribution of oil body cells as possible protection against microherbivores, as indicated by the herbivore damage avoiding the oil body cells, from the Middle Devonian (Givetian Stage) of the Cairo Quarry, Catskill Delta, Green County, New York, USA. Photos modified from Labandeira et al. (2014). **(A)** Thallus showing a sparse distribution of oil body cells and DT1 hole feeding; UCMP-20150.21. **(B)** Thallus exhibiting surface feeding (DT29) and herbivore avoidance of oil body cells; UCMP-250155.07. **(C)** Thallus showing a denser distribution of oil body cells and DT1 hole feeding; UCMP-250150.23. **D, E**: Cretaceous dicot leaves with trichomes or glands. Photos: VEM. **(D)** A leaf of morphotype HC81 (Urticaceae or Cannabaceae) showing capitate trichomes, each represented by an amber dot, from DMNS locality 2203, Williston Basin in Slope County, North Dakota, USA; latest Cretaceous (Maastrichtian Stage, ca. 66.5 Ma); specimen DMNS-19512 (see also Johnson 2002; Labandeira et al. 2002). **(E)** A leaf of an unidentified morphotype, most likely in the family Celastraceae, showing marginal tooth glands represented by a pink dot; from DMNS locality 3725, Kaiparowits Formation in southern Utah; Late Cretaceous, 76.6–74.5 Ma; specimen DMNS-43725 (see also Miller et al. 2013; Roberts et al. 2013). Glands in pink at the intersection of marginal veins and margin sinus are indicated by blue arrows. Repository abbreviations: DMNS, Denver Museum of Nature and Science, Colorado, USA; UCMP, University of California Museum of Paleontology, Berkeley, California, USA.

and chemical defenses against small, typically invertebrate herbivores, as well as against insect oviposition (Gutschick 1999; Werker 2000; Andres and Connor 2003; Dalin and Björkman 2003; Handley et al. 2005; Hanley et al. 2007; Pott et al. 2008, 2012). Trichomes, by their presence, can physically impede the movement or attachment of arthropod herbivores. Some trichomes (called glandular trichomes) also can release stored chemicals to physically or chemically deter invertebrate herbivores (Duffey 1986; Doss et al. 1987; QuIRing et al. 1992; Agren and Schemske 1993; Westerbergh and Nyberg 1995; Eisner et al. 1998; Haddad and Hicks 2000; Traw and Dawson 2002; Dalin and Björkman 2003; Handley et al. 2005). Possible Devonian trichomes or structures representing trichome bases are found on *Longostachys latisporophyllus* from the Middle Devonian of China (Cai and Chen 1996), *Arachnoxylon minor* from the Middle Devonian of the USA (Stein et al. 1983), *Lilingostrobus chaloneri* from the Late Devonian of China (Gerrienne et al. 2018), and various species of *Cosmochlaina* from the Early Devonian of Britain (Edwards 1986). There is no evidence that these early trichome-like features delivered chemical defenses or that they had a defensive role. Their role may have been purely mechanical.

Spines are known, albeit rare, from fossil plants of the Early Devonian (Gensel et al. 1969; Gensel and Berry 2001; Liu et al. 2017). Spines (sharp, pointed leaves or parts of leaves), thorns (sharp, pointed outgrowths of branches), and prickles (sharp, pointed outgrowths from the epidermis or subepidermal or cortical tissues) are all united under the general term spinescence (Grubb 1992; Gutschick 1999) and are often all termed "spines." The Early Devonian Trout Valley Formation in Maine, USA, yields at least three plant genera with axes bearing sparsely distributed, short (ca. 2 mm long), stout spines (Gensel et al. 1969; Andrews et al. 1977): *Kaulangiophyton akantha*, *Psilophyton microspinosum*, and *Drepanophycus* sp. The same site also contains *Thursophyton*, a form genus characterized by sterile axes with small delicate spines or spine-like leaves (Andrews et al. 1977). The Late Devonian Leigutai Member of the Wutong Formation in China bears *Cosmosperma polyloba*, with both stems and fronds densely covered in tiny (0.2 to 0.3 mm long) conical prickles, and *Telangiopsis* sp., with tiny conical prickles only on stems (D. Wang et al. 2014, 2016; Liu et al. 2017). Other Devonian plants, small tree-like lycopsids with secondary growth in the trunk, have spiny margins on their leaves (Gensel and Berry 2001).

Various purposes have been proposed for spines (Nobel et al. 1975; Grubb 1992). However, it is accepted for the most part that spines in extant plants play a defensive role against large, typically vertebrate, herbivores (Cooper and Owen-Smith 1986; Milewski et al. 1991; Myers 1991; Grubb 1992; Gowda

1996; Obeso 1997; Cooper and Ginnett 1998; Young and Okello 1998; Lev-Yadun 2001; Wilson and Kerley 2003; Young et al. 2003; Cash and Fulbright 2005; Hanley et al. 2007). Because of their relatively large size, spines are unsuited for defense against smaller arthropod herbivores (Cooper and Owen-Smith 1986; Potter and Kimmerer 1988). The earliest terrestrial vertebrate fossils are from the Late Devonian (Ward et al. 2006), although trackway evidence suggests that vertebrates at least could have made short forays onto land by the Middle Devonian (Niedźwiedzki et al. 2010; Ahlberg 2018); however, herbivorous vertebrates are unknown until the Late Carboniferous (Sues and Reisz 1998; Dunn et al. 2003a; Reisz and Fröbisch 2014). The spines and prickles in these Devonian plants are therefore unlikely to have defended against vertebrate herbivores and may have simply functioned as supporting structures by providing mutual support in plants growing in thickets or stands (Liu et al. 2017). However, the small size of the spines and prickles in these plants, compared to the larger and more robust versions in modern plants, may also have allowed them to effectively defend against arthropod rather than vertebrate herbivores (M. Wang et al. 2014; Liu et al. 2017).

One occurrence of phytoliths is known from the late Middle Devonian (387 to 382 Ma) (Young 1993) Aztec Siltstone of Antarctica (Carter 1999). These phytoliths are found isolated from any other plant organs, but based on associated spores, they most likely came from the arborescent plant of *Archaeopteris* (Helby and McElroy 1969; White and Frazier 1986; Carter 1999). Phytoliths are mineral crystals and deposits in plant tissues (Hanley et al. 2007). Although mineral deposits embedded in plant tissue can provide structural support (McNaughton et al. 1985), such structures also are known to be an effective defense against invertebrate herbivores (Grime et al. 1968; Vicari and Bazely 1993; Hudgins et al. 2003; Korth et al. 2006; Massey et al. 2007). In the Devonian flora, phytoliths may have provided structural support or defense against arthropod herbivores.

Second Herbivore Expansion: Mid-Carboniferous

The second expansion of insect herbivory, which occurred during the late Paleozoic, is characterized by increased abundance and diversity of all aspects of plant–herbivorous insect interactions, including damage inflicted by a variety of insect functional feeding groups that involve herbivorous insect taxa, herbivorized plant taxa, and insect-damaged plant organs and tissues (Labandeira 2006a; 2013, 2019). This phase corresponds with the expansion of winged insects, many of whom are herbivores, especially piercing-and-sucking paleodictyopteroids, as well as other lineages with more diverse feeding behaviors, such as archaeorthopteroids, caloneurodeans, and early holometabolans, that were responsible for much of the herbivory in the first expansion (Labandeira 2006a, 2019). This expansion includes six of the seven major functional feeding groups seen in modern herbivorous insect groups, lacking only leaf mining (Labandeira 2006a). New plant organs are also being utilized by herbivorous insects, with the first record of insect damage on wood (specifically, the first record of insect damage on live, nonfungal wood), roots, and seeds, as well as much more abundant and diverse damage on leaves (Cichan and Taylor 1982; Scott and Taylor 1983; Labandeira 2006a; Naugolnykh and Ponomarenko 2010; Feng 2018). The insect damage during this phase is typically characterized by—and distinguished from abiological damage based on—wound reaction tissue (Stephenson and Scott 1992) and stereotypy of damage (Labandeira 2006b), but also other criteria (Xu et al. 2018). Plant defenses also change noticeably in this phase; specifically, there is a significantly increased abundance and diversity of trichomes (plate 8.1D; plate 8.2A–E).

Despite occasional records from the Devonian and Early Carboniferous, trichomes, like insect damage, are not reported in abundance on leaves until the Late Carboniferous. Carboniferous pteridosperms, commonly known as seed ferns, have the most herbivore-mediated damage in Euramerica during this time (Beck and Labandeira 1998; Labandeira 2006a; Xu et al. 2018). Pteridosperms exhibit a wide variety of trichomes, including unicellular unbranched trichomes, multicellular unbranched trichomes, trichomes with various branching patterns, and multiple types of capitate glandular trichomes (Cleal and Shute 1992; Krings et al. 2002, 2003a, 2004; Zodrow 2007; Zodrow and Mastalerz 2018). These trichome types may have served a variety of functions, including defense against invertebrate herbivores, but two types found on the Late Carboniferous (Pennsylvanian, 304–299 Ma) seed fern *Blanzyopteris praedentata* are particularly likely to have served a defensive function. The first type are large, stiff, hooked trichomes that bear a striking resemblance to trichomes on extant plants that will mechanically trap arthropod herbivores (Krings et al. 2003a). The second are touch-sensitive capitate glandular trichomes that are similar to "explosive" trichomes in extant plants. When an insect brushes against such glands, they rapidly eject a sticky exudate to trap the insect and prevent it from herbivorizing the plant (Krings et al. 2002, 2003a).

A regional flora with notably high insect damage during this phase is the Permian glossopterid flora in Gondwana (Plumstead 1963; Holmes 1995;

Plate 8.2. Physical defenses of land plants in the fossil record. A–C: Leaves of *Annularia carinata* (Equisetales: Calamitaceae), displaying linear to slightly curved foliar trichomes originating from the midveins of leaves; from the Midland Basin, NMNH locality 40013; Jack County, Texas, USA; latest Carboniferous (latest Pennsylvanian, Gzhelian Stage); specimen USNM-696247a. (**A**) Stem at a node showing the radiate arrangement of leaves. (**B**) Four leaves showing long trichomes originating from midvein region. (**C**) Enlargement of area outlined in **B**, showing insertion, disposition, and course of trichomes, four of which are indicated by blue arrows. D, E: Leaves of *Wielandiella villosa* (Bennettitales: Williamsoniaceae); from the Yanliao Biota, Daohugou 1 locality, Ningcheng County, Inner Mongolia, China; latest Middle Jurassic (Callovian Stage), showing spine-like trichomes along the midrib. (**D**) Upwardly oriented, spine-like trichomes originating along the midrib; CNU-PLA-NN-2006540P. (**E**) Midrib area, showing dark hued trichomes; CNU-PLA-NN-2006902. Repository abbreviations: CNU, Capital Normal University paleontological collections, Beijing, China; USNM, National Museum of Natural History, Washington, DC, USA. Photos modified from Xu et al. (2018).

Adami-Rodrigues and Iannuzzi 2001; Labandeira 2006a; Prevec et al. 2009; Souza Pinheiro et al. 2012; Cariglino 2018). As with other pteridosperms, glossopterids were well defended. The only record of Permian phytoliths, described from the Weller Coal Measures of Antarctica, most likely pertain to *Glossopteris* (Carter 1999). Glossopterids also frequently bore glands, trichomes, and glandular trichomes, all of which can provide a mechanical or chemical defense against arthropod herbivores (Pigg and Trivett 1994; Slater et al. 2012; Cariglino 2018), although these trichome types are not ubiquitous on all glossopterids (Cariglino 2018). Finally, many glossopterids, as well as other plant groups, had spines, albeit these were primarily recorded in climbing plants and were more likely used for mechanical attachment rather than for defense against herbivores (Li and Taylor 1998; Dunn et al. 2003a; Krings et al. 2003b; Burnham 2009).

It may seem counterintuitive that these heavily defended plants experienced high levels of herbivory. However, this pattern also is seen in recent floras because plants often produce structural and chemical defenses in response to herbivory (Hanley et al. 2007; War et al. 2012; Schuman and Baldwin 2016). Trichome growth in particular often is induced by herbivory; new leaves on heavily herbivorized plants will have denser trichome growth than older leaves (Agrawal 1999; War et al. 2012). Alternatively, it is possible that particular groups of plants such as pteridosperms were heavily herbivorized, but only those specific taxa or individual leaves with abundant defenses (e.g., *Blanzyopteris praedentata*) had less insect damage. More detailed comparison of insect damage and plant defenses at the species or individual leaf level would be necessary to untangle these two possibilities.

This phase also includes the earliest record of amber, or fossilized tree resin (Jones and Murchison 1963; Lyons et al. 1982; Bray and Anderson 2009). Terpenoid-rich resin is a sticky, viscous liquid that exudes from trees and shrubs to seal wounds and to defend against pathogens and herbivores (McKellar et al. 2011; Labandeira 2014a). The earliest occurrence of amber comprises tiny fragments of the resin duct system in the trunks of medullosan seed ferns, typically described as "resin rodlets," from the Late Carboniferous (Pennsylvanian, 320 Ma) Tradewater Formation in the USA (Bray and Anderson 2009). Similar resin rodlets have been found at a number of other, somewhat younger, sites of Late Carboniferous age (Labandeira 2014a; Xu et al. 2018), for example, in the leaves of *Macroneuropteris* from the Williamson Drive flora in Texas from the latest Carboniferous (latest Pennsylvanian, 303–298 Ma) (plate 8.3A–C; Xu et al. 2018). The amber is fairly close in age to the earliest damage from insect borings into wood (rather than into hard, nonwoody tissues like those in *Prototaxites*) (Labandeira 2006a), which was iden-

Plate 8.3. More chemical defenses of land plants in the fossil record. **A–C**: Leaves of *Macroneuropteris scheuchzeri* (Medullosales: Neurodontopteridaceae), displaying resin tubules embedded in foliage that are part of the plant's resin duct system, from the Midland Basin, NMNH locality 40013 in Jack County, Texas, USA; latest Carboniferous (latest Pennsylvanian, Gzhelian Stage). **(A)** Leaf with margin at right showing the distribution of tissue embedded tubules oriented upper left to lower right; NMNH-606154a. **(B)** Enlargement of foliage showing tubules, some of which are indicated by blue arrows; note the different trajectories of tubules and leaf venation; NMNH-606298a. **(C)** Same specimen as in **B**, with some tubules indicated by blue arrows and piercing-and-sucking mark in center. Repository abbreviations: NMNH, National Museum of Natural History, Washington, DC, USA. Photos modified from Xu et al. (2018).

tified in the 326 Ma lyginopterid *Trivenia arkansana* from the Fayetteville Formation in the USA (Dunn et al. 2003b; Labandeira 2007a, 2014b; Feng 2018). However, there are not enough data to assess whether the origin of amber is connected to the origin of wood boring.

Third Herbivore Expansion: Middle Triassic

The Permo–Triassic extinction, as well as floral turnover during the Permo–Triassic ecological crisis, caused a significant reduction in plant–insect interactions (Labandeira 2006a). After this global extinction (Benton 2015), the diversity and abundance of plants, arthropod herbivores, and plant–arthropod interactions all remained low until the Middle Triassic (Labandeira 2005, 2006b; Krassilov and Karasev 2008; Shcherbakov 2008; Labandeira and Currano 2013; Kustatscher et al. 2014; Nicholson et al. 2015; Wappler et al. 2015; Labandeira et al. 2016; Pinheiro et al. 2016; Kustatscher et al. 2018). As plants rebounded and diversified during the Triassic, so did their herbivorous insect hosts, resulting in a phase of plant–arthropod interactions characterized by different taxa than in previous phases (Labandeira 2006a; Labandeira and Currano 2013; Labandeira et al. 2018). This co-occurring rebound of taxa is most dramatic in the Late Triassic Molteno Formation of South Africa (Labandeira et al. 2018). Arthropod herbivores are now dominated by neopterous pterygote insects, including hemipteroids, orthopteroids, and early holometabolan lineages, but particularly beetles (Coleoptera), which herbivorize many of the newly emergent crown gymnosperms (Labandeira 2006a). This phase also includes the earliest example leaf mining on entire floras (Labandeira et al. 2018), thereby increasing the diversity of functional feeding groups to encompass all of the seven major herbivore feeding strategies seen in extant interactions (Scott et al. 2004; Labandeira 2006a; Labandeira et al. 2018). The expansion of insect herbivory onto these new plant hosts was quickly matched by an expansion of defensive strategies by these plants.

After this expansion, some of the most heavily herbivorized plants are broad-leaved conifers with few to no adaptations to deterring arthropod herbivores (Ding et al. 2015; Labandeira et al. 2018). These broad-leaved plants lacked trichomes, thick cuticles, and any of the physical features that correlate with chemical defenses. Indeed, these taxa produced thin, delicate, deciduous leaves, and seem to represent the earliest record of the accommodationist strategy of defense against herbivores (Labandeira 2007b). Each individual leaf was short-lived, and it is thought that little investment of energy was required to produce them, based on the observation in modern ecosystems that many of such "cheap" leaves are consumed by arthropod herbivores (Labandeira 2007b; Fürstenberg-Hägg et al. 2013; Ding et al. 2015). The accommodationist strategy is also used by plants today to mitigate herbivory (Wilf et al. 2001; Pott et al. 2012).

Other heavily herbivorized plants, such as bennettitaleans (Strullu-Derrien et al. 2012; Labandeira and Currano 2013; Na et al. 2018), use a more direct defensive strategy to deter herbivore attacks (Ding et al. 2015). These most commonly include fibrous enveloping bracts or scale leaves to protect female and male reproductive organs, although some arthropods could most likely circumvent these defenses to pollinate the plants (Pott et al. 2012; Liu et al. 2019). Less commonly, the leaves were protected with trichomes and sclerophylly (Pott et al. 2012; Liu et al. 2019). Sclerophylly is difficult to define precisely, and even more difficult to quantify, but generally this condition includes a number of features, such as thick cuticle and a high cell wall to cytoplasm ratio that makes the leaf thicker, tougher, or harder (Turner 1994; Hanley et al. 2007). Sclerophylly serves a number of ecological purposes, such as protecting the leaf against water loss, heat, cold, intense UV light, and herbivory (Nobel et al. 1975; Mitrakos 1980; Chabot and Hicks 1982; Turner 1994; Lamont et al. 2002; Groom et al. 2004; Jordan et al. 2005; Hanley et al. 2007). Sclerophylly makes a leaf more difficult to eat and more difficult to digest. Generally speaking, herbivores—large and small, vertebrate and invertebrate—will preferentially avoid sclerophyllous foliage when softer foliage is available (Van Soest 1982; Bernays 1986; Grubb 1986; Björkman and Anderson 1990; Robbins 1993; Choong 1996; Hochuli 1996; Pérez-Barbería and Gordon 1998; Steinbauer et al. 1998; Forsyth et al. 2005; Hanley et al. 2007). Bennettitalean as well as cycadalean foliage from the Middle–Late Triassic Madygen flora in Kyrgyzstan have a number of sclerophyllous features and preserve trichome bases, which mark the position of trichomes in the living plant, that are interpreted to provide a mechanical defense against herbivorous arthropods (Moisan et al. 2011). Other features of the fossil site suggest a humid rather than an arid environment, indicating that the sclerophyllous leaves were a defense against herbivory rather than an adaptation to an arid environment (Moisan et al. 2011). In other sites, including the Early Cretaceous Springhill Formation of Argentina, sclerophyllous features in bennettitalean leaves are interpreted as adaptations to arid environments (Villar de Seoane 2001), or to physiological drought, which can occur even in non-arid conditions (Pott and McLoughlin 2009, 2014). Bennettitalean foliage from Middle Jurassic Daohugou beds in China preserve actual trichomes (plate 8.2D–E), rather than just the trichome bases as in the Kyrgyzstan leaves, which comprise two morphologies—soft flexuous hairs and long stiff hairs—both of which are interpreted as providing a defense against herbivores (Pott et al. 2012, 2015). Aside from trichomes and sclerophylly, bennettitaleans and other phylogenetically derived gymnosperms from various Jurassic sites in Australia have spines, specifically minute marginal and apical spines on their leaves, to defend against insect attacks (McLoughlin et al. 2015). None of these defenses were entirely effective, though, as all of these taxa also display arthropod damage (Moisan et al. 2011; Pott et al. 2012; McLoughlin et al. 2015). However, they exhibit less damage than less well-defended coeval taxa (Pott et al. 2012) and may have been defending the foliage from larger insect herbivores not represented in the leaf damage.

There is one record of phytoliths from the Triassic, namely from the *Dicroidium* flora in the Lashly Formation in Antarctica (Carter 1999). This flora is diverse (White and Frazier 1986), and the phytoliths cannot be assigned reliably to any specific group (Carter 1999). However, they may have been used as a defense against herbivores (Hanley et al. 2007).

The Mesozoic also provides a potential example of an unusual defense against herbivores: insect mimicry. In general, mimicry goes beyond mere camouflage, whereby materials are accumulated in order to disguise the body, and instead includes augmented morphology, behavior, and frequently even physiology that assume patterns, shapes, habits, and sometimes chemical cues approximating the organism being mimicked that are necessary to blend into the surroundings. Mimicry is diverse and frequent among insects, and it is practiced by butterflies, praying mantises, and stick insects. Plant mimicry among insects is ancient, with the oldest record from the middle Permian (*Permotettigonia gallica*) (Garrouste et al. 2016), but it is uncommonly documented in the fossil record (Papier et al. 1997; Nel et al. 2008; Wedmann 2010; Y. Wang et al. 2012; M. Wang et al. 2014; Garrouste et al. 2016; Chen et al. 2018; Liu et al. 2018). In the Late Jurassic, the hangingfly *Juracimbrophlebia ginkgofolia* mimicked the leaves of the ginkgophyte *Yimaia capituliformis*, which is interpreted as allowing the hangingfly to ambush and eat herbivorous insects (Y. Wang et al. 2012). Consequently, this insect's mimicry helped to defend the plant against herbivores.

The third herbivore expansion also corresponds to a major expansion of amber (see Barthel et al., chap. 5, for a more detailed discussion of the amber fossil record), which may have occurred in response to the expansion of clades of wood-boring beetles (Labandeira 2006a, 2014a). Aside from amber, there is very little direct evidence of chemical defenses among gymnosperms in the Mesozoic, although molecular clock estimates suggest that plants had the chemical components necessary for chemical defense by this time interval (Becerra 2003), and some earlier Paleozoic plants had structural traits that are associated with their use (Krings et al. 2002; Pott et al. 2012). Many extant gymnosperms use a wide variety of chemical defenses against insect attacks (Mithöfer and Boland 2012), although there is no direct fossil evidence of defensive chemical compounds until the Cenozoic. The Miocene Clarkia flora in the USA is particularly well known for preserving remnants of original terpenoids, which most likely had a defensive function (e.g., Otto et al. 2003). It would logically follow that the expansion of insect herbivory onto the ancestors of these gymnosperms was matched by an expansion of chemical defenses in many earlier lineages. Nevertheless, currently, we have no direct evidence for support.

Fourth Herbivore Expansion: Late Early Cretaceous

The fourth and final herbivore expansion corresponds to the diversification of angiosperms and involves a complex pattern of changes from gymnosperm to angiosperm insect herbivory (and pollination) behavior (Labandeira 2006a; Peris et al. 2017). First, some insect groups with gymnosperm hosts survived with a minimal loss of diversity by laterally transferring to angiosperm hosts. Second, other insect groups with gymnosperm hosts were unable to make the switch to angiosperms and went extinct. Third, several insect groups with gymnosperm hosts were unable to transition to angiosperm hosts but nonetheless survived and exist today only as relicts on gymnosperms. Fourth, new angiosperm-associated insect groups originated that lacked a previous history of association with gymnosperms. The overwhelming majority of these four major ecological–evolutionary changes occurred within the 26-million-year-long Aptian–Albian Gap, and together, these transfers, originations, and losses of insect groups totaled almost no change in the diversity of insects associated with plants (Labandeira 2014b). This complex, four-part pattern of ecological change transpired for all seven major functional feeding groups, all of which are still present after the expansion of angiosperms and represented by a very diverse array of more specific feeding methods (Labandeira 2006a, 2014b). Shortly after the initiation of herbivory on angiosperms, the angiosperms show a variety of defensive features.

Angiosperms, like gymnosperms, tend to diverge into two strategies: accommodationist and defensive. Accommodationist angiosperms possess large, thin, deciduous leaves without features to deter herbivores, whereas defensive leaves are small, tough leaves often protected by structural or chemical features. This split has been well documented in the late Paleocene Fort Union Formation and the early Eocene Wasatch Formation, both located in Wyoming, USA, as having a corresponding pattern of low insect damage on thick, well-defended leaves and high insect damage on thin, poorly defended leaves (Wilf et al. 2001; Labandeira 2007b).

This expansion of insect herbivory onto angiosperms is also approximately matched by an increase in phytolith-bearing taxa. Phytoliths are known in angiosperms from the Early Cretaceous onwards (e.g., Wu et al. 2018), but phytolith-rich taxa primarily emerged in the Late Cretaceous and continued to diversify and expand into the Cenozoic (Katz 2015). Although phytoliths are typically thought to wear down the teeth of large grazing mammals as part of a defensive strategy (Bouchenak-Khelladi et al. 2009; Strömberg and McInerney 2011; Katz 2015), modern experiments have consistently failed to demonstrate that phytoliths damage teeth or deter large mammalian herbivores (Strömberg et al. 2016). In any case, the Late Cretaceous phytolith expansion does not correspond to an expansion of large mammalian herbivores (Katz 2015). By contrast, phytoliths are an effective defense against arthropod and small vertebrate grazers (Grime et al. 1968; Vicari and Bazely 1993; Hanley et al. 2007; Massey et al. 2007), suggesting this phytolith expansion may rather have been a response to the coeval expansion of arthropod herbivores onto angiosperms.

Trichomes of various morphologies, including glandular trichomes and secretory cells, are known from a variety of Cretaceous angiosperms (Crane et al. 1994; Mohr et al. 2008; e.g., Friis et al. 2010) (plate 8.1D, E). However, some trichomes are known from the Early Cretaceous, prior to fourth expansion of arthropod herbivores (e.g., Crane et al. 1994). Similarly, many of the earliest branching angiosperms have trichomes and secretory cells, suggesting that angiosperms had these features prior to the mid-Cretaceous herbivore expansion (Carpenter 2006). Without a detailed review of all occurrences of angiosperm trichomes and glands throughout the Cretaceous, which is beyond the scope of this chapter, it is not clear if there is an angiosperm trichome expansion that matches the fourth herbivore expansion.

Aside from glandular trichomes and secretory cells, there are several other lines of evidence to suggest fossil angiosperms also utilized chemical defenses. The most abundant such evidence is angiosperm amber, which first originated in significant amounts after the K–Pg extinction (Labandeira 2014a; Seyfulla et al. 2018). The lag between angiosperm herbivorization and angiosperm amber production may be due to the sparseness of woody angiosperms during the Cretaceous (Labandeira 2014a). In the Neo-gene, there is a major expansion in angiosperm resin that correlates with a shift in wood-boring insects, particularly beetles, from gymnosperms to angiosperms (Labandeira 2014a). Some of these angiosperm ambers still preserve remnants of their original defensive chemicals (Dutta et al. 2017; McCoy et al. 2017; Paruya et al. 2018). Starting in the Eocene, some angio-sperms also preserve fossilized remnants of latex. Latex, like amber, is a de-fensive exudate from plants that may deter arthropod herbivores through physical effects, such as gumming up insect mouthparts or chemical toxic-ity (Hagel et al. 2008; Konno 2011; Mithöfer and Boland 2012). Latex rarely fossilizes because, unlike resin, it does not polymerize (Labandeira 2014a); however, rubber-rich latex can fossilize through a natural vulcanization pro-cess (Mahlberg and Störr 1989). As with gymnosperms, many extant angio-sperms utilize a wide variety of chemical defenses (Mithöfer and Boland 2012; War et al. 2012), and the actual diversity and abundance of chemical defenses in ancient angiosperms most likely was much more elevated than is currently apparent from the fossil record.

The Cenozoic record of angiosperm–insect interactions also includes an interaction that is rarely preserved in the fossil record: extrafloral nectaries. Extrafloral nectaries are plant structures outside the flower that produce nectar and attract invertebrate predators, such as ants, that then defend the plant against herbivores (Bentley 1977; Heil and McKey 2003; Bronstein et al. 2006; Weber and Keeler 2013). Extrafloral nectaries are known from ca. 4000 extant plant species (Weber and Keeler 2013) and have been identified in the fossil record (Pemberton 1992). For example, *Populus crassa* leaves from the 34-million-year-old Florissant fossil site bear structures with a striking similarity to extant extrafloral nectaries, suggesting they represent such fossil flower structures (Pemberton 1992).

Future Perspectives

The fossil record of plant defenses against insect attacks primarily comprises structural defenses, and occasional structural features indicative of chemical defenses (e.g., Krings et al. 2002; Pott et al. 2012). This record alone is suffi-cient to show the general pattern of plant responses to the major expansions of arthropod herbivory. Nonetheless, even for structural defenses, more re-search is needed to understand smaller-scale patterns. For example, as insect damage on plants waxes and wanes with ecological factors such as diversification and extinction episodes (Labandeira and Currano 2013), do plant defenses change in concert? Similarly, investigations that are more detailed are needed to assess the interplay of insect damage and plant defenses in individual assemblages to complement those studies that already exist.

Moreover, the fossil record of plant chemical defenses against arthropod herbivores is almost unknown. In older Paleozoic and Mesozoic fossils, it may not be possible to gather direct evidence of plant chemical defenses. However, there is ample evidence that original defensive chemicals can be preserved in Cenozoic samples (Rieseberg and Soltis 1987; Otto et al. 2003, 2005; Simoneit et al. 2003; Edwards et al. 2014; Dutta et al. 2017; McCoy et al. 2017; Paruya et al. 2018). Additional chemical investigations of Cenozoic samples would contribute to an understanding of the evolutionary patterns of plant chemical defenses against arthropod herbivores.

Conclusions

There have been four major expansions of arthropod herbivory throughout geological time, each of which was matched by an expansion of plant defensive strategies.

1. The first expansion in the Late Devonian, shortly after arthropods became terrestrial, involves minimal herbivory and minimal defenses.
2. The second expansion in the mid-Carboniferous is characterized by increased abundance and diversity in both insect feeding strategies and in plant defensive strategies, notably trichomes.
3. The third expansion in the Middle Triassic is characterized by increased diversity of insect feeding strategies and an expansion of insect herbivory onto new host plants, which quickly evolved a variety of strategies to accommodate or defend against herbivory.
4. The fourth expansion in the late Early Cretaceous is characterized by insect herbivorizing of angiosperms and angiosperms using a variety of methods to accommodate or defend against herbivores. A significant innovation in plant defenses is the diversification of phytolith-bearing taxa.

More research to better understand smaller-scale patterns of plant defenses against arthropod herbivory and the fossil record of plant chemical defenses will help to clarify further our understanding of plant–insect interactions through geological time.

ACKNOWLEDGMENTS

The authors thank Finnegan Marsh and Carole Gee for producing the plates from rough drafts, and Lefang Xiao from Capital Normal University in Beijing, who photographed *Wielandiella villosa* for plate 8.2. Thank you to Ian Miller and Gussie Maccracken for access to specimens and for helpful discussion about the leaves with glands preserved in plate 8.1D and E, as well as to the Grand Staircase Escalante National Monument, the source of these two specimens. The chapter was also improved by the comments from Christian Pott and one anonymous reviewer. This work was funded by the Deutsche Forschungsgemeinschaft (DFG, German Research Foundation), Project number 396637283 to Jes Rust for VEM (University of Bonn). This is contribution number 22 of the DFG Research Unit 2685, "The Limits of the Fossil Record: Analytical and Experimental Approaches to Fossilization," as well as contribution 375 of the Evolution of Terrestrial Ecosystems consortium at the National Museum of Natural History, in Washington, DC.

WORKS CITED

Adami-Rodrigues, K., and Iannuzzi, R. 2001. Late Paleozoic terrestrial arthropod faunal and floral successions in the Paraná Basin: A preliminary synthesis. *Acta Geologica Leopoldensia*, 24: 165–179.

Agrawal, A.A. 1999. Induced responses to herbivory in wild radish: Effects on several herbivores and plant fitness. *Ecology*, 80: 1713–1723.

Agren, J., and Schemske, D.W. 1993. The cost of defense against herbivores: An experimental study of trichome production in *Brassica rapa*. *The American Naturalist*, 141: 338–350.

Ahlberg, P.E. 2018. Follow the footprints and mind the gaps: A new look at the origin of tetrapods. *Earth and Environmental Science Transactions of the Royal Society of Edinburgh*, 109: 115–137. doi:10.1017/S1755691018000695.

Andres, M.R., and Connor, E.F. 2003. The community-wide and guild-specific effects of pubescence on the folivorous insects of manzanitas *Arctostaphylos* spp. *Ecological Entomology*, 28: 383–396.

Andrews, H.N., Kasper, A.E., Forbes, W.H., Gensel, P.G., and Chaloner, W.G. 1977. Early Devonian flora of the Trout Valley Formation of northern Maine. *Review of Palaeobotany and Palynology*, 23: 255–285.

Banks, H.P., and Colthart, B.J. 1993. Plant–animal–fungal interactions in Early Devonian Trimerophytes from Gaspé, Canada. *American Journal of Botany*, 80: 992–1001.

Baron, C., and Zambryski, P.C. 1995. The plant response in pathogenesis, symbiosis, and wounding: Variations on a common theme? *Annual Review of Genetics*, 29: 107–129.

Becerra, J.X. 2003. Synchronous coadaptation in an ancient case of herbivory. *Proceedings of the National Academy of Sciences of the United States of America*, 100: 12804–12807.

Beck, A.L., and Labandeira, C.C. 1998. Early Permian insect folivory on a gigantopterid-dominated riparian flora from north-central Texas. *Palaeogeography, Palaeoclimatology, Palaeoecology,* 142: 139–173.

Bentley, B.L. 1977. Extrafloral nectaries and protection by pugnacious bodyguards. *Annual Review of Ecology and Systematics,* 8: 407–427.

Benton, M.J. 2015. *When Life Nearly Died: The Greatest Mass Extinction of All Time,* revised edition. Thames & Hudson, London.

Bernays, E.A. 1986. Diet-induced head allometry among foliage-chewing insects and its importance for graminivores. *Science,* 231: 495–497.

Björkman, C., and Anderson, D.B. 1990. Trade-off among antiherbivore defences in a South American blackberry (*Rubus bogotensis*). *Oecologia,* 85: 247–249.

Bouchenak-Khelladi, Y., Anthony Verboom, G., Hodkinson, T.R., Salamin, N., Francois, O., Ni Chonghaile, G., and Savolainen, V. 2009. The origins and diversification of C_4 grasses and savanna-adapted ungulates. *Global Change Biology,* 15: 2397–2417.

Boucot, A.J., and Poinar, G.O., Jr. 2010. *Fossil Behavior Compendium.* CRC Press, Boca Raton.

Bray, P.S., and Anderson, K.B. 2009. Identification of Carboniferous (320 million years old) class Ic amber. *Science,* 326: 132–134.

Bronstein, J.L., Alarcón, R., and Geber, M. 2006. The evolution of plant–insect mutualisms. *The New Phytologist,* 172: 412–428.

Bryant, J.P., Provenza, F.D., Pastor, J., Reichardt, P.B., Clausen, T.P., and Du Toit, J.T. 1991. Interactions between woody plants and browsing mammals mediated by secondary metabolites. *Annual Review of Ecology and Systematics,* 22: 431–446.

Burnham, R.J. 2009. An overview of the fossil record of climbers: Bejucos, sogas, trepadoras, lianas, cipós, and vines. *Revista Brasileira de Paleontologia,* 12: 149–160.

Cai, C., and Chen, L. 1996. On a Chinese Givetian lycopod, *Longostachys latisporophyllus* Zhu, Hu and Feng, emend.: Its morphology, anatomy and reconstruction. *Palaeontographica Abteilung B,* 238: 1–43.

Cariglino, B. 2018. Patterns of insect-mediated damage in a Permian *Glossopteris* flora from Patagonia (Argentina). *Palaeogeography, Palaeoclimatology, Palaeoecology,* 507: 39–51.

Carpenter, K.J. 2006. Specialized structures in the leaf epidermis of basal angiosperms: Morphology, distribution, and homology. *American Journal of Botany,* 93: 665–681.

Carter, J.A. 1999. Late Devonian, Permian and Triassic phytoliths from Antarctica. *Micropaleontology,* 45: 56–61.

Cash, V.W., and Fulbright, T.E. 2005. Nutrient enrichment, tannins, and thorns: Effects on browsing of shrub seedlings. *The Journal of Wildlife Management,* 69: 782–793.

Chabot, B.F., and Hicks, D.J. 1982. The ecology of leaf life spans. *Annual Review of Ecology and Systematics,* 13: 229–259.

Chen, S., Yin, X., Lin, X., Shih, C., Zhang, R., Gao, T., and Ren, D. 2018. Stick insect in Burmese amber reveals an early evolution of lateral lamellae in the Mesozoic. *Proceedings of the Royal Society B,* 285: 20180425.

Choong, M.F. 1996. What makes a leaf tough and how this affects the pattern of *Castanopsis fissa* leaf consumption by caterpillars. *Functional Ecology,* 10: 668–674.

Cichan, M.A., and Taylor, T.N. 1982. Wood-borings in *Premnoxylon*: Plant–animal interactions in the Carboniferous. *Palaeogeography, Palaeoclimatology, Palaeoecology*, 39: 123–127.

Cleal, C.J., and Shute, C.H. 1992. Epidermal features of some Carboniferous neuropteroid fronds. *Review of Palaeobotany and Palynology*, 71: 191–206.

Collinson, M.E., and van Bergen, P.F. 2004. Evolution of angiosperm fruit and seed dispersal biology and ecophysiology: Morphological, anatomical and chemical evidence from fossils, pp. 343–377. In: Hemsley, A.R., and Poole, I., eds., *The Evolution of Plant Physiology*. Academic Press, Oxford.

Condamine, F.L., Clapham, M.E., and Kergoat, G.J. 2016. Global patterns of insect diversification: Towards a reconciliation of fossil and molecular evidence? *Scientific Reports*, 6: 19208.

Cooper, S.M., and Ginnett, T.F. 1998. Spines protect plants against browsing by small climbing mammals. *Oecologia*, 113: 219–221.

Cooper, S.M., and Owen-Smith, N. 1986. Effects of plant spinescence on large mammalian herbivores. *Oecologia*, 68: 446–455.

Crane, P.R., Friis, E.M., and Pedersen, K.R. 1994. Palaeobotanical evidence on the early radiation of magnoliid angiosperms, pp. 51–72. In: Endress, P.K., and Friis, E.M., eds., *Early Evolution of Flowers*. Springer, Vienna.

Dalin, P., and Björkman, C. 2003. Adult beetle grazing induces willow trichome defence against subsequent larval feeding. *Oecologia*, 134: 112–118.

Ding, Q., Labandeira, C.C., Meng, Q., and Ren, D. 2015. Insect herbivory, plant–host specialization and tissue partitioning on mid-Mesozoic broadleaved conifers of northeastern China. *Palaeogeography, Palaeoclimatology, Palaeoecology*, 440: 259–273.

Doss, R.P., Shanks, C.H., Jr., Chamberlain, J.D., and Garth, J.K.L. 1987. Role of leaf hairs in resistance of a clone of beach strawberry, *Fragaria chiloensis*, to feeding by adult black vine weevil, *Otiorhynchus sulcatus* (Coleoptera: Curculionidae). *Environmental Entomology*, 16: 764–768.

Duffey, S.S. 1986. Plant glandular trichomes: Their partial role in defence against insects, pp. 152–172. In: Juniper, B., and Southwood, R., eds., *Insects and the Plant Surface*. Edward Arnold, London.

Dunn, M.T., Krings, M., Mapes, G., Rothwell, G.W., Mapes, R.H., and Keqin, S. 2003a. *Medullosa steinii* sp. nov., a seed fern vine from the Upper Mississippian. *Review of Palaeobotany and Palynology*, 124: 307–324.

Dunn, M.T., Rothwell, G.W., and Mapes, G. 2003b. On Paleozoic plants from marine strata: *Trivena arkansana* (Lyginopteridaceae) gen. et sp. nov., a lyginopterid from the Fayetteville Formation (middle Chesterian/Upper Mississippian) of Arkansas, USA. *American Journal of Botany*, 90: 1239–1252.

Dutta, S., Mehrotra, R.C., Paul, S., Tiwari, R.P., Bhattacharya, S., Srivastava, G., Ralte, V.Z., and Zoramthara, C. 2017. Remarkable preservation of terpenoids and record of volatile signalling in plant–animal interactions from Miocene amber. *Scientific Reports*, 7: 10940.

Edwards, D. 1986. Dispersed cuticles of putative non-vascular plants from the Lower Devonian of Britain. *Botanical Journal of the Linnean Society*, 93: 259–275.

Edwards, D., and Selden, P.A. 1992. The development of early terrestrial ecosystems. *Botanical Journal of Scotland*, 46: 337–366.

Edwards, D., Selden, P.A., Richardson, J.B., and Axe, L. 1995. Coprolites as evidence for plant–animal interaction in Siluro–Devonian terrestrial ecosystems. *Nature,* 377: 329.

Edwards, N.P., Manning, P.L., Bergmann, U., Larson, P.L., Van Dongen, B.E., Sellers, W.I., Webb, S.M., Sokaras, D., Alonso-Mori, R., Ignatyev, K., and Barden, H.E. 2014. Leaf metallome preserved over 50 million years. *Metallomics,* 6: 774–782.

Edwards, P.J., and Wratten, S.D. 1983. Wound induced defences in plants and their consequences for patterns of insect grazing. *Oecologia,* 59: 88–93.

Eisner, T., Eisner, M., and Hoebeke, E.R. 1998. When defense backfires: Detrimental effect of a plant's protective trichomes on an insect beneficial to the plant. *Proceedings of the National Academy of Sciences of the United States of America,* 95: 4410–4414.

Feng, Z. 2018. Paleozoic fossil records of plant–insect interaction: A window into the deep time terrestrial ecosystems. *Science and Technology Review,* 36: 36–41.

Forsyth, D.M., Richardson, S.J., and Menchenton, K. 2005. Foliar fibre predicts diet selection by invasive Red Deer *Cervus elaphus scoticus* in a temperate New Zealand forest. *Functional Ecology,* 19: 495–504.

Friis, E.M., Pedersen, K.R., and Crane, P.R. 2010. Cretaceous diversification of angiosperms in the western part of the Iberian Peninsula. *Review of Palaeobotany and Palynology,* 162: 341–361.

Fürstenberg-Hägg, J., Zagrobelny, M., and Bak, S. 2013. Plant defense against insect herbivores. *International Journal of Molecular Sciences,* 14: 10242–10297.

Garrouste, R., Hugel, S., Jacquelin, L., Rostan, P., Steyer, J.S., Desutter-Grandcolas, L., and Nel, A. 2016. Insect mimicry of plants dates back to the Permian. *Nature Communications,* 7: 13735.

Gensel, P.G., and Berry, C.M. 2001. Early lycophyte evolution. *American Fern Journal,* 91: 74–98.

Gensel, P., Kasper, A., and Andrews, H.N. 1969. *Kaulangiophyton,* a new genus of plants from the Devonian of Maine. *Bulletin of the Torrey Botanical Club,* 96: 265–276.

Gerrienne, P., Cascales-Minana, B., Prestianni, C., Steemans, P., and [Li, C.-S.]. 2018. *Lilingostrobus chaloneri* gen. et sp. nov., a Late Devonian woody lycopsid from Hunan, China. *PLOS ONE,* 13: e0198287.

Gowda, J.H. 1996. Spines of *Acacia tortilis:* What do they defend and how? *Oikos,* 77: 279–284.

Grime, J.P., MacPherson-Stewart, S.F., and Dearman, R.S. 1968. An investigation of leaf palatability using the snail *Cepaea nemoralis* L. *The Journal of Ecology,* 56: 405–420.

Groom, P.K., Lamont, B.B., Leighton, S., Leighton, P., and Burrows, C. 2004. Heat damage in sclerophylls is influenced by their leaf properties and plant environment. *Écoscience,* 11: 94–101.

Grubb, P.J. 1986. Sclerophylls, pachyphylls, and pycnophylls: The nature and significance of hard leaf surface, pp. 137–150. In: Juniper, B., and Southwood, R., eds., *Insects and the Plant Surface.* Edward Arnold, London.

Grubb, P.J. 1992. A positive distrust in simplicity—Lessons from plant defences and from competition among plants and among animals. *Journal of Ecology,* 80: 585–610.

Gutschick, V.P. 1999. Biotic and abiotic consequences of differences in leaf structure. *The New Phytologist,* 143: 3–18.

Habgood, K.S., Hass, H., and Kerp, H. 2003. Evidence for an early terrestrial food web: Coprolites from the Early Devonian Rhynie chert. *Earth and Environmental Science Transactions of the Royal Society of Edinburgh,* 94: 371–389.

Haddad, N.M., and Hicks, W.M. 2000. Host pubescence and the behavior and performance of the butterfly *Papilio troilus* (Lepidoptera: Papilionidae). *Environmental Entomology,* 29: 299–303.

Hagel, J.M., Yeung, E.C., and Facchini, P.J. 2008. Got milk? The secret life of laticifers. *Trends in Plant Science,* 13: 631–639.

Handley, R., Ekbom, B., and Agren, J. 2005. Variation in trichome density and resistance against a specialist insect herbivore in natural populations of *Arabidopsis thaliana. Ecological Entomology,* 30: 284–292.

Hanley, M.E., Lamont, B.B., Fairbanks, M.M., and Rafferty, C.M. 2007. Plant structural traits and their role in anti-herbivore defence. *Perspectives in Plant Ecology, Evolution and Systematics,* 8: 157–178.

He, X., Sun, Y., and Zhu, R.-L. 2013. The oil bodies of liverworts: Unique and important organelles in land plants. *Critical Reviews in Plant Sciences,* 32: 293–302.

Heil, M., and McKey, D. 2003. Protective ant–plant interactions as model systems in ecological and evolutionary research. *Annual Review of Ecology, Evolution, and Systematics,* 34: 425–553.

Helby, R.J., and McElroy, C.T. 1969. Microfloras from the Devonian and Triassic of the Beacon Group, Antarctica. *New Zealand Journal of Geology and Geophysics,* 12: 376–382.

Hernick, L.V., Landing, E., and Bartowski, K.E. 2008. Earth's oldest liverworts—*Metzgeriothallus sharonae* sp. nov. from the Middle Devonian (Givetian) of eastern New York, USA. *Review of Palaeobotany and Palynology,* 148: 154–162.

Hochuli, D.F. 1996. The ecology of plant/insect interactions: Implications of digestive strategy for feeding by phytophagous insects. *Oikos,* 75: 133–141.

Holmes, W.B.K. 1995. The late Permian megafossil flora from Cooyal, New South Wales, Australia, pp. 123–152. In: Pant D.D., Nautiyal D.D., Bhatnagar A.N., Bose M.D., and Khare P.K., eds., *Proceedings of the International Conference on Global Environment and Diversification of Plants through Geological Time.* Society of Indian Plant Taxonomists, Allahabad.

Howe, G.A., and Jander, G. 2008. Plant immunity to insect herbivores. *Annual Review of Plant Biology,* 59: 41–66.

Hudgins, J.W., Krekling, T., and Franceschi, V.R. 2003. Distribution of calcium oxalate crystals in the secondary phloem of conifers: A constitutive defense mechanism? *New Phytologist,* 159: 677–690.

Johnson, K.R. 2002. Megaflora of the Hell Creek and lower Fort Union Formations in the western Dakotas: Vegetational response to climate change, the Cretaceous–Tertiary boundary event, and rapid marine transgression. *Geological Society of America Special Paper,* 361: 329–391.

Jones, J.M., and Murchison, D.G. 1963. The occurrence of resinite in bituminous coals. *Economic Geology and the Bulletin of the Society of Economic Geologists,* 58: 263–273.

Jordan, G.J., Dillon, R.A., and Weston, P.H. 2005. Solar radiation as a factor in the evolution of scleromorphic leaf anatomy in Proteaceae. *American Journal of Botany,* 92: 789–796.

Katz, O. 2015. Silica phytoliths in angiosperms: Phylogeny and early evolutionary history. *New Phytologist,* 208: 642–646.

Kevan, P.G., Chaloner, W.G., and Savile, D.B.O. 1975. Interrelationships of early terrestrial arthropods and plants. *Palaeontology,* 18: 391 417.

Konno, K. 2011. Plant latex and other exudates as plant defense systems: Roles of various defense chemicals and proteins contained therein. *Phytochemistry,* 72: 1510–1530.

Korth, K.L., Doege, S.J., Park, S.-H., Goggin, F.L., Wang, Q., Gomez, S.K., Liu, G., Jia, L., and Nakata, P.A. 2006. *Medicago truncatula* mutants demonstrate the role of plant calcium oxalate crystals as an effective defense against chewing insects. *Plant Physiology,* 141: 188–195.

Krassilov, V., and Karasev, E. 2008. First evidence of plant–arthropod interaction at the Permian–Triassic boundary in the Volga Basin, European Russia. *Alavesia,* 2: 247–252.

Krings, M., and Harper, C.J. 2019. Fungal intruders of enigmatic propagule clusters occurring in microbial mats from the Lower Devonian Rhynie chert. *Paläontologische Zeitschrift,* 93: 135–149.

Krings, M., Kellogg, D.W., Kerp, H., and Taylor, T.N. 2003a. Trichomes of the seed fern *Blanzyopteris praedentata*: Implications for plant–insect interactions in the Late Carboniferous. *Botanical Journal of the Linnean Society,* 141: 133–149.

Krings, M., Kerp, H., Taylor, T.N., and Taylor, E.L. 2003b. How Paleozoic vines and lianas got off the ground: On scrambling and climbing Carboniferous–early Permian pteridosperms. *Botanical Review,* 69: 204–224.

Krings, M., Taylor, T.N., and Kellogg, D.W. 2002. Touch-sensitive glandular trichomes: A mode of defence against herbivorous arthropods in the Carboniferous. *Evolutionary Ecology Research,* 4: 779–786.

Krings, M., Taylor, T.N., and Taylor, E.L. 2004. Structural diversity and spatial arrangement of trichomes in a Carboniferous seed fern, pp. 61–69. In: Srivastava P.C., ed., *Vistas in Palaeobotany and Plant Morphology: Evolutionary and Environmental Perspectives*. Professor D.D. Pant Memorial Volume, U.P. Offset, Lucknow.

Kustatscher, E., Ash, S.R., Karasev, E., Pott, C., Vajda, V., Yu, J., and McLoughlin, S. 2018. Flora of the Late Triassic, pp. 545–622. In: Tanner, L., ed., *The Late Triassic World*. Springer, Cham, Switzerland.

Kustatscher, E., Franz, M., Heunisch, C., Reich, M., and Wappler, T. 2014. Floodplain habitats of braided river systems: Depositional environment, flora and fauna of the Solling Formation (Buntsandstein, Lower Triassic) from Bremke and Fürstenberg (Germany). *Palaeobiodiversity and Palaeoenvironments,* 94: 237–270.

Labandeira, C.C. 1998. Early history of arthropod and vascular plant associations. *Annual Review of Earth and Planetary Sciences,* 26: 329–377.

Labandeira, C.C. 2005. The fossil record of insect extinction: New approaches and future directions. *American Entomologist,* 51: 14–29.

Labandeira, C.C. 2006a. The four phases of plant–arthropod associations in deep time. *Geologica Acta,* 4: 409–438.

Labandeira, C.C. 2006b. Silurian to Triassic plant and insect clades and their associations: New data, a review, and interpretations. *Arthropod Systematics and Phylogeny,* 64: 53–94.

Labandeira, C.C. 2007a. The origin of herbivory on land: Initial patterns of plant tissue consumption by arthropods. *Insect Science,* 14: 259–275.

Labandeira, C.C. 2007b. Assessing the fossil record of plant–insect associations: Ichnodata versus body-fossil data, pp. 9–26. In: Bromley, R.G., Buatois, L.A., Mángano, G., Genise, J.F., and Melchor, R.N., eds., *Sediment–Organism Interactions: A Multifaceted Ichnology. SEPM Special Publication 88.* Society for Sedimentary Geology, Tulsa, Oklahoma.

Labandeira, C.C. 2013. Deep-time patterns of tissue consumption by terrestrial arthropod herbivores. *Die Naturwissenschaften,* 100: 355–364.

Labandeira, C.C. 2014a. Amber. Reading and writing of the fossil record: Preservational pathways to exceptional fossilization. *Paleontological Society Papers,* 20: 163–216.

Labandeira, C.C. 2014b. Why did terrestrial insect diversity not increase during the angiosperm radiation? Mid-Mesozoic, plant-associated insect lineages harbor some clues, pp. 261–299. In: Pontarotti, P., ed., *Evolutionary Biology: Genome Evolution, Speciation, Coevolution and the Origin of Life.* Springer, Cham, Switzerland.

Labandeira, C.C. 2019. The fossil record of insect mouthparts: Innovation, functional convergence and associations with other organisms, p. 567–671. In: Krenn, H., ed., *Insect Mouthparts—Form, Function, Development and Performance.* Zoological Monographs, vol. 5. Springer Nature.

Labandeira, C.C., Anderson, J.M., and Anderson, H.M. 2018. Expansion of arthropod herbivory in Late Triassic South Africa: The Molteno Biota, Aasvoëlberg 411 site and developmental biology of a gall, pp. 623–719. In: Tanner, L., ed., *The Late Triassic World.* Springer, Cham, Switzerland.

Labandeira, C.C., and Currano, E.D. 2013. The fossil record of plant–insect dynamics. *Annual Review of Earth and Planetary Sciences,* 41: 287–311.

Labandeira, C.C., Dilcher, D.L., Davis, D.R., and Wagner, D.L. 1994. Ninety-seven million years of angiosperm–insect association: Paleobiological insights into the meaning of coevolution. *Proceedings of the National Academy of Sciences of the United States of America,* 91: 12278–12282.

Labandeira, C.C., Johnson, K.R., and Lang, P. 2002. A preliminary assessment of insect herbivory across the Cretaceous/Tertiary boundary: Extinction and minimal rebound, pp. 296–327. In: Hartman, J.H., Johnson, K.R., and Nichols, D.J., eds., *The Hell Creek Formation and the Cretaceous–Tertiary Boundary in the Northern Great Plains: An Integrated Continental Record at the End of the Cretaceous. Geological Society of America Special Paper,* 361.

Labandeira, C.C., Kustatscher, E., and Wappler, T. 2016. Floral assemblages and patterns of insect herbivory during the Permian to Triassic of Northeastern Italy. *PLOS ONE,* 11: e0165205.

Labandeira, C.C., Tremblay, S.L., Bartowski, K.E., and VanAller Hernick, L. 2014. Middle Devonian liverwort herbivory and antiherbivore defence. *New Phytologist,* 202: 247–258.

Lamont, B.B., Groom, P.K., and Cowling, R.M. 2002. High leaf mass per area of related species assemblages may reflect low rainfall and carbon isotope discrimination rather than low phosphorus and nitrogen concentrations. *Functional Ecology*, 16: 403–412.

Levin, D.A. 1973. The role of trichomes in plant defense. *Quarterly Review of Biology*, 48: 3–15.

Lev-Yadun, S. 2001. Aposematic (warning) coloration associated with thorns in higher plants. *Journal of Theoretical Biology*, 210: 385–388.

Li, H., and Taylor, D.W. 1998. *Aculeovinea yunguiensis* gen. et sp. nov. (Gigantopteridales), a new taxon of gigantopterid stem from the Upper Permian of Guizhou Province, China. *International Journal of Plant Sciences*, 159: 1023–1033.

Liu, L., Wang, D., Meng, M., and Xue, J. 2017. Further study of Late Devonian seed plant *Cosmosperma polyloba*: Its reconstruction and evolutionary significance. *BMC Evolutionary Biology*, 17: 149.

Liu, X., Shi, G., Xia, F., Lu, X., Wang, B., and Engel, M.S. 2018. Liverwort mimesis in a Cretaceous lacewing larva. *Current Biology*, 28: 1475–1481.

Liu, Z.J., Hou, Y.M., and Wang, X. 2019. *Zhangwuia*: An enigmatic organ with a bennettitalean appearance and enclosed ovules. *Earth and Environmental Science Transactions of the Royal Society of Edinburgh*, 108: 419–428.

Lyons, P.C., Finkelman, R.B., Thompson, C.L., Brown, F.W., and Hatcher, P.G. 1982. Properties, origin and nomenclature of rodlets of the inertinite maceral group in coals of the central Appalachian basin, USA. *International Journal of Coal Geology*, 1: 313–346.

Mahlberg, P.G., and Störr, M. 1989. Fossil rubber in brown coal deposits: An overview. *Zeitschrift für Geologische Wissenschaften*, 17: 475–488.

Manetas, Y. 2003. The importance of being hairy: The adverse effects of hair removal on stem photosynthesis of *Verbascum speciosum* are due to solar UV-B radiation. *New Phytologist*, 158: 503–508.

Mark, D.F., Rice, C.M., Fallick, A.E., Trewin, N.H., Lee, M.R., Boyce, A., and Lee, J.K.W. 2011. ^{40}Ar/^{39}Ar dating of hydrothermal activity, biota and gold mineralization in the Rhynie hot-spring system, Aberdeenshire, Scotland. *Geochimica et Cosmochimica Acta*, 75: 555–569.

Massey, F.P., Ennos, A.R., and Hartley, S.E. 2007. Herbivore specific induction of silica-based plant defences. *Oecologia*, 152: 677–683.

McCoy, V.E., Boom, A., Solórzano Kraemer, M.M., and Gabbott, S.E. 2017. The chemistry of American and African amber, copal, and resin from the genus *Hymenaea*. *Organic Geochemistry*, 113: 43–54.

McGregor, D.C. 1973. Lower and Middle Devonian spores of eastern Gaspé, Canada. II. Biostratigraphy. *Palaeontographica Abteilung B*, 142: 1–77.

McKellar, R.C., Wolfe, A.P., Muehlenbachs, K., Tappert, R., Engel, M.S., Cheng, T., and Sanchez-Azofeifa, G.A. 2011. Insect outbreaks produce distinctive carbon isotope signatures in defensive resins and fossiliferous ambers. *Proceedings of the Royal Society B*, 278: 3219–3224.

McLoughlin, S., Martin, S.K., and Beattie, R. 2015. The record of Australian Jurassic plant–arthropod interactions. *Gondwana Research*, 27: 940–959.

McNaughton, S.J., Tarrants, J.L., McNaughton, M.M., and Davis, R.D. 1985. Silica as a defense against herbivory and a growth promotor in African grasses. *Ecology*, 66: 528–535.

Milewski, A.V., Young, T.P., and Madden, D. 1991. Thorns as induced defenses: Experimental evidence. *Oecologia,* 86: 70–75.

Miller, I.M., Johnson, K.R., Kline, D.E., Titus, A.L., and Loewen, M.A. 2013. A Late Campanian flora from the Kaiparowits, pp. 107–131. In: Titus, A., and Loewen, M., eds., *At the Top of the Grand Staircase: The Late Cretaceous of Southern Utah.* Indiana University Press, Indianapolis.

Misof, B., Liu, S., Meusemann, K., Peters, R.S., Donath, A., Mayer, C., Frandsen, P.B., et al. 2014. Phylogenomics resolves the timing and pattern of insect evolution. *Science,* 346: 763–767.

Mithöfer, A., and Boland, W. 2012. Plant defense against herbivores: Chemical aspects. *Annual Review of Plant Biology,* 63: 431–450.

Mitrakos, K. 1980. A theory for Mediterranean plant life [evergreen sclerophyllous shrubs, climatic stresses, Mediterranean climate]. *Acta Oecologica–Oecologia Plantarum,* 1: 245–252.

Mohr, B.A.R., Bernardes-de-Oliveira, M.E.C., and Taylor, D.W. 2008. *Pluricarpellatia,* a nymphaealean angiosperm from the Lower Cretaceous of northern Gondwana (Crato Formation, Brazil). *Taxon,* 57: 1147–1158.

Moisan, P., Voigt, S., Pott, C., Buchwitz, M., Schneider, J.W., and Kerp, H. 2011. Cycadalean and bennettitalean foliage from the Triassic Madygen Lagerstätte (SW Kyrgyzstan, Central Asia). *Review of Palaeobotany and Palynology,* 164: 93–108.

Myers, J. 1991. Thorns, spines, prickles, and hairs: Are they stimulated by herbivory and do they deter herbivores, pp. 325–344. In: Tallamy, D.W., and Raupp, M.J., eds., *Phytochemical Induction by Herbivores.* Wiley, New York.

Na, Y.L., Sun, C.L., Wang, H., Dilcher, D.L., Yang, Z.Y., Li, T., Li, Y.F. 2018. Insect herbivory and plant defense on ginkgoalean and bennettitalean leaves of the Middle Jurassic Daohugou Flora from Northeast China and their paleoclimatic implications. *Palaeoworld,* 27: 202–210.

Naugolnykh, S.V., and Ponomarenko, A.G. 2010. Possible traces of feeding by beetles in coniferophyte wood from the Kazanian of the Kama River Basin. *Paleontological Journal,* 44: 468–474.

Nel, A., Prokop, J., and Ross, A.J. 2008. New genus of leaf-mimicking katydids (Orthoptera: Tettigoniidae) from the Late Eocene–Early Oligocene of France and England. *Comptes Rendus Palevol,* 7: 211–216.

Nicholson, D.B., Mayhew, P.J., and Ross, A.J. 2015. Changes to the fossil record of insects through fifteen years of discovery. *PLOS ONE,* 10: e0128554.

Niedźwiedzki, G., Szrek, P., Narkiewicz, K., Narkiewicz, M., and Ahlberg, P.E. 2010. Tetrapod trackways from the early Middle Devonian period of Poland. *Nature,* 463: 43–48.

Nobel, P.S., Zaragoza, L.J., and Smith, W.K. 1975. Relation between mesophyll surface area, photosynthetic rate, and illumination level during development for leaves of *Plectranthus parviflorus* Henckel. *Plant Physiology,* 55: 1067–1070.

Obeso, J.R. 1997. The induction of spinescence in European holly leaves by browsing ungulates. *Plant Ecology,* 129: 149–156.

Otto, A., Simoneit, B.R.T., and Rember, W.C. 2003. Resin compounds from the seed cones of three fossil conifer species from the Miocene Clarkia flora, Emerald Creek, Idaho, USA, and from related extant species. *Review of Palaeobotany and Palynology,* 126: 225–241.

Otto, A., Simoneit, B.R.T., and Rember, W.C. 2005. Conifer and angiosperm biomarkers in clay sediments and fossil plants from the Miocene Clarkia Formation, Idaho, USA. *Organic Geochemistry,* 36: 907–922.

Papier, F., Nel, A., Grauvogel-Stamm, L., and Gall, J.C. 1997. La plus ancienne sauterelle Tettigoniidae, Orthoptera (Trias, NE France): Mimétisme ou exaptation? *Paleontological Journal,* 71: 71–77.

Paruya, D.K., Chakraborty, T., Bera, S., and Dutta, S. 2018. Amber embalms essential oils: A rare preservation of monoterpenoids in fossil resins from eastern Himalaya. *Palaios,* 33: 218–227.

Pemberton, R.W. 1992. Fossil extrafloral nectaries, evidence for the ant–guard antiherbivore defense in an Oligocene *Populus. American Journal of Botany,* 79: 1242–1246.

Pérez-Barbería, F.J., and Gordon, I.J. 1998. Factors affecting food comminution during chewing in ruminants: A review. *Biological Journal of the Linnean Society,* 63: 233–256.

Peris, D., Pérez-de la Fuente, R., Peñalver, E., Delclòs, X., Barrón, E., and Labandeira, C.C. 2017. False blister beetles and the expansion of gymnosperm-insect pollination modes before angiosperm dominance. *Current Biology,* 27. doi:10.1016/jcub.2017.02.009.

Pigg, K.B., and Trivett, M.L. 1994. Evolution of the glossopterid gymnosperms from Permian Gondwana. *Journal of Plant Research,* 107: 461–477.

Pinheiro, E.R., Iannuzzi, R., and Duarte, L.D. 2016. Insect herbivory fluctuations through geological time. *Ecology,* 97: 2501–2510.

Plumstead, E.P. 1963. The influence of plants and environment on the developing animal life of Karroo times. *South African Journal of Science,* 59: 147–152.

Pott, C., Labandeira, C.C., Krings, M., and Kerp, H. 2008. Fossil insect eggs and ovipositional damage on bennettitalean leaf cuticles from the Carnian (Upper Triassic) of Austria. *Journal of Paleontology,* 82: 778–789.

Pott, C., and McLoughlin, S. 2009. Bennettitalean foliage in the Rhaetian–Bajocian (latest Triassic–Middle Jurassic) floras of Scania, southern Sweden. *Review of Palaeobotany and Palynology,* 158: 117–166.

Pott, C., and McLoughlin, S. 2014. Divaricate growth habit in Williamsoniaceae (Bennettitales): Unravelling the ecology of a key Mesozoic plant group. *Palaeobiodiversity and Palaeoenvironments,* 94: 307–325.

Pott, C., McLoughlin, S., Wu, S., and Friis, E.M. 2012. Trichomes on the leaves of *Anomozamites villosus* sp. nov. (Bennettitales) from the Daohugou beds (Middle Jurassic), Inner Mongolia, China: Mechanical defence against herbivorous arthropods. *Review of Palaeobotany and Palynology,* 169: 48–60.

Pott, C., Wang, X., and Zheng, X. 2015. *Wielandiella villosa* comb. nov. from the Middle Jurassic of Daohugou, China: More evidence for divaricate plant architecture in Williamsoniaceae. *Botanica Pacifica,* 42: 137–148.

Potter, D.A., and Kimmerer, T.W. 1988. Do holly leaf spines really deter herbivory? *Oecologia,* 75: 216–221.

Prevec, R., Labandeira, C.C., Neveling, J., Gastaldo, R.A., Looy, C.V., and Bamford, M. 2009. Portrait of a Gondwanan ecosystem: A new late Permian fossil locality from KwaZulu-Natal, South Africa. *Review of Palaeobotany and Palynology,* 156: 454–493.

QulRing, D.T., Timmins, P.R., and Park, S.J. 1992. Effect of variations in hooked tri-chome densities of *Phaseolus vulgaris* on longevity of *Liriomyza trifolii* (Diptera: Agromyzidae) adults. *Environmental Entomology,* 21: 1357–1361.

Reisz, R.R., and Fröbisch, J. 2014. The oldest caseid synapsid from the Late Pennsyl-vanian of Kansas, and the evolution of herbivory in terrestrial vertebrates. *PLOS ONE,* 9: e94518.

Rieseberg, L.H., and Soltis, D.E. 1987. Flavonoids of fossil Miocene *Platanus* and its extant relatives. *Biochemical Systematics and Ecology,* 15: 109–112.

Robbins, C.T. 1993. *Wildlife Nutrition and Feeding.* Elsevier, New York.

Roberts, E.M., Sampson, S.D., Deino, A.L., Bowring, S.A., and Buchwaldt, R. 2013. The Kaiparowits Formation: A remarkable record of Late Cretaceous terrestrial envi-ronments, ecosystems, and evolution in western North America, pp. 85–106. In: Titus, A., and Loewen, M., eds., *At the Top of the Grand Staircase: The Late Creta-ceous of Southern* Utah. Indiana University Press, Indianapolis.

Schuman, M.C., and Baldwin, I.T. 2016. The layers of plant responses to insect herbi-vores. *Annual Review of Entomology,* 61: 373–394.

Scott, A.C. 1991. Evidence for plant–arthropod interactions in the fossil record. *Geol-ogy Today,* 7: 58–61.

Scott, A.C. 1992. Trace fossils of plant–arthropod interactions. *Short Courses in Paleon-tology,* 5: 197–223.

Scott, A.C., Anderson, J.M., and Anderson, H.M. 2004. Evidence of plant–insect inter-actions in the Upper Triassic Molteno Formation of South Africa. *Journal of the Geological Society,* 161: 401–410.

Scott, A.C., Stephenson, J., and Chaloner, W.G. 1992. Interaction and coevolution of plants and arthropods during the Palaeozoic and Mesozoic. *Philosophical Trans-actions of the Royal Society of London B,* 335: 129–165.

Scott, A.C., and Taylor, T.N. 1983. Plant/animal interactions during the Upper Carbon-iferous. *Botanical Review,* 49: 259–307.

Seilacher, A. 1970. Begriff und bedeutung der Fossil-Lagerstätten. Neues Jahrbuch für Geologie und Paläontologie, 1970: 34–39.

Seyfullah, L.J., Beimforde, C., Dal Corso, J., Perrichot, V., Rikkinen, J., and Schmidt, A.R. 2018. Production and preservation of resins—Past and present. *Biological Reviews,* 93: 1684–1714.

Shcherbakov, D.E. 2008. Insect recovery after the Permian/Triassic crisis. *Alavesia,* 2: 125–131.

Shear, W.A., and Kukalová-Peck, J. 1990. The ecology of Paleozoic terrestrial arthro-pods: The fossil evidence. *Canadian Journal of Zoology,* 68: 1807–1834.

Simoneit, B.R.T., Otto, A., and Wilde, V. 2003. Novel phenolic biomarker triterpenoids of fossil laticifers in Eocene brown coal from Geiseltal, Germany. *Organic Geo-chemistry,* 34: 121–129.

Slater, B.J., McLoughlin, S., and Hilton, J. 2012. Animal–plant interactions in a Middle Permian permineralised peat of the Bainmedart Coal Measures, Prince Charles Mountains, Antarctica. *Palaeogeography, Palaeoclimatology, Palaeoecology,* 363: 109–126.

Souza Pinheiro, E.R., de, Iannuzzi, R., and Tybusch, G.P. 2012. Specificity of leaf dam-age in the Permian "*Glossopteris* Flora": A quantitative approach. *Review of Palae-obotany and Palynology,* 174: 113–121.

Stein, W.E., Jr., Wight, D.C., and Beck, C.B. 1983. *Arachnoxylon* from the Middle Devonian of southwestern Virginia. *Canadian Journal of Botany,* 61: 1283–1299.

Steinbauer, M.J., Clarke, A.R., and Madden, J.L. 1998. Oviposition preference of a *Eucalyptus* herbivore and the importance of leaf age on interspecific host choice. *Ecological Entomology,* 23: 201–206.

Stephenson, J., and Scott, A.C. 1992. The geological history of insect-related plant damage. *Terra Nova,* 4: 542–552.

Strömberg, C.A.E., Di Stilio, V.S., and Song, Z. 2016. Functions of phytoliths in vascular plants: An evolutionary perspective. *Functional Ecology,* 30: 1286–1297.

Strömberg, C.A.E., and McInerney, F.A. 2011. The Neogene transition from C_3 to C_4 grasslands in North America: Assemblage analysis of fossil phytoliths. *Paleobiology,* 37: 50–71.

Strullu-Derrien, C., McLoughlin, S., Philippe, M., Mørk, A., and Strullu, D.G. 2012. Arthropod interactions with bennettitalean roots in a Triassic permineralized peat from Hopen, Svalbard Archipelago (Arctic). *Palaeogeography, Palaeoclimatology, Palaeoecology,* 348: 45–58.

Sues, H.D., and Reisz, R.R. 1998. Origins and early evolution of herbivory in tetrapods. *Trends in Ecology and Evolution,* 13: 141–145.

Taylor, E.L., Taylor, T.N., and Krings, M. 2009. *Paleobotany: The Biology and Evolution of Fossil Plants.* Academic Press, San Diego.

Trant, C.A., and Gensel, P.G. 1985. Branching in *Psilophyton*: A new species from the Lower Devonian of New Brunswick, Canada. *American Journal of Botany,* 72: 1256–1273.

Traw, B.M., and Dawson, T.E. 2002. Differential induction of trichomes by three herbivores of black mustard. *Oecologia,* 131: 526–532.

Turner, I.M. 1994. Sclerophylly: Primarily protective? *Functional Ecology,* 8: 669–675.

Van Soest, P.J. 1982. *Nutritional Ecology of the Ruminant.* Durham and Downey, Portland, Oregon.

Vicari, M., and Bazely, D.R. 1993. Do grasses fight back? The case for antiherbivore defences. *Trends in Ecology and Evolution,* 8: 137–141.

Villar de Seoane, L. 2001. Cuticular study of Bennettitales from the Springhill Formation, Lower Cretaceous of Patagonia, Argentina. *Cretaceous Research,* 22: 461–479.

Walling, L.L. 2000. The myriad plant responses to herbivores. *Journal of Plant Growth Regulation,* 19: 195–216.

Wang, D., Liu, L., Meng, M., Xue, J., Liu, T., and Guo, Y. 2014. *Cosmosperma polyloba* gen. et sp. nov., a seed plant from the Upper Devonian of South China. *Die Naturwissenschaften,* 101: 615–622.

Wang, D.-M., Meng, M.-C., and Guo, Y. 2016. Pollen organ *Telangiopsis* sp. of Late Devonian seed plant and associated vegetative frond. *PLOS ONE,* 11: e0147984.

Wang, M., Béthoux, O., Bradler, S., Jacques, F.M., Cui, Y., and Ren, D. 2014. Under cover at pre-angiosperm times: A cloaked phasmatodean insect from the Early Cretaceous Jehol biota. *PLOS ONE,* 9: e91290.

Wang, Y., Labandeira, C.C., Shih, C., Ding, Q., Wang, C., Zhao, Y., and Ren, D. 2012. Jurassic mimicry between a hangingfly and a ginkgo from China. *Proceedings of the National Academy of Sciences of the United States of America,* 109: 20514–20519.

Wappler, T. 2010. Insect herbivory close to the Oligocene–Miocene transition—A quantitative analysis. *Palaeogeography, Palaeoclimatology, Palaeoecology*, 292: 540–550.

Wappler, T., Currano, E.D., Wilf, P., Rust, J., and Labandeira, C.C. 2009. No post-Cretaceous ecosystem depression in European forests? Rich insect-feeding damage on diverse middle Palaeocene plants, Menat, France. *Proceedings of the Royal Society B: Biological Sciences*, 276: 4271–4277.

Wappler, T., Kustatscher, E., and Dellantonio, E. 2015. Plant–insect interactions from Middle Triassic (late Ladinian) of Monte Agnello (Dolomites, N-Italy)—Initial pattern and response to abiotic environmental perturbations. *PeerJ*, 3: e921.

Wappler, T., Labandeira, C.C., Rust, J., Frankenhäuser, H., and Wilde, V. 2012. Testing for the effects and consequences of mid Paleogene climate change on insect herbivory. *PLOS ONE*, 7: e40744.

War, A.R., Paulraj, M.G., Ahmad, T., Buhroo, A.A., Hussain, B., Ignacimuthu, S., and Sharma, H.C. 2012. Mechanisms of plant defense against insect herbivores. *Plant Signaling and Behavior*, 7: 1306–1320.

Ward, P., Labandeira, C.C., Laurin, M., and Berner, R.A. 2006. Confirmation of Romer's Gap as a low oxygen interval constraining the timing of initial arthropod and vertebrate terrestrialization. *Proceedings of the National Academy of Sciences of the United States of America*, 103: 16818–16822.

Weber, M.G., and Keeler, K.H. 2013. The phylogenetic distribution of extrafloral nectaries in plants. *Annals of Botany*, 111: 1251–1261.

Wedmann, S. 2010. A brief review of the fossil history of plant masquerade by insects. *Palaeontographica Abteilung B*, 283: 175–182.

Werker, E. 2000. Trichome diversity and development. *Advances in Botanical Research*, 31: 1–35.

Westerbergh, A., and Nyberg, A.-B. 1995. Selective grazing of hairless *Silene dioica* plants by land gastropods. *Oikos*, 73: 289–298.

White, M.E., and Frazier, J. 1986. *Greening of Gondwana*. Reeds, New South Wales.

Wilf, P., Labandeira, C.C., Johnson, K.R., Coley, P.D., and Cutter, A.D. 2001. Insect herbivory, plant defense, and early Cenozoic climate change. *Proceedings of the National Academy of Sciences of the United States of America*, 98: 6221–6226.

Wilson, S.L., and Kerley, G.I.H. 2003. The effect of plant spinescence on the foraging efficiency of bushbuck and boergoats: Browsers of similar body size. *Journal of Arid Environments*, 55: 150–158.

Wu, Y., You, H.-L., and Li, X.-Q. 2018. Dinosaur–associated *Poaceae* epidermis and phytoliths from the Early Cretaceous of China. *National Science Review*, 5: 721–727.

Xu, Q.Q., Jin, J., and Labandeira, C.C. 2018. Williamson Drive: Herbivory of a north-central Texas flora of latest Pennsylvanian age shows discrete component community structure, expansion of piercing and sucking, and plant counterdefenses. *Review of Palaeobotany and Palynology*, 251: 28–72.

Young, G.C. 1993. Middle Palaeozoic macrovertebrate biostratigraphy of eastern Gondwana, pp. 208–251. In: Long, J.A., ed., *Palaeozoic Vertebrate Biostratigraphy and Biogeography*. Belhaven Press, London.

Young, T.P., and Okello, B.D. 1998. Relaxation of an induced defense after exclusion of herbivores: Spines on *Acacia drepanolobium*. *Oecologia*, 115: 508–513.

Young, T.P., Stanton, M.L., and Christian, C.E. 2003. Effects of natural and simulated herbivory on spine lengths of *Acacia drepanolobium* in Kenya. *Oikos,* 101: 171–179.

Zodrow, E.L. 2007. Reconstructed tree fern *Alethopteris zeilleri* (Carboniferous, Medullosales). *International Journal of Coal Geology,* 69: 68–89.

Zodrow, E.L., and Mastalerz, M. 2018. A study of functional groups of trichomes of *Odontopteris cantabrica*: Implications for molecular taxonomy (Seed fern, Carboniferous, Canada). *International Journal of Coal Geology,* 198: 77–87.

CHAPTER 9

Color in Living and Fossil Plants

The Search for Biological Pigments in the Paleobotanical Record

CAROLE T. GEE AND VICTORIA E. MCCOY

A B S T R A C T | Although the riot of colors found in living plants has been well studied, colors in fossil plants have received little attention. In the extant flora, plant color is produced by biological pigments, also known as biochromes, as well as by structural color and bioluminescence. In the paleobotanical record, only a handful of cases of pigmentary color are known, although more is understood about colors resulting from diagenesis. Here we review color in the modern flora and survey it briefly in the paleobotanical record in order to assess what we know about biochromes in fossil plants. Living plants are colored green, yellow, orange, red, blue, violet, ultraviolet (UV), brown, gray, pink, and white. They can also appear metallic blue, silver, and iridescent. Hues in fossil plants, however, are generally limited to the browns and blacks of carbonized plant matter, although there are sometimes vivid shades found in fossil wood that are produced by geological processes. While it should theoretically be possible to detect a variety of extant plant pigments in the rock record, the study of plant biochromes in the fossil record is rare and mostly restricted to chlorophyll derivatives and, to a lesser extent, to carotenoids and flavonoids. Green Cenozoic leaves and a distinctly pink-colored Jurassic alga are among the few intriguing examples of fossil plants with pigmentation. Nevertheless, in view of the many new analytical techniques and methods currently available, and in conjunction with high-resolution imaging, future forays into the color research of fossil plants have the potential to be very fruitful. |

Introduction

Color! What a deep and mysterious language, the language of dreams.
PAUL GAUGUIN

I found that I could say things with color and shapes that I couldn't
say any other way.
GEORGIA O'KEEFFE

The plant world is awash in color. Conifer forests shimmer in refreshing greens and deep browns, shortgrass prairies glow golden in summer, deciduous leaves radiate in crimsons and yellows in the fall, and many desert plants sport a cool blue-green color all year round. Despite their intense and multifaceted effects on us, the colors in plants evolved not to evoke human emotions but to facilitate plant growth and to promote the plant's physical well-being, reproduction, and survival. Many botanical colors are the product of metabolic pathways that are only found in plants.

Plant colors span the visible color spectrum—from red, orange, and yellow to green, blue, indigo, and violet. They also include white, pink, tan, and brown, in addition to a metallic silver and UV.* Nearly all hues are produced by plants as pigments, although a few cases of structural color and bioluminescence have been described. In contrast, many more cases of structural color have been studied in animals, in addition to pigmentary colors and bioluminescence.

Pigments are chemical compounds in plant tissues that reflect the wavelengths of visible light that produce a certain color, while also absorbing all other wavelengths. In the case of chlorophyll, for example, the pigment reflects green wavelengths and absorbs reds and blues, bouncing verdant color into the color receptors in our eyes. Structural colors are not produced by pigments, but by the physical or structural modification of surfaces through periodically arranged materials. These modifications are micro- or nanostructures that interfere with visible light and result in glowing, dazzling, metallic, or shiny hues. The iridescent wing scales of butterflies, for instance, come from structural color, as does the black coloration of some insect bodies. In plants, the study of structural colors is still in its infancy, but it offers much promise for further investigations (Lee 2007; Glover and Whitney 2010). Some plant colors are produced in their entirety from structural color,

---

* UV color can be seen by many birds and insects, and also by some fish, amphibians, reptiles, marsupials, and mammals, such as rats, bats, and reindeer (Hunt et al. 2009). Some humans—children and young adults—can see UV under special conditions, and others can see near-UV color (Lynch and Livingston 2001; Dash and Dash 2009).

such as the metallic blue berries of *Elaeocarpus angustifolius*, the blue quandong, which is native to eastern Australia and New Caledonia (plate 9.1A, B). Alternatively, plant color can result from a combination of structural color and pigment, such as in the glittery blue-colored lower lip of the flower of *Orchis speculum*, or mirror orchid, in the Mediterranean (plate 9.1C).

Another rare source of color in plants is bioluminescence, which is a bright, short-lived color of cold blue light. The best-known examples of bioluminescence occur in animals, such as in fireflies, glow worms, and deep-sea fish. This biochemical emission is generated in animals by the oxidation of a class of compounds called luciferins. Bioluminescence in plants is produced only by some marine dinoflagellates (Valiadi and Iglesias-Rodriguez 2013). During a "red tide," or algal bloom in the sea, minute dinoflagellate algae, measuring only 30 μm to 1 mm, individually flash a bright blue light when agitated in the water (plate 9.1D). This leads to spectacular light shows in a wave-agitated sea or on the edges of waves pounding on a beach. In contrast to the luciferins in animals, the bioluminescence in dinoflagellates derive from a tetrapyrrole, a derivative of chlorophyll (Morse and Mittag 2000).

However dramatic the flashy bioluminescence of marine algae or the electric blue structural color of quandong berries, it is actually pigmentary color that is the basis of virtually all plant coloration. Chlorophyll is the major pigment in nearly all plants, and there are five forms of chlorophyll. Chlorophyll *a* is the universal form and is found in cyanobacteria, algae, and higher plants. Chlorophyll *b* is produced in green algae and higher plants; chlorophyll *c* in diatoms, dinoflagellates, and brown algae; and chlorophyll *d* only in red algae. Recently, a new type of chlorophyll, chlorophyll *f*, was found in the cyanobacteria that form the limestone mounds of stromatolites in Shark Bay, Western Australia (Chen et al. 2010).

In addition to chlorophyll type, plants can be sorted out from one another taxonomically on the basis of accessory pigments. Cyanobacteria, which were formerly referred to as blue-green algae, have the accessory pigment phycocyanobilin, which imparts a bluish tinge. Thus, the combination of the green chlorophyll and blue phycocyanobilin pigments gives cyanobacteria their very distinctive blue-green color (plate 9.1E).

Algae are traditionally sorted by the presence of accessory pigments (Bold and Wynne 1978). While green algae contain chlorophylls *a* and *b*, which are responsible for their verdant color, red algae also have phycoerythrin, which produces a red hue, and phycocyanin, which is a light blue color. In the red alga *Porphyra*, which is known in the Japanese culinary world as nori, the deep purple color of the fresh seaweed is produced by the combination of these red and blue pigments (Kohata et al. 2010). In the case

Plate 9.1. Structural color, bioluminescence, and the pigmentary colors of chlorophylls, anthocyanins, and carotenoids in plants. Cluster of dark blue berries (**A**) and a close-up (**B**) of *Elaeocarpus angustifolius* (commonly known as the blue quandong), showing its shiny indigo surface produced exclusively by structural color, at the Brisbane Botanic Gardens Mt. Coot-tha, Australia. Photos: CTG. (**C**) The sparkly blue "mirror" of the mirror orchid, *Orchis speculum* subspecies *speculum*, which results from an anthocyanin coupled with structural color. Photo: Esculapio, Creative Commons Attribution–ShareAlike 2.5 Generic license. (**D**) Blue-colored light from bioluminescent dinoflagellates during a red tide. Photo: Catalano82, Creative Commons Attribution 2.0 Generic License. (**E**) Marine stromatolite at Shark Bay, Australia, containing chlorophylls *a* and *f*, as well as the bluish pigment phycocyanobilin, releasing here numerous oxygen bubbles as a byproduct of photosynthesis. Photo courtesy of Louise Woo. (**F**) The spores of *Equisetum* are unusual in containing chlorophyll. They are typically 50 μm in diameter. Photo: CTG. (**G**) Anthocyanin (reds) and carotenoid (yellows) pigments in a grape leaf, which show up in the fall after the breakdown of chlorophyll (greens), in the Ahr Valley near Bonn, Germany, in early November. Photo: CTG.

of brown algae or marine kelps, the brown to olive-green colors come from fucoxanthin. A range of yellowish to brownish colors are produced in microscopic algae by a potpourri of pigments: the dinoflagellates of bioluminescence fame (see above) also contain beta-carotene and xanthophylls, which are responsible for their golden-brown colors, while the siliceous diatoms produce fucoxanthin, carotenoids, and xanthophylls, which make up their yellowish-green to dark brown colors.

While biologically produced pigments, or biochromes, in living plants have received much study, little is known about the original color of plants in the fossil record. Very few studies have searched for and confirmed the presence of pigments in fossil plants, let alone elucidated color patterns in plants. Color pigments and patterns in flowers and fruits would be instrumental in offering us insight into plant–animal interactions, for example. Similarly, differential coloration on leaves can indicate feeding traces due to wound tissue and leaf regeneration, which would also facilitate a deeper understanding of insect herbivory and strategic biochemical defenses in the leaf (McCoy et al., chap. 8).

In this chapter, we examine color in living plants and approach the question of the fossilization of plant color and pigments in the paleobotanical record. We will start by first reviewing the purpose, occurrence, and chemical structure of pigments in living plants as a basis for identifying biochromes in fossil eukaryotic plants. We will then narrow the focus to color in the leaves and wood of living plants and survey the coloration of these plant organs in the paleobotanical record in order to distinguish between fossil plant colors as either deriving from original pigmentation or through diagenetic processes. We will also review the evidence for structural color in living plants. In closing, we will offer suggestions for future studies for the detection of original coloration in fossil plants.

Major Plant Pigments in Living Plants

The chief function of color should be to serve expression
as well as possible.
    HENRI MATISSE

The pigmentary colors in living plants are produced by four major pathways: the porphyrin pathway (chlorophyll greens), isoprene pathway (carotenoid yellows, oranges, and reds), phenylpropanoid/flavonoid pathway (yellows, oranges, reds, tans, browns), and the betalain pathway (beet-red colors). A pathway may produce pigments of a limited color range, such as

Plate 9.2. Colors in leaves. (**A**) Molecular structure of chlorophyll *a*. (**B**) Variegated fern frond of the Cretan brake, *Pteris cretica*, with pale yellow pinnules rimmed with green, against a background of kelly green-colored *Selaginella*, at the Phipps Conservatory in Pittsburgh, Pennsylvania, USA. Photo: VEM. (**C**) The fresh, light to medium green colors of various ferns, again at the Phipps Conservatory. Photo: VEM. (**D**) Blue-grayish green fronds of the cycad *Encephalartos arenarius* at the Huntington Gardens, California, USA. Photo: CTG. (**E**) Dark forest green conifers on Latemar mountain in the Dolomites, South Tyrol, Italy. Photo: CTG. (**F**) *Plectranthus* (*Solenostemon*) *scutellarioides* "Wizard Coral Sunrise," commonly known as coleus, is prized for its deep red and pink colors, which are produced by anthocyanins, and the striking color patterns in its leaves. Photo: Aleksandrs Balodis, Creative Commons Attribution–ShareAlike 4.0 International license.

the chlorophyll greens arising from the porphyrin pathway or, alternatively, a wide range of pigmentary compounds that produce yellows, oranges, reds, tans, and browns, such as the phenylpropanoid/flavonoid pathway.

Chlorophyll is by far the most important pigment in plants, for it is responsible for the capture of sunlight and its subsequent conversion to chemical energy for growth and maintenance. Chlorophyll occurs in leaves, stems, flowers, fruits, and in other organs colored green, sometimes even in spores (plate 9.1F; plate 9.2). It is present in virtually all plants, with the exception of saprophytes—plants, fungi, or microorganisms that live on dead or decaying organic matter—as well as occurring in algae, prokaryotes such as cyanobacteria, and other photosynthesizing bacteria like prochlorophytes. The five types of chlorophyll that were mentioned earlier, which are designated as chlorophyll *a*, *b*, *c*, *d*, and *f*, differ from one another in their light absorbance, particularly in the infrared range (e.g., Airs et al. 2014).

In the plant cell, chlorophyll is sequestered in special plastid organelles called chloroplasts. The chlorophyll molecule itself centers around a Mg^{2+} cation complexed in a large ring structure (plate 9.2A) with a long hydrocarbon tail. It is structurally similar to heme of hemoglobin in humans, which, however, represents an Fe^{2+}/Fe^{3+} complex.

Carotenoid pigments absorb in the blue and indigo range, and reflect yellows, oranges, and reds (plate 9.1G; plate 9.3). As accessory pigments, they facilitate absorbance of light energy for use in photosynthesis, protect chlorophyll from photodamage, and therefore occur primarily in leaves. The glowing yellow and orange colors in leaves (plate 9.1G) are produced in the autumn when the chloroplasts and chlorophyll molecules break down and the carotenoid pigments (xanthophylls) are unmasked (Lee et al. 2003). Carotenoids are also found in roots, such as carrots and yams, and in flowers, producing colors ranging from a pale yellow to a bright orange to a deep red. They also occur in a huge variety of fruits with these colors, such as lemons, pumpkins, tomatoes, and rosehips.

In the plant cell, carotenoids are stored in plastids called chromoplasts or are found in the double membranes of chloroplasts. Chemically, carotenoids can be divided into two major classes, carotenes and xanthophylls. Carotenes consist of only carbon and hydrogen, while xanthophylls include oxygen (plate 9.3A). Perhaps the best-known carotene is the beta-carotene found in brightly colored carrots. The most common xanthophyll in plants is lutein, which is found in fruits and vegetables such as corn and orange bell peppers, but also in kiwis, grapes, and dark leafy greens like spinach. An abundance of the xanthophyll eschscholtzxanthin produces a bright orange color in the California poppy, *Eschscholzia californica* (plate 9.3B),

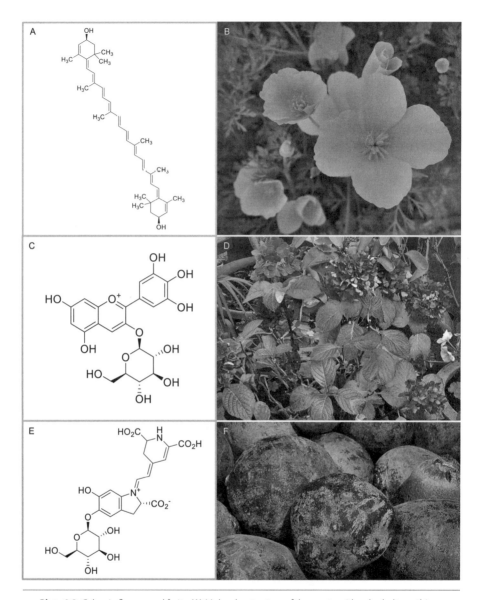

Plate 9.3. Colors in flowers and fruits. (**A**) Molecular structure of the carotenoid eschscholtzxanthin. (**B**) The bright orange color in the California poppy, *Eschscholzia californica*, is produced by carotenoids (including abundant eschscholtzxanthin) and a ridged ultrastructural surface. Photo: CTG. (**C**) Molecular structure of the anthocyanin delphinidin-3-glucoside. (**D**) *Hydrangea* is unusual because the red-violet color of its flower heads and autumnal crimson of its leaves come from a sole pigment, delphinidin-3-glucoside. Photo: CTG. (**E**) Molecular structure of the betalain pigment called betanin, which is the primary pigment in beetroots. (**F**) Even when raw, unwashed, and unpeeled, the beet-red color of *Beta vulgaris* is unmistakable. Photo: CTG.

while fucoxanthin is responsible for the yellowish-brown color of kelp and diatoms, as mentioned earlier.

Flavonoid pigments show up as ivory or cream, yellow, red, blue, and in the UV spectrum. Many food plants are high in flavonoids, and these include parsley, onions, blueberries, strawberries, black tea, green tea, oolong tea, bananas, lemons, oranges, grapefruit, red wine, and dark chocolate, to name a few (Bhagwat and Haytowitz 2015). The phenyl-propanoid/flavonoid pathway gives rise to anthocyanins (plate 9.3C), which are responsible for most of the orange-reds, pink-reds, blues, and purples in flowers (plate 9.3D), fruits, and leaves. Anthocyanins are found in red to purple fruits such as berries, currants, Concord and muscadine grapes, cherries, and eggplants, as well as in leafy vegetables like red cabbage, grains like black rice, and edible roots and tubers (Khoo et al. 2017). Anthocyanin pigments are usually contained in the large vacuole in plant cells. While flavonoids are ubiquitous in the plant kingdom, anthocyanins are patchier in their occurrence, showing up in some bryophytes, some ferns and gymnosperms, and most angiosperms (Lee 2007).

Betalains are another class of pigments (plate 9.3E) that have yellow or red to violet color effects. Typical of red betalain color is the deep wine-red hue of beetroots (plate 9.3F) or the crimson red of *Amaranthus* flowers, while yellow-colored betalains are found in cactus and *Portulaca* flowers (Lee 2007). Betalains differ fundamentally in structure from the other types of red or yellow pigments in that betalains are produced by joining two nitrogen-containing molecules derived from the amino acid tyrosine with an indole structure. Curiously, betalain pigments are restricted to the Caryophyllales, an order that notably lacks anthocyanins.

Colors in Living and Fossil Plants

I prefer living in colour.
DAVID HOCKNEY

I'm obsessed with color—never saw one I didn't like.
DALE CHIHULY

In the preceding section, the discussion focused on the major pigments in plants that produce color. Now a closer look will be made at plant organs that are most commonly found in the rock record—leaves and wood—in regard to their pigmentation in living plants. We will also carry out a brief survey of colors in fossil leaves and wood, as well as a review of the few studies on ancient biochromes in the paleobotanical record.

Leaves

They give us the greens of summers
Makes you think all the world's a sunny day
PAUL SIMON, "KODACHROME"

It's not easy being green.
KERMIT THE FROG

Color in Living Leaves

While chlorophyll green is the quintessential color of plants, a closer look at leaves reveals a wondrous array of colors and patterns. These colors range from a pale yellowish hue (plate 9.2B) to various fresh, vivid greens (plate 9.2C), and a blue-gray green (plate 9.2D) to a dark forest green (plate 9.2E). Two-toned leaves are commonly green around their margins and a different color in the center of the leaf blade (plate 9.2B, F). Variation in leaf color comes from the other molecules involved in photosynthesis, such as the carotenoids, flavonoids, anthocyanins, and betalains. These accessory pigments form complexes with the chlorophyll to supplement the wavelengths that are being absorbed, thereby refining the color of the leaf.

Some leaves also include shades of red, pink, or purple among the green (plate 9.2F). These reddish hues are provided by the anthocyanins, or by betalains in the order Caryophyllales. The spectacular red and purple coloration in deciduous leaves in the fall is also due to anthocyanins (plate 9.1G). These autumnal anthocyanins are synthesized anew once roughly 50% of the chlorophyll in the leaf has been lost (Lee et al. 2003; Archetti et al. 2009). The colors in fall leaves can also vary with factors such as chlorophyll concentration—a higher concentration can give leaves a purple to brown appearance. The pH value also plays a major role, for a basic pH in the cell vacuoles results in blue colors, whereas a more acidic pH produces reddish colors.

In contrast, the yellow colors in deciduous leaves (plate 9.1G) are commonly found during the growth seasons and are thus already in place once the autumn comes around. As chlorophyll breaks down into colorless metabolites in the fall, the yellows are unmasked with diminishing of the green pigments (Archetti et al. 2009). Reddish brown shades can occur in autumnal foliage at the start of the season from the combined presence of anthocyanins and chlorophyll (Lee et al. 2003).

Color in Fossil Leaves

Unlike the green color in living leaves, the most common color found among fossil leaves (and flowers) is brown, ranging from a light tan to beige to nut brown (plate 9.5A–D), or sometimes even black. The lack of green, bright yellow, and red colors implies a lack of chlorophyll, carotenoids, flavonoids, anthocyanins, or betalains. Instead, the shades of brown in fossil leaves and flowers are derived from the degraded organic tissue of the leaf, through a fossilization process called carbonization.

When a senescing leaf tumbles off a tree, it falls on the soil or on other leaves in the forest litter. Once pigments such as chlorophyll, carotenoids, flavonoids, and anthocyanins have been lost and the leaf has dried out, the resulting color of the leaf is a dusty brown. This brown color is primarily produced by the lignin, cellulose, and hemicellulose that make up the bulk of the cell walls of the leaf tissue (Elvidge 1990). The cellular structures in the dead leaf then begin to break down in their chemical composition with the help of various fungi. The process is helped by bacteria and some invertebrates, such as slugs, snails, and springtails, as well as by earthworms when the decay becomes advanced.

Through decomposition and decay, the oxygen, hydrogen, and nitrogen compounds are degraded in the plant matter until only a flat film of carbonaceous matter is left. If the carbon film finds its way into anaerobic conditions, where it does not come in contact with oxygen, such as under water in a pond, lake, or river, the carbonaceous film of the leaf will not necessarily rot away. With time and the right conditions, this leaf will become fossilized. The darker shade of brown in the fossil, the more carbon it contains.

When the internal tissues of a fossil leaf are completely degraded into a thin smear of organic matter, the organic matter is sometimes "sandwiched" between waxy films representing the original upper and lower surfaces of the once-living leaf. These waxy layers, called the cuticle, often bear the imprint of the outlines of the original epidermal cells of the leaf. Since the structure and patterns of these epidermal cells commonly hold the key to identification of the leaf, laboratory methods such as cuticle preparation have been developed to clear out the black organic matter to see the cell patterns of the cuticle (Kerp and Krings 1999). To macerate cuticle, a fossil leaf is carefully placed into an oxidizing mixture of Schulze's solution or reagent, which consists of saturated aqueous potassium chlorate ($KClO_3$) and concentrated nitric acid (HNO_3). This bleaching agent removes the dark, collapsed, carbonaceous matter from inside the leaf and lightens the waxy cuticle until the cell patterns are evident. The optimal color of a prepared leaf cuticle is a light brown or raw sienna color.

Stunning Exceptions: Green Fossil Leaves

There are, surprisingly, a few fossil floras that have yielded leaves with a vivid green color to them when freshly excavated. These are the Eocene lignite flora in Geiseltal near Halle in eastern Germany (Weigelt and Noack 1931; Dilcher et al. 1970), the Miocene Succor Creek flora in Oregon, USA (Niklas and Giannasi 1977; Giannasi and Niklas 1977), and the Miocene Clarkia flora in northern Idaho, USA (Niklas and Brown 1981). Chlorophyll derivatives were found in the fossil leaf impressions on the surface of the brown coal from Geiseltal (Dilcher et al. 1970) and in the preserved fossil leaf tissues of three species of dicot leaves from the Succor Creek Flora (Niklas and Giannasi 1977; Giannasi and Niklas 1977). The chlorophyll derivatives of pheophorbides (fig. 9.1A) and pristane/phytane, along with flavonoids and carotenoids, were extracted in a different set of four dicot leaf species in the Clarkia flora (Niklas and Brown 1981). Ultrastructural study using transmission electron microscopy confirmed the existence of well-preserved chloroplasts, which would have contained the chlorophyll, in the 15-million-year-old leaves from both Miocene localities (Niklas et al. 1978; Niklas and Brown 1981).

It has been reported that the green color of these fossil leaves from these localities fades almost immediately after excavation and is difficult to stabilize. Other accounts state that these green fossil leaves are not the same color as living leaves, but are a vivid green or slightly darker. In the single photo of a green leaf from the Clarkia flora that we could find in the literature (Taylor et al. 1993: fig. 1.27), one fossil leaf is dark, nearly black in color, but does have a greenish cast to it. It should be mentioned, though, that the rock on which the fossil leaf is found has the same greenish cast, which means that, in this case, the green coloration is not specific to the fossil leaf. Interestingly, on the same rock in this photo, there are also reddish colors in the leaf tissues that do not appear on the rock surface, suggesting the preservation of red pigments in the fossil leaves and supporting informal accounts

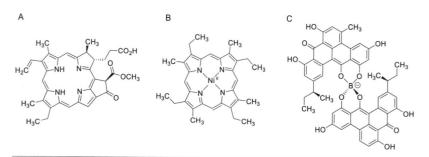

Figure 9.1. Molecular structure of pigments or derivatives that have been found in fossil plants. (A) Pheophorbide *a*. (B) Etioporphyrin I nickel, an example of a common porphyrin. (C) Example of a borolithochrome.

of fossil leaves with autumnal red hues at the Clarkia site (e.g., Steinthors-dottir and Coxall 2018).

Our discovery of green fossil dicot leaves from the Eocene Geiseltal flora some decades after excavation was thus unexpected and, for us, sensational. In this case, the embedding rock is beige to dark brown in color and the leaf tissues occur in shades of green, implying the preservation of chlorophyll in the leaves themselves. One fossil dicot leaf appears as a bluish grass green (plate 9.6A), which is not a green color commonly found in most living leaves.

The unusual green colors exhibited by the Eocene leaves from the Geiseltal is certainly due to the degradation of the pigment complexes in the leaf, as had been suggested by Dilcher et al. (1970). Slight changes in green hues are produced when the proteins degrade from the chlorophyll–protein complex (Milne et al. 2015), and more extreme changes result from the breakdown of chlorophyll itself into various degradation products, which range from dark green, to grayish-brownish green, to reddish, to colorless (Kräutler 2019). Furthermore, as chlorophyll breaks down, the color of the leaves will be more strongly influenced by any remaining leaf pigments, such as carotenoids (e.g., Lee et al. 2003; Archetti et al. 2009), as well as the natural brown colors of the cell walls produced by lignin, cellulose, and hemicellulose (Elvidge 1990).

Wood

I cannot pretend to feel impartial about the colours.
I rejoice with the brilliant ones, and am genuinely sorry
for the poor browns.
WINSTON CHURCHILL

When thinking about the colors of plants, most people immediately focus on the bright greens, reds, oranges, and pinks of leaves, flowers, and fruits. However, plants are also characterized by equally beautiful but more subtle shades of brown and gray: the stark brown and gray bark of trees in winter, the delicate patterns in wood grain, and the dry brown leaves that crunch appealingly underfoot during an autumnal walk. All of these more subdued colors derive from lignin (Elvidge 1990; Hon and Minemura 2000; Zhang et al. 2018).

Wood is composed of three major components: cellulose, hemicellulose, and lignin. Cellulose is a polysaccharide consisting of chains of glucose monomers held together by hydrogen bonding, whereas hemicellulose is a smaller branched carbohydrate that can be made of different monosaccharides. Lignin, on the other hand, is a complex, heterogeneous organic polymer (plate 9.4A) that is composed of combinations of three basic monolignol compounds (Freudenberg and Nash 1968). Although lignin is ubiquitous in cell walls in all vascular plants, it has no defined structure, its composition is highly variable, and its exact chemical structure in wood is unknown (Boerjan et al. 2003; Vanholme et al. 2010; Heitner et al. 2016).

Colors in Living Wood

Woods are quite variable in color, ranging from yellow to reddish brown and from beige to pale brown or darker brown (plate 9.4B–K). There are even woods that are black, such as ebony and African blackwood. While it is commonly said that wood gets its color from lignin, in a way, this is an oversimplification that can lead to misunderstandings. Lignin in its purest chemical form is colorless, white, or pale yellow. When occurring in wood, lignin is also described as being colorless (Hon and Minemura 2000; Wang et al. 2016). Nevertheless, wood does get its color from lignin because most of the color of wood comes from chromophores—color-bearing compounds and functional groups—in the lignin (Hon and Minemura 2000; Lee 2007), in addition to extractable compounds such as tannins or pigments. Many chromophores can contribute to the color of lignin; most are known from the analysis of extracted lignin and may have been produced during the extraction process (e.g., Ajao et al. 2018). However, some that have been identified in situ in wood include stilbene carboxaldehyde, stilbene methanol, syringyl stilbene, coniferaldehyde, o-quinone, and p-quinone (Agarwal and Atalla 2000). In contrast, hemicellulose and cellulose are colorless, white, or pale yellow, suggesting that they contain very few chromophores and contribute little to wood color (Korntner et al. 2015).

Colors in Fossil Wood

Wood can be fossilized in several ways: as a carbonaceous compression, as charcoal, or as a permineralization or mineral replacement.* In the first case, fossil wood compressions are formed by carbonization in the same way most fossil leaves are preserved. Carbonaceous fossil wood compressions are thus dark brown to black in color, reflecting the earthy dark colors of carbon. Fossil charcoal is produced by wildfire, in which the wood has been charred, and is black in color with a silky luster.

On the other hand, fossil wood that is preserved through permineralization, whether by silica (e.g., Gee and Liesegang, chap. 6; Liesegang et al.,

---

* Recently, it has been suggested that "wood mineralization" should be used instead as a general descriptive term (Mustoe 2017).

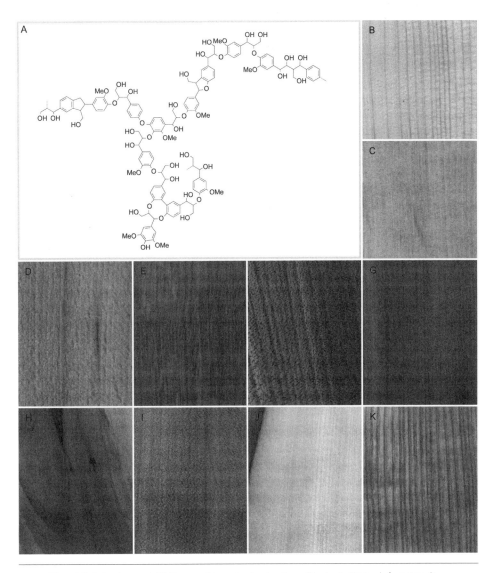

Plate 9.4. Colors in wood. (**A**) One possible molecular structure of lignin. **B–K**: Woods from Scandinavia. Photos: Anonimski, Creative Commons CC0 1.0 Universal Public Domain Dedication. (**B**) *Picea abies* (Norway spruce). (**C**) *Juniperus communis* (common juniper). (**D**) *Fagus sylvatica* (beech). (**E**) *Quercus robur* (pedunculate oak). (**F**) *Ulmus glabra* (wych elm). (**G**) *Prunus avium* (wild cherry). (**H**) *Pyrus communis* (pear). (**I**) *Acer platanoides* (Norway maple). (**J**) *Tilia cordata* (small-leaved lime). (**K**) *Fraxinus excelsior* (ash).

Plate 9.5. Brown colors in fossil plants attributable to carbonization of organic matter. **A–D**: 23-million-year-old fossil plants from the late Oligocene of Rott near Bonn, Germany. Photos: Georg Oleschinski. (**A**) Leaves of *Tremophyllum tenerrimum*, a member of the Ulmaceae. (**B**) An unknown fossil flower in which the intensity of brown colors differs between the delicate petals and the sturdier sepals and stalk. (**C**) Another unidentified flower in which the brown colors are darker and more uniform; nevertheless, darker stripes can be observed on the petals. (**D**) Fossil palm frond. **E, F**: 150-million-year-old wood from the Morrison Formation in northeast Utah, USA. Photos: CTG. (**E**) Fossil log in the field. The light beige, orange, and brown colors were likely produced by traces of iron precipitates. (**F**) Thin section in radial section through the fossil wood. The dark brown horizontal bars are resin plugs that have likely also undergone carbonization since the Late Jurassic.

chap. 7) or by other minerals (Mustoe 2018), can take on a huge range of colors. Many silicified logs and wood, for example, appear in various earth tones ranging from light beige, to orange, to brown colors on the outside (plate 9.5E). Brown tones are even evident on the microscopic level, that is, in the cell walls of the wood when viewed in thin section (plate 9.5F). In cases like this, brown colors in the cell walls of silicified wood likely come from remnants of the original organic matter. Fossil resin preserved in the silicified wood cells can appear as a darker brown (plate 9.5F), likely reflecting the concentration and degradation of a different suite of organic compounds than those found in the cell walls.

There are, however, other examples of silicified wood that are brighter and flashier. In these cases, these colors are not derived from the organic material in wood, but are produced diagenetically. One rare example is the precipitation of precious opal in fossil wood (plate 9.6B). More commonly, however, bright colors in fossil wood can be produced by small quantities of trace metals that enter the ancient wood from the embedding sediment during fossilization. These colors in silicified wood can consist of a multitude of shades of red, orange, yellow, green, blue, purple, brown, beige, black, and white, or a mixture of these colors (plate 9.6C) (Daniels and Dayvault 2006; Yoon and Kim 2008; Mustoe and Acosta 2016; Mustoe et al. 2019). Through the permineralization process, silicified wood loses all of its biological coloration. In the absence of colors produced during diagenesis, it will take up the color of the silica (Mustoe and Acosta 2016).

In silicified wood, trace metals, most commonly iron, produce a range of reds, oranges, yellows, and greens, and the specific color depends on the metal, the oxidation or reduction state, the particle size, and interactions with other colors (Mustoe and Acosta 2016). Sometimes a bright green can be produced by the trace element chromium.

Blue and purple are less common in silicified wood; blue typically comes from blue-tinted chalcedony, and purple from blue chalcedony with red-tinting trace metals (Daniels and Dayvault 2006; Mustoe and Acosta 2016; Mustoe et al. 2019). The coloration of brown, beige, and black silicified wood is still not fully understood. As mentioned earlier, brown and beige colors may come from some remnant organic matter and may occur in combination with iron and minute traces of titanium, vanadium, manganium, or copper. Black coloration does not yet have a well-supported explanation (Mustoe and Acosta 2016).

Some of the brightest and most spectacular silicified woods are rainbow-colored. Silicified wood with an explosion of colors is commonly found in the so-called Rainbow Forest of the Petrified Forest National Park in Ari-

Plate 9.6. Bright colors in fossil plants. (**A**) Fossil dicot leaf that is still green, presumably due to chlorophyll derivatives present even after 45 million years, collected from the Eocene lignites in Geiseltal but now stored in the collections of the Museum für Naturkunde Berlin. Photo: VEM. (**B**) Precious opal-replaced fossil wood from the middle Miocene Virgin Valley Formation, Nevada, USA. Photo: James St. John, Creative Commons Attribution 2.0 Generic license. (**C**) Tumbled bits of silicified wood from the Ginkgo Petrified Forest State Park, showing diagenetic colors ranging from red to pink, orange to light and dark brown, and even black and white. Photo: Georg Oleschinki. (**D**) The center of a 230-million-year-old tree trunk with a rainbow of diagenetic colors from the Petrified Forest National Park, Arizona, USA. Photo courtesy of Sidney R. Ash.

zona, USA (plate 9.6D). As with most other colored silicified wood, it is iron-oxide minerals that are responsible for these delightful rainbow colors (Mustoe and Acosta 2016).

Other Plant Colors in the Fossil Record

Recently, a sensational discovery was made of the oldest organic pigment—intact porphyrins (fig. 9.1B), the molecular fossils of chlorophylls—which came from 1.1-billion-year-old black shales in Mauritania, West Africa (Gueneli et al. 2018); this pigment was produced by prokaryotic cyanobacteria, not true eukaryotic plants. Cyanobacterial pigments are of the utmost importance for the evolution of early earth and the history of life. It was the chlorophyll pigments in cyanobacteria 3.5 billion years ago that enabled them to photosynthesize and release oxygen gas, thus filling earth's atmosphere with enough O_2 to trigger the Great Oxygenation Event some 2.4 billion years ago and the evolution of multicellular organisms as early as 2.3 billion years ago (Schirrmeister et al. 2013). Biological diversification and the proliferation of more complex, multicellular organisms 800 million years ago followed.

Another recent spectacular discovery was that of an entire class of previously unknown plant pigments in a calcareous red alga from the Jurassic of Great Britain and France (Wolkenstein et al. 2010). This fossil red alga, *Solenopora jurassica*, is known for its striking pink coloration. Concentric internal banding in the algal body, coupled with a rosy hue, has led to its nickname, the "Beetroot stone." The pink color comes from boron-containing pigments, termed borolithochromes (fig. 9.1C), which were extracted from the fossil alga preserved as a rock-hard limestone. Further biogeochemical work has linked these unusual boron-containing pigments to a living *Clostridium* bacterium, suggesting that the pink-colored pigments in the fossil alga may have been produced by an ancient bacterium (Wolkenstein et al. 2015).

Biological Structural Color in Plants

Now, that this Whiteness is a Mixture of the severally colour'd rays, falling confusedly on the paper, I see no reason to doubt of.
ISAAC NEWTON

Biological structural color is produced by nano- or microscale structures that scatter visible (white) light, thereby changing its color, an effect which may or may not occur in combination with pigments (Sun et al. 2013). Structural colors are very common in extant animals and produce some of the purest colors, including the brightest brights (Seago et al. 2009; Sun et al. 2013), the whitest whites (Lafait et al. 2010; Burresi et al. 2014), and the blackest blacks (McCoy et al. 2019). Structural colors have also been identified in fossil animals, including a dinosaur with iridescent plumage (Hu et al. 2018), brightly colored beetles (McNamara et al. 2012a, 2012b), and moths with a subtle golden green shine (McNamara et al. 2011).

Structural colors are less well-studied in living plants (Glover and Whitney 2010; Sun et al. 2013), and, as far as we know, not at all studied in fossil plants. However, living plants produce structural color using the same range of structures as animals (Glover and Whitney 2010): incoherent scattering, in which small, widely spaced particles scatter light in a random way; multilayer reflectors, layered structures alternating tissues with different refractive indices; diffraction gratings, which have a regular, ordered, periodic structure; photonic crystals, which are ordered nanostructures in one, two, or three dimensions; and microlens arrays, which are conical structures that focus light onto an underlying pigment, making the color richer and decreasing white highlights (Vogelmann 1993).

Most structural colors known in plants produce blue coloration, ranging from the dull blue tinge in various conifer needles due to incoherent scattering of light by surface waxy deposits (Vogelmann 1993; Glover and Whitney 2010) to the deep metallic blue iridescence of the fruits of *Elaeocarpus angustifolius* (plate 9.1A, B) and *Delarbrea michiana* due to multilayer reflectors (Lee 1991; Lee et al. 2000). Most commonly, multilayer reflectors are found on the leaves of a variety of shade-growing plants, including *Selaginella willdenowii*, *S. uncinata*, *Danaea nodosa*, *Diplazium tomentosum*, *Lindsaea lucida*, *Trichomanes elegans*, *Phyllagathis rotundifolia*, and *Begonia pavonina*, and produce a distinctive blueish-green iridescence (Hébant and Lee 1984; Graham et al. 1993; Gould and Lee 1996; Glover and Whitney 2010).

Diffraction gratings and photonic crystals in plants are only known to have effects in the UV spectrum (Glover and Whitney 2010): diffraction gratings in plants primarily produce UV iridescence in flowers, most likely to signal pollinators capable of seeing UV light (Whitney et al. 2009a, 2009b), and photonic crystals, so far known only in edelweiss flowers (*Leontopodium nivale* subsp. *alpinum*), absorb all UV light to act as a sunblock (Vigneron et al. 2005).

Microlens arrays in plants are found in a number of leaves and flowers, and they involve elongated conical epidermal cells that focus light onto underlying pigment, resulting in a deeper color than would be produced by pigment alone (Vogelmann 1993; Gorton and Vogelmann 1996; Gkikas et al. 2015). One dazzling example is the glowing orange or yellow color of the California poppy, *Eschscholzia californica* (plate 9.3B), which is produced by the carotenoid eschscholtzxanthin (plate 9.3A) enhanced by the densely thickened, deeply ridged, prism-like surface of the epidermal cells of the petals (Wilts et al. 2018). In general, while microlens arrays certainly influence plant color, they may also have other major functions, such as increasing the efficiency of photosynthesis or providing a textured surface that is easier for pollinators to grip (Kay et al. 1981; Vogelmann 1993).

The preservation of structural color in the fossil record depends upon the preservation of the tissues that make up the structure: the multilayered reflectors are produced either in the cell walls of epidermal cells, and include cellulose layers (Hébant and Lee 1984; Lee 1991; Graham et al. 1993; Gould and Lee 1996; Lee et al. 2000; Lee 2007), or in the protoplast as stacked layers of iridophores (Graham et al. 1993; Gould and Lee 1996; Lee 2007); diffraction gratings and microlens arrays are specific morphologies of epidermal cells in flowers (Whitney et al. 2009b); and photonic crystals are found in hairs on edelweiss flower bracts (Vigneron et al. 2005). Visible color patterns in the fossil are not necessary for preserved structural color.

Future Research Directions

If one could only catch that true color of nature—the very thought of it drives me mad.

    ANDREW WYETH

Despite the large strides that have been made in the understanding of the biochemistry, physiology, ecological strategies, and evolutionary advantages of colors in living plants, as well as in the advances in detecting colors in fossil animals, the study of color in ancient plants is still woefully lagging behind. Theoretically, it should be possible to find all or at least some of the colors in the modern flora in the paleobotanical record. However, there are only a very few studies on pigmentation in fossil plants and, as far as we know, none on structural colors, although the coloration of fossil plants produced through mineralization is becoming better understood, particularly in silicified wood (Mustoe and Acosta 2016). This means, on the other hand, that this field of study is wide open for forays of investigations into fossil plant color. With the high-resolution analytical methods and tools available today (Sander and Gee, chap. 1), for instance, Raman spectroscopy (Geisler and Menneken, chap. 4), scanning electron microscopy (SEM) (Liesegang et al., chap. 7), high-resolution X-ray microcomputed tomography (micro-CT) (e.g., Gee 2013), synchrotron rapid scanning X-ray fluorescence (SRS-XRF) elemental mapping, and X-ray absorption spectroscopy (Bergmann et al. 2012; Edwards et al. 2014; Manning et al. 2019), the extraction and analysis of data from fossil material should get easier and become more precise.

The obvious starting point for color pigment studies in paleobotany is to begin with fossil plants that have a distinctive color that cannot be attributed to diagenesis. For example, the extensive study into the pink pigmentation of the "Beetroot stone," *Solenopora jurassica*, is an ideal case (Wolkenstein et al. 2010, 2015; see above, "Other Plant Colors in the Fossil Record"). The unusual boron-containing pigments in the Jurassic alga of *Solenopora*, and, hence, the chemical link to a pink pigment-producing bacterium, is still rather enigmatic at this time, but it opens up a fascinating, previously unknown dimension to the interactions of plants and microbes in deep time.

The fossil green dicot leaves from the Eocene Geiseltal in Germany and the two Miocene sites of Succor Creek and Clarkia in the USA were also obvious starting points for chlorophyll extraction and analysis in the 1970s (Dilcher et al. 1970; Giannasi and Niklas 1977; Niklas and Giannasi 1977; Niklas and Brown 1981). These studies proved to be successful in extracting and identifying chlorophyll derivatives (see above, "Stunning Exceptions: Green Fossil Leaves"), as well as giving rise to further investigations into the ultrastructure of chloroplasts and other organelles in the 15-million-year-old leaf cells (Niklas and Brown 1981).

In future studies, biogeochemical work could focus on other fossil plant material thought to contain biochromes. In addition to the chemical extraction of pigments, SRS-XRF elemental mapping and X-ray absorption spectroscopy could be applied to fossil plants to identify metals that may indicate the presence of certain pigments, for it has been shown in the fossil plants from the Eocene Green River Formation that the chemical content of well-preserved fossil leaves is derived from the original biochemistry of the closely related living leaves (Edwards et al. 2014). In animals, for example, the study of metallomes—the distribution of free metal ions in cells and tissues—in fossil animals and, in particular, the identification of chemical markers for the black pigment melanin have allowed for the reconstruction of black and white coloration in feathers in the body of a 120-million-year-old bird from the Cretaceous of China (Wogelius et al. 2011; Edwards et al. 2016).

The discovery, elucidation, and analysis of structural color in fossil plants would also be a potentially rewarding area of research, given the current success of finding structural color in fossil animals (see above, "Biological Structural Color in Plants"). The micrometer- and nanometer-level study of surfaces producing structural color in fossil plants can be achieved with today's possibilities of high-resolution imaging using micro-CT or SEM, including environmental SEM (ESEM) or variable pressure SEM (VPSEM) (Liesegang et al., chap. 7), which do not require specimen coating or destructive sampling (McCoy, chap. 10). Continued technical improvements in imaging capability and the utilization of nondestructive methods will surely help lead the way to the study of structural color in paleobotanical specimens.

Conclusions

The vibrant world of plant color is well known in the living flora, mostly through a plethora of biochrome pigments, such as chlorophylls, carotenoids, and anthocyanins, but also from structural color and, to a much lesser extent, bioluminescence. Color in fossil plants generally reflects the degradation of the carbon-rich plant tissues into the browns of coalified leaves and flowers, the silky black luster of charcoalified wood, and the earth tones of fossil wood, although some silicified wood becomes brightly colored through the inclusion of minute quantities of trace minerals during diagenesis.

Plant pigments in the fossil record are rare. The oldest organic pigment, molecular fossils of chlorophylls, was extracted from 1.1-billion-year-old black shales in Mauritania, West Africa, and presumed to have originated from cyanobacteria. Pink pigment found in a calcareous red alga from the Jurassic is now thought to have been produced by a bacterium. Green fossil leaves from the Cenozoic provide the first examples of pigments—chlorophyll derivatives—extracted from eukaryotic plants. Yet, given the many new techniques and methods now available, coupled with new possibilities for high-resolution imaging, as well as organic and inorganic chemical analysis, future forays into fossil plant color research offer much potential.

ACKNOWLEDGMENTS

The authors sincerely thank the two reviewers of this chapter, George Mustoe (Western Washington University, USA) and Christa Müller (University of Bonn, Germany), for their helpful comments, as well as handling editor Martin Sander. This research was supported by the Deutsche Forschungsgemeinschaft (DFG, German Research Foundation), Project number 39676817 to CTG and Project number 396637283 to Jes Rust for VEM (both University of Bonn). This is contribution number 23 of the DFG Research Unit 2685, "The Limits of the Fossil Record: Analytical and Experimental Approaches to Fossilization."

WORKS CITED

Agarwal, U.P., and Atalla, R.H. 2000. Using Raman spectroscopy to identify chromophores in lignin–lignocellulosics, pp. 250–264. In: ACS Symposium Series 742. *Lignin: Historical, Biological, and Materials Perspectives*. American Chemical Society.

Airs, R.L., Temperton, B., Sambles, C., Farnham, G., Skill, S.C., and Llewellyn, C.A. 2014. Chlorophyll *f* and chlorophyll *d* are produced in the cyanobacterium *Chlorogloeopsis fritschii* when cultured under natural light and near-infrared radiation. *FEBS Letters*, 588: 3770–3777.

Ajao, O., Jeaidi, J., Benali, M., Restrepo, A.M., El Mehdi, N., and Boumghar, Y. 2018. Quantification and variability analysis of lignin optical properties for colour—Dependent industrial applications. *Molecules*, 23: 377–398.

Archetti, M., Döring, T.F., Hagen, S.B., Hughes, N.M., Leather, S.R., Lee, D.W, Lev-Yadun, S., Manetas, Y., Ougham, H.J., Schaberg, P.G., and Thomas, H. 2009. Unravelling the evolution of autumn colours: An interdisciplinary approach. *Trends in Ecology & Evolution*, 24: 166–173.

Bergmann, U., Manning, P.L., and Wogelius, R.A. 2012. Chemical mapping of paleontological and archeological artifacts with synchrotron X-rays. *Annual Reviews of Analytical Chemistry*, 5: 361–389.

Bhagwat, S., and Haytowitz, D.B. 2015. USDA database for the flavonoid content of selected foods. Release 3.2. https://data.nal.usda.gov/dataset/usda-database-flavo noid-content-selected-foods-release-32-november-2015/resource/b6ae8bff.

Boerjan, W., Ralph, J., and Baucher, M. 2003. Lignin biosynthesis. *Annual Review of Plant Biology*, 54: 519–546.

Bold, H.C., and Wynne, M.J. 1978. *Introduction to the Algae*. Prentice-Hall, Inc., Englewood Cliffs, New Jersey, USA.

Burresi, M., Cortese, L., Pattelli, L., Kolle, M., Vukusic, P., Wiersma, D.S., Steiner, U., and Vignolini, S. 2014. Bright-white beetle scales optimise multiple scattering of light. *Scientific Reports*, 4: 6075. doi:10.1038/srep06075.

Chen, M., Schliep, M., Willows, R.D., Cai, Z.L., Neilan, B.A., and Scheer, H. 2010. A red-shifted chlorophyll. *Science*, 329: 1318–1319.

Daniels, F.J., and Dayvault, R.D. 2006. *Ancient Forests: A Closer Look at Fossil Wood*. Western Colorado Publishing Company Grand Junction, Colorado, USA.

Dash, M.C., and Dash, S.P. 2009. *Fundamentals of Ecology*, 3rd ed. The McGraw-Hill Companies, New Delhi, India.

Dilcher, D.L., Pavlick, R.J., and Mitchell, J. 1970. Chlorophyll derivatives in middle Eocene sediments. *Science*, 168: 1447–1449.

Edwards, N.P., Manning, P.L., Bergmann, U., Larson, P.L., van Dongen, B.E., Sellers, W.I., Webb, S.M., Sokaras, D., Alonso-Mori, R., Ignatyev, K., Barden, H.E., van Veelen, A., Anné, J., Egerton, V.M., and Wogelius, R.A. 2014. Leaf metallome preserved over 50 million years. *Metallomics*, 6: 774–782.

Edwards, N.P., van Veelen, A., Anné, J., Manning, P.L., Bergmann, U., Sellers, W.I., Egerton, V.M., Sokaras, D., Alonso-Mori, R., Wakamatsu, K., Ito, S., and Wogelius, R.A. 2016. Elemental characterisation of melanin in feathers via synchrotron X-ray imaging and absorption spectroscopy. *Scientific Reports*, 6: 34002. doi:10.1038/srep34002.

Elvidge, C.D. 1990. Visible and near infrared reflectance characteristics of dry plant materials. *International Journal of Remote Sensing*, 11: 1775–1795.

Freudenberg, K., and Nash, A.C., eds. 1968. *Constitution and Biosynthesis of Lignin*. Springer-Verlag, Berlin.

Gee, C.T. 2013. Applying microCT and 3D visualization to Jurassic silicified conifer seed cones: A virtual advantage over thin-sectioning. *Applications in Plant Sciences*, 1: 1300039. doi:10.3732/apps.1300039.

Giannasi, D.E., and Niklas, K.J. 1977. Flavonoid and other chemical constituents of fossil Miocene *Celtis* and *Ulmus* (Succor Creek Flora). *Science*, 197: 765–767.

Gkikas, D., Argiropoulos, A., and Rhizopoulou, S. 2015. Epidermal focusing of light and modelling of reflectance in floral-petals with conically shaped epidermal cells. *Flora: Morphology, Distribution, Functional Ecology of Plants*, 212: 38–45.

Glover, B.J., and Whitney, H.M. 2010. Structural colour and iridescence in plants: The poorly studied relations of pigment colour. *Annals of Botany*, 105: 505–511.

Gorton, H.L., and Vogelmann, T.C. 1996. Effects of epidermal cell shape and pigmentation on optical properties of *Antirrhinum* petals at visible and ultraviolet wavelengths. *Plant Physiology*, 112: 879–888.

Gould, K.S., and Lee, D.W. 1996. Physical and ultrastructural basis of blue leaf iridescence in four Malaysian understory plants. *American Journal of Botany*, 83: 45–50.

Graham, R.M., Lee, D.W., and Norstog, K. 1993. Physical and ultrastructural basis of blue leaf iridescence in two neotropical ferns. *American Journal of Botany*, 80: 198–203.

Gueneli, N., McKenna, A.M., Ohkouchi, N., Boreham, C.J., Beghin, J., Javaux, E.J., and Brocks, J.J. 2018. 1.1-billion-year-old porphyrins establish a marine ecosystem dominated by bacterial primary producers. *Proceedings of the National Academy of Sciences of the United States*, 115: E6978–E6986.

Han, Z., Li, B., Mu, Z., Yang, M., Niu, S., Zhang, J., and Ren, L. 2015. An ingenious super light trapping surface templated from butterfly wing scales. *Nanoscale Research Letters*, 10: 1052.

Hébant, C., and Lee, D.W. 1984. Ultrastructural basis and developmental control of blue iridescence in *Selaginella* leaves. *American Journal of Botany*, 71: 216–219.

Heitner, C., Dimmel, D., and Schmidt, J. 2016. *Lignin and Lignans: Advances in Chemistry*. CRC Press, Taylor & Francis, Boca Raton, Florida, USA.

Hon, D.N.S., and Minemura, N. 2000. Color and discoloration, pp. 385–442. In: Hon, D.N.S., and Shiraishi, N., eds. *Wood and Cellulosic Chemistry*. Marcel Dekker Inc., New York.

Hu, D., Clarke, J.A., Eliason, C.M., Qiu, R., Li, Q., Shawkey, M.D., Zhao, C., D'Alba, L., Jiang, J., and Xu, X. 2018. A bony-crested Jurassic dinosaur with evidence of iridescent plumage highlights complexity in early paravian evolution. *Nature Communications*, 9: 217.

Hunt, D.M., Carvalho, L.S., Cowing, J.A., and Davies, W.L. 2009. Evolution and spectral tuning of visual pigments in birds and mammals. *Philosophical Transactions of the Royal Society B: Biological Sciences*, 364: 2941–2955.

Kay, Q.O.N., Daoud, H.S., and Stirton, C.H. 1981. Pigment distribution, light reflection and cell structure in petals. *Botanical Journal of the Linnean Society*, 83: 57–83.

Kerp, H., and Krings, M. 1999. Light microscope of cuticles, pp. 52–56. In: Jones, T.P., and Rowe, N.P., eds. *Fossil Plants and Spores: Modern Techniques*. Geological Society of London, UK.

Khoo, H.E., Azlan, A., Tang, S.T., and Lim, S.M. 2017. Anthocyanidins and anthocyanins: Colored pigments as food, pharmaceutical ingredients, and the potential health benefits. *Food Nutrition Research*, 61: 1361779. doi:10.1080/16546628.2017.1361779.

Kohata, S., Matsunaga, N., Hamabe, Y., Yumihara, K., and Sumi, T. 2010. Photo stability of mixture of violet pigments phycoerythrin and phycocyanin extracted without separation from discolored nori seaweed. *Food Science and Technology Research* 16: 617–620.

Korntner, P., Hosoya, T., Dietz, T., Eibinger, K., Reiter, H., Spitzbart, M., Koder, T., Borgards, A., Kreiner, W., Mahler, A.K., Winter, H., Groiss, Y., French, A.D., Henniges, U., Potthast, A., and Rosenau, T. 2015. Chromophores in lignin-free cellulosic materials belong to three compound classes. Chromophores in cellulosics, XII. *Cellulose*, 22: 1053–1062.

Kräutler, B. 2019. Chlorophyll breakdown—How chemistry has helped to decipher a striking biological enigma. *Synlett*, 30: 263–274.

Lafait, J., Andraud, C., Berthier, S., Boulenguez, J., Callet, P., Dumazet, S., Rassart, M., and Vigneron, J.P. 2010. Modeling the vivid white color of the beetle *Calothyrza margaritifera*. *Materials Science and Engineering B*, 169: 16–22.

Lee, D. 2007. *Nature's Palette: The Science of Plant Color*. University of Chicago Press, Chicago and London.

Lee, D.W. 1991. Ultrastructural basis and function of iridescent blue colour of fruits in *Elaeocarpus*. *Nature*, 349: 260.

Lee, D.W., O'Keefe, J., Holbrook, N.M., and Feild, T.S. 2003. Pigment dynamics and autumn leaf senescence in a New England deciduous forest, eastern USA. *Ecological Research*, 18: 677–694.

Lee, D.W., Taylor, G.T., and Irvine A.K. 2000. Structural fruit coloration in *Delarbrea michieana* (Araliaceae). *International Journal of Plant Sciences*, 161: 297–300.

Lynch, D.K., and Livingston, W.C. 2001. *Color and Light in Nature*, 2nd ed. Cambridge University Press, Cambridge, UK.

Manning, P.L., Edwards, N.P., Bergmann, U., Anné, J., Sellers, W.I., van Veelen, A., Sokaras, D., Egerton, V.M., Alonso-Mori, R., Ignatyev, K., van Dongen, B.E., Wakamatsu, K., Ito, S., and Knoll, F., and Wogelius, R.A. 2019. Pheomelanin pigment remnants mapped in fossils of an extinct mammal. *Nature Communications*, 10: 2250. doi:10.1038/s41467-019-10087-2.

McCoy, D.E., McCoy, V.E., Mandsberg, N.K., Shneidman, A.V., Aizenberg, J., Prum, R.O., and Haig, D. 2019. Structurally assisted super black in colourful peacock spiders. *Proceedings of the Royal Society B*, 286: 20190589. doi:10.1098/rspb.2019.0589.

McNamara, M.E., Briggs, D.E.G., and Orr, P.J. 2012a. The controls on the preservation of structural color in fossil insects. *Palaios*, 27: 443–454.

McNamara, M.E., Briggs, D.E.G., Orr, P.J., Noh, H., and Cao, H. 2012b. The original colours of fossil beetles. *Proceedings of the Royal Society B*, 279: 1114–1121.

McNamara, M.E., Briggs, D.E.G., Orr, P.J., Wedmann, S., Noh, H., and Cao, H. 2011. Fossilized biophotonic nanostructures reveal the original colors of 47-million-year-old moths. *PLOS Biology*, 9: e1001200.

Milne, B.F., Toker, Y., Rubio, A., and Nielsen, S.B. 2015. Unraveling the intrinsic color of chlorophyll. *Angewandte Chemie*, 54: 2170–2173.

Morse, D., and Mittag, M. 2000. Dinoflagellate luciferin-binding protein. *Methods in Enzymology*, 305: 285–276.

Mustoe, G.E. 2017. Wood petrification: A new view of permineralization and replacement. *Geosciences*, 17: 1–17.

Mustoe, G.E. 2018. Mineralogy of non-silicified wood. *Geosciences*, 8: 1–31.

Mustoe, G.E., and Acosta, M. 2016. Origin of petrified wood color. *Geosciences*, 6: 25.

Mustoe, G.E., Viney, M., and Mills, J. 2019. Mineralogy of Eocene fossil wood from the "Blue Forest" locality, southwestern Wyoming, United States. *Geosciences*, 9: 35.

Niklas, K.J., and Brown, R.M., Jr. 1981. Ultrastructural and paleobiochemical correlations among fossil leaf tissues from the St. Maries River (Clarkia) Area, northern Idaho, USA. *American Journal of Botany*, 68: 332–341.

Niklas, K.J., Brown, R.M., Jr., Santos, R., and Vian, B. 1978. Ultrastructure and cytochemistry of Miocene angiosperm leaf tissues. *Proceedings of the National Academy of Sciences of the United States*, 75: 3263–3267.

Niklas, K.J., and Giannasi, D.E. 1977. Flavonoids and other chemical constituents of fossil Miocene *Zelkova* (Ulmaceae). *Science*, 196: 877–878.

Schirrmeister, B.E., de Vos, J.M., Antonelli, A., and Bagheri, H.C. 2013. Evolution of multicellularity coincided with increased diversification of cyanobacteria and the Great Oxidation Event. *Proceedings of the National Academy of Sciences of the United States*, 110: 1791–1796.

Seago, A.E., Brady, P., Vigneron, J.P., and Schultz, T.D. 2009. Gold bugs and beyond: A review of iridescence and structural colour mechanisms in beetles (Coleoptera). *Journal of the Royal Society Interface*, 6: S165–S184.

Steinthorsdottir, M., and Coxall, H.K. 2018. The Clarkia flora: 16-million-year-old plants offer a window into the past. *Deposits Magazine*. https://depositsmag.com/2019 /03/11/the-clarkia-flora-16-million-year-old-plants-offer-a-window-into-the-past.

Sun, J., Bhushan, B., and Tong, J. 2013. Structural coloration in nature. *RSC Advances*, 3: 14862–14889.

Taylor, T.N., Taylor, E.L., and Krings, M. 1993. *Paleobotany: The Biology and Evolution of Fossil Plants*, 2nd ed. Academic Press, San Diego, USA.

Valiadi, M., and Iglesias-Rodriguez, D. 2013. Understanding bioluminescence in dinoflagellates—How far have we come? *Microorganisms*, 1: 3–25.

Vanholme, R., Demedts, B., Morreel, K., Ralph, J., and Boerjan, W. 2010. Lignin biosynthesis and structure. *Plant Physiology*, 153: 895–905.

Vigneron, J.P., Rassart, M., Vértesy, Z., Kertész, K., Sarrazin, M., Biró, L.P., Ertz, D., and Lousse, V. 2005. Optical structure and function of the white filamentary hair covering the edelweiss bracts. *Physical Review E*, 71: 011906. doi:10.1103/PhysRevE .71.011906.

Vogelmann, T.C. 1993. Plant tissue optics. *Annual Review of Plant Physiology and Plant Molecular Biology*, 44: 231–251.

Wang, J., Deng, Y., Qian, Y., Qiu, X., Ren, Y., and Yang, D. 2016. Reduction of lignin color via one-step UV irradiation. *Green Chemistry*, 18: 695–699.

Weigelt, J., and Noack, K. 1931. Über Reste von Blattfarbstoffen aus der Geiseltal-Braunkohle (Mitteleozän). *Nova Acta Leopoldina*, 1: 1–96.

Whitney, H.M., Kolle, M., Alvarez-Fernandez, R., Steiner, U., and Glover, B.J. 2009a. Contributions of iridescence to floral patterning. *Communicative & Integrative Biology*, 2: 230–232.

Whitney, H.M., Kolle, M., Andrew, P., Chittka, L., Steiner, U., and Glover, B.J. 2009b. Floral iridescence, produced by diffractive optics, acts as a cue for animal pollinators. *Science*, 323: 130–133.

Wilts, B.D., Rudall, P.J., Moyroud, E., Gregory, T., Ogawa, Y., Vignolini, S., Steiner, S., and Glover, B.J. 2018. Ultrastructure and optics of the prism-like petal epidermal cells of *Eschscholzia californica* (California poppy). *The New Phytologist*, 219: 1124–1133.

Wogelius, R.A., Manning, P.L., Barden, H.E., Edwards, N.P., Webb, S.M., Sellers, W.I., Taylor, K.G., Larson, P.L., Dodson, P., You, H. Da-qing, L., and Bergmann, U. 2011. Trace Metals as biomarkers for eumelanin pigment in the fossil record. *Science*, 333: 1622–1626.

Wolkenstein, K., Gross, J.H., and Falk, H. 2010. Boron-containing organic pigments from a Jurassic red alga. *Proceedings of the National Academy of Sciences of the United States*, 107: 19374–19378.

Wolkenstein, K., Sun, H., Falk, H., and Griesinger, C. 2015. Structure and absolute configuration of Jurassic polyketide-derived spiroborate pigments obtained from microgram quantities. *Journal of the American Chemical Society*, 137: 13460–13463.

Yoon, C.J., and Kim, K.W. 2008. Anatomical descriptions of silicified woods from Madagascar and Indonesia by scanning electron microscopy. *Micron*, 39: 825–831.

Zhang, P., Wei, Y., Liu, Y., Gao, J., Chen, Y., and Fan, Y. 2018. Heat-induced discoloration of chromophore structures in *Eucalyptus* lignin. *Materials*, 11: E1686. doi:10.3390/ma11091686.

CHAPTER **10**

The Future of Fossilization

VICTORIA E. MCCOY

To know your future you must know your past.
GEORGE SANTAYANA

Taphonomy, which includes the study of fossilization, is an ancient science, that started in about AD 77 with Pliny the Elder, who hypothesized about the fossilization of insects in amber—insects are trapped in liquid amber and then entombed (or fossilized) as the amber hardens (Bostock and Riley 1855)—and has continued to the present day to play many important roles within the field of paleobiology (Behrensmeyer et al. 2000). Paleobiology relies upon an accurate reading of the fossil record, and an accurate reading of the fossil record depends upon a deep and complete understanding of both the process of fossilization and the material nature of fossils (Sander and Gee, chap. 1). In recent years, there has been a renaissance in the field of taphonomy, built around three realizations: (1) soft tissues, such as skin and muscle, can fossilize along with hard tissues (Conway Morris 1979; Briggs 2014); (2) the early stages of soft tissue fossilization can be observed in laboratory time (Briggs et al. 1993; Briggs and Kear 1993; Briggs and Mc-Mahon 2016); and (3) these fossil soft tissues can include unaltered or only slightly altered remnants of original biological information, often in the form of original biomolecules (Schweitzer et al. 2005; Briggs and Summons 2014). Together, these three realizations mean that fossil organisms can be studied and interpreted as living organisms, within a biological framework. Furthermore, the process of soft tissue fossilization can be studied in controlled laboratory experiments to determine how original materials change throughout this process. Finally, the application of modern analytical methods can help reveal the material nature of fossils and extract relevant biological information. The future of fossilization is to build upon and unify these three research directions to develop a fundamental theory of fossilization and untangle the complex information in fossils.

Fossils as Living Organisms

Fossilization, both the process and the science of studying it, does not exist in isolation. As a process, it acts upon a living (or recently dead) organism, to preserve, obscure, alter, or destroy original biological information from that living organism (Parry et al. 2018). The science of studying fossilization (a subset of taphonomy) acts in the service of paleobiology to contribute to the accurate interpretation of fossils as remains of living organisms. To this end, it is necessary to frame the goals of fossilization research within broader paleobiological questions (Purnell et al. 2018).

Color is one of the main features that distinguishes fossils from living organisms, so perhaps it is not surprising that research into the fossilization of color has already moved in the direction of answering broad paleontological questions. After a few initial papers primarily demonstrating that color, both pigmentary and structural, can be reconstructed for extinct organisms (Vinther et al. 2008; McNamara et al. 2011a), more recent research has focused on interpreting the role of color patterns, for example, for camouflage, countershading, or display (McNamara et al. 2011b; Vinther et al. 2016; Smithwick et al. 2017; Gee and McCoy, chap, 9). Similar trends can be seen in other areas as well. For example, fossilized reproductive tissues can inform about reproductive behavior (Yang et al. 2018; Yang and Canoville, chap. 3), and fossilized plant chemical defenses can inform ancient plant–insect interactions (McCoy et al., chap. 8). Some ground-truthing analyses, both experimental and analytical, will always be necessary to ensure fossils are being correctly interpreted, but these ground-truthing analyses should all work toward a long-term goal of answering specific questions about extinct organisms.

Controlled Laboratory Experiments

Fossilization experiments were initially very simple, fish-in-a-jar-type, experiments meant to demonstrate that early patterns of decay and mineralization can be observed in a laboratory setting (Briggs and Kear 1993; Briggs 1995), a necessary demonstration prior to more complicated experiments (Briggs and McMahon 2016). More recently, the emphasis has been placed on controlled fossilization experiments, with the goal of isolating and constraining the effects of one (or a few) variables (Briggs and McMahon 2016; Purnell et al. 2018; Gee and Liesegang, chap. 6). The key to such experiments is simplifying the complex process of fossilization, so that the most important components can be tested in a laboratory, without oversimplifying to the point where the results no longer reveal information about real-world processes (Briggs and McMahon 2016; Parry et al. 2018). Certain fossilization processes such as silicification and phosphatization, which may be easier to simplify, have been the primary focus of fossilization experiments.

Moving forward, fossilization experiments will cover broader ranges of study, including more complex, overlooked, or unusual processes of fossilization. For example, fossilization in amber is both an extremely complex and traditionally understudied process, although a few recent papers have begun to address it (Solorzano Kraemer et al. 2018; McCoy et al. 2018; McCoy et al. 2019; Barthel et al., chap. 5). Similarly, the fossilization of plant tissues and biomolecules, including color or the complex chemistry of plants, is very understudied from a controlled experiment perspective (McCoy et al., chap. 8, Gee and McCoy, chap. 9; but see Iniesto et al. 2018).

Moreover, methodological advances for controlling an experimental environment and analyzing complex systems will allow fossilization experiments to assess wider ranges of variables, and test both the effects and interactions of these variables. One particular variable of interest is the effect of microbial activity; although this can be summarized as one variable, it includes a complex set of organic and inorganic interactions (Briggs and McMahon 2016). These interactions are generally thought to be important for fossilization but are typically only considered in experiments in fairly simple ways. However, there are many biological approaches for growing, controlling, and analyzing complex microbial assemblages in a laboratory setting that could be applied to fossilization experiments (e.g., Brown et al. 2019). Similarly, other types of complex variables or variable interactions, such as sediment and seawater chemistry, are also interesting for fossilization experiments, and protocols could be developed in collaboration with organic chemists or geochemists.

Modern Analytical Methods

One of the primary research directions within paleobiology has always been understanding the material composition of fossils. The recent realization that some fossils contain remnants of original biological information made this research direction even more important and fundamental (Schweitzer et al. 2005; Briggs 2014; Briggs and Summons 2014; Wiersma et al., chap. 2). However, the more the material nature of fossils is investigated, the more complex fossils are seen to be. Even that most iconic and well-studied fossil, the dinosaur bone, has a complex composition that is not well understood (and that seems to be an exceptional host for preserved remnants of original soft tissue and biomolecules) (Wiersma et al., chap. 2). The application of advanced analytical methods has already proven to be very useful in understanding the material nature of fossils and specifically in extracting original biological information from fossils (Edwards et al. 2014; Wiemann et al. 2018; Liesegang et al., chap. 7), and this trend will certainly continue in the future.

Typically, analytical methods are not applied to fossils until they have become a little mainstream, meaning that paleobiology is missing out on the most cutting-edge techniques. Collaborations with biologists, chemists, pharmacists, physicists, and other experts in complimentary fields will help paleobiologists get access to the newest and most advanced analytical methods. This would also lead to research focused on developing new methods specifically to address the challenges offered by fossils: low concentrations of altered components in dirty samples (originally complex composition, possibly further influenced by more recent contamination). Such methods would be useful for both experiments and fossil analyses.

The current trend is toward nondestructive techniques that can be applied to the rarest and most exceptional fossils. Our research unit in particular is focused around Raman spectroscopy as a powerful yet nondestructive tool for analyzing fossils (Geisler and Menneken, chap. 4). Moving forward, the focus will be refining the protocols for such information-dense, nondestructive techniques to maximize their effectiveness on a wide range of fossils.

Conclusions: A Unified Theory

The ideal future research project on fossilization would unify these three approaches: modern analytical techniques to determine the material nature of a fossil, fossilization experiments to interpret the fossil composition and determine what components retain biological information, and finally developing these results within a biological framework to gain a better understanding of an extinct organism. Such a research project would be complex and interdisciplinary, and it would benefit from collaborators from a wide range of fields to access the most up-to-date analytical methods and detailed knowledge of biological and geological processes. Moreover, one key component for unifying various lines of research is developing standard protocols for experiments and analyses. Standard experimental protocols, and standard procedures for gathering and analyzing data from experiments and from fossils, will allow the data from multiple experiments to be collated and quantitatively analyzed, which is necessary for identifying a fundamental model and unified theory of fossilization (Briggs and McMahon 2016; Purnell et al. 2018).

Our DFG Research Unit 2685 on "The Limits of the Fossil Record: Analytical and Experimental Approaches to Fossilization" includes investigators from a wide range of fields carrying out experimental and analytical projects, and it is working on developing standard protocols and addressing paleobiological hypotheses. Together we are taking the first steps along this path leading to the future of fossilization.

ACKNOWLEDGMENTS

The author was supported by the Deutsche Forschungsgemeinschaft (DFG, German Research Foundation), Project number 396637283 to Jes Rust for VEM. Much of the research mentioned herein was also supported by this and other projects within the DFG Research Unit 2685. This is contribution number 24 of the DFG Research Unit 2685, "The Limits of the Fossil Record: Analytical and Experimental Approaches to Fossilization."

WORKS CITED

Behrensmeyer, A.K., Kidwell, S.M., and Gastaldo, R.A. 2000. Taphonomy and paleobiology. *Paleobiology*, 26: 103–147.

Bostock, J., and Riley, H. 1855. *Pliny the Elder, The Natural History.* Taylor and Francis, London.

Briggs, D.E.G. 1995. Experimental taphonomy. *Palaios*, 10: 539–550.

Briggs, D.E.G. 2014. Konservat-Lagerstätten 40 years on: The exceptional becomes mainstream. *The Paleontological Society Papers*, 20: 1–14.

Briggs, D.E.G., and Kear, A.J. 1993. Fossilization of soft tissue in the laboratory. *Science*, 259: 1439–1442.

Briggs, D.E.G., Kear, A.J., Martill, D.M., and Wilby, P.R. 1993. Phosphatization of soft-tissue in experiments and fossils. *Journal of the Geological Society*, 150: 1035–1038.

Briggs, D.E.G., and McMahon, S. 2016. The role of experiments in investigating the taphonomy of exceptional preservation. *Palaeontology,* 59: 1–11.

Briggs, D.E.G., and Summons, R.E. 2014. Ancient biomolecules: Their origins, fossilization, and role in revealing the history of life. *BioEssays,* 36: 482–490.

Brown, J.L., Johnston, W., Delaney, C., Short, B., Butcher, M.C., Young, T., Butcher, J., Riggio, M., Culshaw, S., and Ramage, G. 2019. Polymicrobial oral biofilm models: Simplifying the complex. *Journal of Medical Microbiology,* 68: 1573–1584.

Conway Morris, S. 1979. The Burgess Shale (Middle Cambrian) fauna. *Annual Review of Ecology and Systematics,* 10: 327–349.

Edwards, N., Manning, P., Bergmann, U., Larson, P., van Dongen, B., Sellers, W., Webb, S., Sokaras, D., Alonso-Mori, R., Ignatyev, K., Barden, H.E., van Veelen, A., Anné, J., Egerton, V.M., and Wogelius, R.A. 2014. Leaf metallome preserved over 50 million years. *Metallomics,* 6: 774–782.

Iniesto, M., Blanco-Moreno, C., Villalba, A., Buscalioni, Á., Guerrero, M., and López-Archilla, A. 2018. Plant tissue decay in long-term experiments with microbial mats. *Geosciences,* 8: 387.

McCoy, V.E., Gabbott, S.E., Penkman, K., Collins, M.J., Presslee, S., Holt, J., Grossman, H., Wang, B., Solorzano Kraemer, M.M., and Delclòs, X. 2019. Ancient amino acids from fossil feathers in amber. *Scientific Reports,* 9: 6420.

McCoy, V.E., Soriano, C., Pegoraro, M., Luo, T., Boom, A., Foxman, B., and Gabbott, S.E. 2018. Unlocking preservation bias in the amber insect fossil record through experimental decay. *PLOS ONE,* 13: e0195482.

McNamara, M.E., Briggs, D.E., Orr, P.J., Noh, H., and Cao, H. 2011a. The original colours of fossil beetles. *Proceedings of the Royal Society B: Biological Sciences,* 279: 1114–1121.

McNamara, M.E., Briggs, D.E., Orr, P.J., Wedmann, S., Noh, H., and Cao, H. 2011b. Fossilized biophotonic nanostructures reveal the original colors of 47-million-year-old moths. *PLOS Biology,* 9: e1001200.

Parry, L.A., Smithwick, F., Nordén, K.K., Saitta, E.T., Lozano-Fernandez, J., Tanner, A.R., Caron, J.B., Edgecombe, G.D., Briggs, D.E., and Vinther, J. 2018. Soft-bodied fossils are not simply rotten carcasses. *BioEssays,* 40: 1700167.

Purnell, M.A., Donoghue, P.J., Gabbott, S.E., McNamara, M., Murdock, D.J., and Sansom, R.S. 2018. Experimental analysis of soft-tissue fossilization—Opening the black box. *Paleontology,* 61: 317–323.

Schweitzer, M.H., Wittmeyer, J.L., Horner, J.R., and Toporski, J.K. 2005. Soft-tissue vessels and cellular preservation in *Tyrannosaurus rex. Science,* 307: 1952–1955.

Smithwick, F.M., Nicholls, R., Cuthill, I.C., and Vinther, J. 2017. Countershading and stripes in the theropod dinosaur *Sinosauropteryx* reveal heterogeneous habitats in the Early Cretaceous Jehol Biota. *Current Biology,* 27: 3337–3343.e2.

Solorzano Kraemer, M.M., Delclòs, X., Clapham, M.E., Arillo, A., Peris, D., Jäger, P., Stebner, F., and Peñalver, E. 2018. Arthropods in modern resins reveal if amber accurately recorded forest arthropod communities. *Proceedings of the National Academy of Sciences of the United States,* 115: 6739–6744.

Vinther, J., Briggs, D.E., Prum, R.O., and Saranathan, V. 2008. The colour of fossil feathers. *Biology Letters,* 4: 522–525.

Vinther, J., Nicholls, R., Lautenschlager, S., Pittman, M., Kaye, T.G., Rayfield, E., Mayr, G., and Cuthill, I.C. 2016. 3D camouflage in an ornithischian dinosaur. *Current Biology*, 26: 2456–2462.

Wiemann, J., Fabbri, M., Yang, T.-R., Stein, K., Sander, P.M., Norell, M.A., and Briggs, D.E. 2018. Fossilization transforms vertebrate hard tissue proteins into *N*-heterocyclic polymers. *Nature Communications*, 9: 4741.

Yang, T.-R., Chen, Y.-H., Wiemann, J., Spiering, B., and Sander, P.M. 2018. Fossil eggshell cuticle elucidates dinosaur nesting ecology. *PeerJ*, 6: e5144.

Contributors

Editors

Carole T. Gee is a paleobotanist and Associate Professor of Paleontology at the University of Bonn in Germany. Her interests include Mesozoic plants, *Araucaria* cones, Cenozoic water lilies, and the fossilization of the most common plant megafossil in terrestrial ecosystems, silicified wood. She is pictured with stromatolites at Shark Bay, Australia.

Institute of Geosciences, Division of Paleontology, University of Bonn, Nussallee 8, 53115 Bonn, Germany; Huntington Botanical Gardens, 1151 Oxford Road, San Marino, California 91108, USA

Victoria E. Mccoy is a paleobotanist and invertebrate paleontologist. Currently a Visiting Assistant Professor of Paleontology at the University of Wisconsin–Milwaukee, her interests include biomolecule fossilization and the fossil record of plant defenses, both physical and chemical, against insect herbivores.

Institute of Geosciences, Division of Paleontology, University of Bonn, Nussallee 8, 53115 Bonn, Germany; currently at the Department of Geosciences, University of Wisconsin–Milwaukee, 3209 N Maryland Ave, Milwaukee, Wisconsin 53211, USA

Authors

GEORG OLESCHINSKI

JESSICA OLDENBURGER

P. Martin Sander is a vertebrate paleontologist and Professor of Paleontology at the University of Bonn. His interests include the biology and evolution of dinosaurs and Mesozoic marine reptiles, as well as the histology and chemistry of preserved organics in fossil bone and teeth from the age of dinosaurs.

Institute of Geosciences, Division of Paleontology, University of Bonn, Nussallee 8, 53115 Bonn, Germany; Dinosaur Institute, Natural History Museum of Los Angeles County, 900 Exposition Boulevard, Los Angeles, California 90007, USA

H. Jonas Barthel is an invertebrate paleontologist and PhD student in paleontology at the University of Bonn. His dissertation is focused on the fossilization of arthropod soft tissues in amber.

Institute of Geosciences, Division of Paleontology, University of Bonn, Nussallee 8, 53115 Bonn, Germany

Aurore Canoville is a paleobiologist and a postdoctoral researcher at North Carolina State University in the United States. Her research focuses on drawing paleobiological inferences from the study of tetrapod biomineralized tissues.

Division of Paleontology, North Carolina Museum of Natural Sciences, 11 W Jones Street, Raleigh, North Carolina 27601, USA; Department of Biological Sciences, North Carolina State University, 100 Brooks Avenue, Raleigh, North Carolina 27607, USA

Thorsten Geisler(-Wierwille) is a mineralogist and a Professor of Geochemistry at the University of Bonn. His main interests include the stability of ceramic materials and silicate glasses in aqueous solutions, which are proposed as nuclear waste forms, Raman spectroscopy of self-irradiated materials, solid solutions, and fossils, and U-Th-Pb dating of U- and Th-bearing minerals.

Institute of Geosciences, Division of Geochemistry and Petrology, University of Bonn, Meckenheimer Allee 169, 53115 Bonn, Germany

STEFFI GÖTZE

ANONYMOUS

Jens Götze is a mineralogist and a Professor of Applied Mineralogy at the Freiberg University of Mining and Technology in Germany. His research interests include investigations of quartz raw materials and SiO_2 minerals, methodological development of cathodoluminescence microscopy and spectroscopy, typomorphic properties of selected minerals, and the synthesis of single crystals and materials with applicable properties.

Institute of Mineralogy, TU Bergakademie Freiberg, Brennhausgasse 14, 09599 Freiberg, Germany

Conrad C. Labandeira is a paleoecologist and Senior Research Geologist and Curator of Fossil Arthropods at the Smithsonian's National Museum of Natural History. His interests include interactions between plants and insects in the fossil record, terrestrial fossil arthropods (and in particular insects), the evolution of insect mouthparts, and fossil insect diversity.

Smithsonian Institution, National Museum of Natural History, Department of Paleobiology, 10th St. & Constitution Ave., Washington, District of Columbia 20013-7012, USA; University of Maryland, Department of Entomology and BEES Program, College Park, Maryland 20742, USA; College of Life Sciences, Capital Normal University, Beijing, 100048, China

SASHIMA LÄBE

ESTHER SCHWARZENBACH

Sashima Läbe is a vertebrate paleontologist and a postdoctoral researcher at the University of Bonn. Her research interests include interpreting sauropod tracks with soil mechanics to determine body mass and sauropod gait analysis based on tracks, as well as the fossilization and permineralization of tetrapod hard tissue.

Institute of Geosciences, Division of Paleontology, University of Bonn, Nussallee 8, 53115 Bonn, Germany

Moritz Liesegang is a mineralogist interested in silica minerals, mineral nucleation and growth, and nanolevel processes. He worked on wood silicification as a postdoctoral researcher at the University of Bonn before moving on to the electron microprobe and X-ray diffraction labs at the Freie Universität in Berlin.

Institute of Geosciences, Division of Paleontology, University of Bonn, Nussallee 8, 53115 Bonn, Germany; currently at the Institute of Geological Sciences, Arbeitsbereich Mineralogie–Petrologie, Freie Universität Berlin, Malteserstrasse 74–100, 12249 Berlin, Germany

Martina Menneken is a geoscientist and postdoctoral researcher at the University of Bonn. Her research focuses on the application of Raman spectroscopy in various petrological and material science issues, including early crustal evolution (that is, the formation of crust) and understanding potential early crustal recycling mechanisms.

Institute of Geosciences, Division of Geochemistry and Petrology, University of Bonn, Meckenheimer Allee 169, 53115 Bonn, Germany

Jes Rust is an invertebrate paleontologist and a Professor of Paleontology at the University of Bonn. His interests include the evolution, phylogeny, and paleobiology of arthropods; the paleobiology of mollusks; the origin, nature, and importance of *Konservat-Lagerstätten*; and theoretical aspects of evolutionary biology, phylogeny, and biological systematics.

Institute of Geosciences, Division of Paleontology, University of Bonn, Nussallee 8, 53115 Bonn, Germany

Frank Tomaschek is a geochemist and mineralogist. As a postdoctoral researcher at the University of Bonn, his interests include geochronology and microanalytical characterization of geomaterials.

Institute of Geosciences, Division of Geochemistry and Petrology, University of Bonn, Meckenheimer Allee 169, 53115 Bonn, Germany

Torsten Wappler is an invertebrate paleontologist and a Curator of Natural History at the Hessisches Landesmuseum Darmstadt in Germany. His research interests include plant–insect associations through the ages and their importance for understanding paleoenvironments, as well as the evolution of insects.

Department of Natural History, Hessisches Landesmuseum Darmstadt, Friedensplatz 1, 64283 Darmstadt, Germany

YOLANDA SCHICKER

CAROLE T. GEE

Kayleigh Wiersma(-Weyand) is a vertebrate paleontologist and PhD student in paleontology at the University of Bonn. Her dissertation concentrates on dinosaur bone histology and soft tissue preservation.

Institute of Geosciences, Division of Paleontology, University of Bonn, Nussallee 8, 53115 Bonn, Germany

Tzu-Rui Yang is a vertebrate paleontologist and Assistant Curator in Vertebrate Paleontology at the National Museum of Natural Sciences in Taiwan. His interests include the reproductive biology of oviraptorid dinosaurs, eggshell parataxonomy, gender identification of fossil bones, eggshell histology, and taxonomic processes of fossil eggshells from chemical perspectives.

Division of Geology, National Museum of Natural Sciences, Guancian Road 1, 40453 Taichung, Taiwan; Department of Earth Sciences, National Cheng Kung University, University Road 1, 70101 Tainan, Taiwan

Index